Nicola Gess
Primitive Thinking

Paradigms

Literature and the Human Sciences

Edited by
Rüdiger Campe · Karen S. Feldman

Editorial Board
Paul Fleming · Eva Geulen · Rüdiger Görner · Barbara Hahn
Daniel Heller-Roazen · Helmut Müller-Sievers
William Rasch · Joseph Vogl · Elisabeth Weber

Volume 13

Nicola Gess
Primitive Thinking

Figuring Alterity in German Modernity

Translated by Erik Butler and Susan L. Solomon

DE GRUYTER

The translation was funded by Geisteswissenschaften International – Translation Funding for Work in the Humanities and Social Sciences from Germany, a joint initiative of the Fritz Thyssen Foundation, the German Federal Foreign Office, the collecting society VG WORT and the Börsenverein des Deutschen Buchhandels (German Publishers & Booksellers Association).

ISBN 978-3-11-152080-3
e-ISBN (PDF) 978-3-11-069509-0
e-ISBN (EPUB) 978-3-11-069515-1
ISSN 2195-2205
DOI https://
doi.org/10.1515/9783110695090

This work is licensed under the Creative Commons Attribution-NonCommercial-NoDerivatives 4.0 International License. For details go to https://creativecommons.org/licenses/by-nc-nd/4.0/.

Creative Commons license terms for re-use do not apply to any content (such as graphs, figures, photos, excerpts, etc.) that is not part of the Open Access publication. These may require obtaining further permission from the rights holder. The obligation to research and clear permission lies solely with the party re-using the material.

Library of Congress Control Number: 2022936057

Bibliographic information published by the Deutsche Nationalbibliothek
The Deutsche Nationalbibliothek lists this publication in the Deutsche Nationalbibliografie; detailed bibliographic data are available on the internet at http://dnb.dnb.de.

© 2024 the author(s), published by Walter de Gruyter GmbH, Berlin/Boston. The book is published open access at www.degruyter.com.
This volume is text- and page-identical with the hardback published in 2022.
First published in German under the title *Primitives Denken Wilde, Kinder und Wahnsinnige in der literarischen Moderne (Müller, Musil, Benn, Benjamin)*. © 2013 Wilhelm Fink Verlag, ein Imprint der Brill Gruppe (Koninklijke Brill NV, Leide, Niederlande; Brill USA Inc.; Boston MA, USA; Brill Asia Pte Ltd, Singapore; Brill Deutschland GmbH, Paderborn, Deutschland).

Cover image: Paul Goesch. Generalarzt (1928). Private Collection (courtesy Moeller Fine Art, New York–Berlin).

www.degruyter.com

Contents

Chapter 1
The Presence of the 'Primitive': An Introduction —— 1
        The 'Primitive' as a Narrative of Origins —— 3
        The 'Primitive' as a Scientific Paradigm —— 5
        The 'Primitive' in the Service of Cultural Critique —— 7
        The 'Primitive' as a Figure of Thought —— 9
        The 'Primitive' as Literary Utopia —— 12
        The Presence of the 'Primitive' in Disenchanted Modernity —— 15
        Two Cultures: The 'Primitive' as *Poème* —— 22
        Literary Primitivism —— 25

Part One: Figures of 'Primitive Thinking'

Chapter 2
The Ethnological Paradigm of the 'Primitive' —— 33
        The Paradigm of the 'Primitive' in Tylor's *Primitive Culture* —— 36
        Analogy, Allochrony, and Survival —— 41
        The 'Primitive' as a Figure of Thought in Early Ethnology —— 43
        Lucien Lévy-Bruhl's Notion of Participatory Thinking —— 51
        The Ethnological *Poème* of the 'Primitive' —— 60

Chapter 3
The Child as 'Primitive' —— 75
        Recapitulating Phylogeny —— 76
        Othering: The 'Bad' Child —— 80
        The Question of Conscious Deception —— 88
        Jean Piaget and the Magical Thinking of Children —— 97
        Between Natural Science, Philology, and Literature: The Methodological Dilemma of Developmental Psychology —— 103

Chapter 4
Psychopathology in the Paradigm of the 'Primitive' —— 110
        The *Poèmes* of Psychology —— 112
        The Analogy of Regression —— 115
        Phylogenetic Regression —— 120
        Ontologization —— 130
        The Schizophrenic Artist —— 136

Part Two: Art, Language, and 'Primitive Thinking'

Chapter 5
The Origins of Art —— 143
        Justifying the Study of Art —— 146
        The Enigma of Creativity —— 157
        Pathology or Heroization: Genius and Madness —— 160
        Normalizing the Artist —— 166
        Art as (Child's) Play —— 167
        Art and Deception —— 170
        Art and Destruction —— 172

Chapter 6
'Primitive Language' – Theories of Metaphor —— 176
        Constructions of 'Primitive Language': The Cratylist Tradition —— 178
        Malinowski and the Magical Power of Language —— 182
        Theories of Metaphor around 1900: Nietzsche, Mauthner, Vischer, Biese, Cassirer —— 184

Part Three: 'Primitive Thinking' in German Literary Modernism

Chapter 7
The "Tropological Nature" of the Poet in Müller and Benn —— 205
        A Biological Reverie —— 205
        The Tropics —— 209
        Tropological Language —— 217
        Poets —— 222
        Returning to Primordial Slime —— 228
        The Body as Hieroglyph of the Archaic —— 231
        The Way to Fascism —— 234

Chapter 8
A Sister in Madness: Figures of 'Primitive Thinking' in Robert Musil —— 236
        Musil's Ethnological Readings —— 236
        Red Parrots or 'Primitive Thinking' —— 239
        On 'Primitive Language' and Its Magic —— 241
        Animal-Humans: From Expedition to (Self-)Experimentation —— 243
        Clarisse – A Sister in Madness (*The Man Without Qualities*) —— 248

Regression —— 275
Psychology as the Springboard for Literature —— 285
Primitivistic Narration —— 291
Another Primitivist Aesthetics? —— 297

Chapter 9
The Dialectical Turn of 'Primitive Thinking': The Child and Gesture in Walter Benjamin —— 303
　　The Child as 'Barbarian' —— 308
　　The Child as 'Primitive' —— 314
　　A Dialectical Turn —— 322
　　The Sovereign Child —— 329
　　Toward the Child's Language of Gesture —— 332
　　A Theory of Gestures in the "Problems in the Sociology of Language" —— 335
　　The Politics of Gestural Language —— 349
　　The *Arcades Project:* The Child as Historiographical Model —— 352

Epilogue —— 358

Bibliography —— 361

Index —— 391

Chapter 1
The Presence of the 'Primitive': An Introduction

This book is about modernity under the spell of the 'primitive.'[1] From proponents of the ideology of progress to critics of civilization, from utopians dreaming of a re-enchanted existence to the supporters and opponents of nascent fascism, the first decades of the twentieth century were profoundly shaped by the phantasm of the 'primitive,' which European modernity in its many variations regarded as both its origin and opposite. This book, however, argues that the 'primitive' must instead be seen as modernity's own product and presence.

Exploring modernity's primitivism, it takes a somewhat different angle than previous research.[2] A central element of the phantasm of the 'primitive,' this book shows, is the notion of 'primitive thought,' a distinct mode of thinking – characterized by turns as magical, mythical, mystical, or prelogical – that forges a fundamentally other relationship to the world. In this respect, modernity's fascination with the 'primitive' may also be described as *epistemological primitivism*. The book also demonstrates that epistemological primitivism of the early

---

[1] *Modernity* is used here first as an epochal designation, that is, as "a shorthand term for modern society, or industrial civilization" (Anthony Giddens, *Conversations with Anthony Giddens: Making Sense of Modernity* [Stanford: Stanford University Press, 1998], 94) in the late nineteenth and early twentieth centuries, which witnessed the emergence of philosophical and aesthetic modernism. It is used, second, as a critical concept, applied by social and cultural theorists of that time when reflecting upon the specificities of their own society (whether in a normative and affirmative sense, or to negative and critical ends). Throughout this study, part of my aim is to show the extent to which this self-understanding of modernity depended on the 'primitive' as its imaginary counterpart. Because this problem is inscribed into the very concept of modernity, its uncritical, naïve use has quite rightly provoked critique (e. g., from a postcolonial perspective by Frederick Cooper, *Colonialism in Question: Theory, Knowledge, History* [Berkeley: University of California Press, 2005] or from a sociological standpoint by Bruno Latour, *We Have Never Been Modern*, trans. Catherine Porter [Cambridge, MA: Harvard University Press, 1993]).

[2] Representative works from literary studies, which I will talk about below in more detail, include Marianna Torgovnick, *Gone Primitive: Savage Intellects, Modern Lives* (Chicago: University of Chicago Press, 1991); Erhard Schüttpelz, *Die Moderne im Spiegel des Primitiven. Weltliteratur und Ethnologie (1870–1960)* (Munich: Fink, 2005); and Sven Werkmeister, *Kulturen jenseits der Schrift. Zur Figur des Primitiven in Ethnologie, Kulturtheorie und Literatur um 1900* (Munich: Fink, 2010). Last to appear, after the publication of the German edition of this book, was Ben Etherington, *Literary Primitivism* (Stanford: Stanford University Press, 2017); and Samuel Spinner, *Jewish Primitivism* (Stanford: Stanford University Press, 2021). Further works from literary studies and adjacent fields such as art history, cultural studies, and anthropology will be referenced below.

twentieth century concerned not only indigenous cultures but also other figures of alterity. As important as the postcolonial perspective on the 'primitive' has been, its focus on the (false) representation of indigenous cultures has tended to obscure two things: First, the constructed nature of the 'primitive,' which should be understood as the product, not the object, of primitivism. And second, the observation that indigenous cultures constitute only one of many figurations of alterity constructed by early twentieth-century primitivist discourse.[3] Other embodiments of the 'primitive' that proved just as important include children and the mentally ill.[4] Thus, in this book I locate primitivism first of all in the human sciences, noting how the emergent disciplines of ethnology, child psychology, and psychopathology conceived of children, indigenous cultures, and the mentally ill as contemporary incarnations of the 'primitive.' The heterogeneity of the three groups makes plain that this conception was not motivated by scholarly interest in any one of these figures in particular. Instead, their shared interest in the 'primitive' (and classification of these groups as 'primitive' in the first place) concerned their supposed representation of a *presence* of an *origin* marked, among other traits, by 'primitive thinking.'

Based on these premises, the book focuses on the relevance of epistemological primitivism for the theory and practice of the arts in early twentieth-century Europe and for German literature in particular. The search for humankind's origins and the widespread fascination with other modes of thinking, both of which were shaped by the paradigm of the 'primitive,' intersected with contemplations on the arts to yield a new perspective on both. Interest turned to art's beginning, which was supposed to shed light on the origins of modern civilization and the essence of humanity.[5] Even more importantly, a new view of contemporary art

---

[3] This runs counter to what most scholarship on literary primitivism has assumed. However, since the publication of the German edition of this book, two studies have appeared that offer a corrective: the exhibition catalog edited by Anselm Franke and Tom Holert, *Neolithische Kindheit: Kunst in einer falschen Gegenwart, ca. 1930* (Zurich: HKW, 2018); and Barbara Wittmann, *Bedeutungsvolle Kritzeleien: Eine Kultur- und Wissensgeschichte der Kinderzeichnung, 1500–1950* (Zurich: Diaphanes, 2018).

[4] These three figures are only the most conspicuous. Others – with a range of different connotations – include women, the working class (cf. Gina M. Rossetti, *Imagining the Primitive in Naturalist and Modernist Literature* [Columbia, MO: University of Missouri Press, 2006]), ethnic minorities (cf. Eva Blome, *Reinheit und Vermischung. Literarisch-kulturelle Entwürfe von "Rasse" und Sexualität [1900–1930]* [Cologne: Böhlau, 2011]; Sieglinde Lemke, *Primitivist Modernism: Black Culture and the Origins of Transatlantic Modernism* [Oxford: Oxford University Press, 1998]), and rural populations.

[5] In respect to the fine arts, see Susanne Leeb's approach in *Die Kunst der Anderen: "Weltkunst" und die anthropologische Konfiguration der Moderne* (Berlin: b-books, 2015).

emerged insofar as it was supposed to have preserved 'primitive thinking,' which was thought to be observable also in children at play, the delusions of the mentally ill and the rituals and myths of indigenous peoples. In these contexts the arts enjoyed a new form of legitimacy. They became a privileged site to which questions of origin were addressed and a starting point for utopian efforts promoting 'primitive thinking' against a disenchanted modernity.

In this introduction I would like to accomplish three goals: The first is to address the question of why the 'primitive' held such an abiding fascination for European modernity. To do so, I approach the category of the 'primitive' as a narrative of origin, an instrument for critiquing civilization, a literary utopia, and a diagnosis of the present (of that time) – four assumptions that I consider to be fundamental to understanding the early twentieth century's primitivist discourse. The second goal is to reconceptualize the 'primitive' as a paradigm, a figure of thought, and a *poème*. Third, and finally, I offer a brief overview of existing research on literary primitivism before moving on to preview the structure of this book.

The 'Primitive' as a Narrative of Origins

"They *are* what we *were*" – Friedrich Schiller's famous dictum in "Über naïve und sentimentalische Dichtung" (1795–1796; "On Naïve and Sentimental Poetry," 1861) summarizes a view that has recurred time and again throughout European history: a primordial condition of human existence may be found among peoples living elsewhere, and also among children, animals, and parts of the natural world.[6] "We were nature just as they," Schiller writes, referring to "plants, minerals, animals, [...] landscapes, [...] children, [...] the customs of country folk, and [...] the primitive world."[7] At a basic level, this figure of thought transfers the foreign of the present into the past of the familiar – what Johannes Fabian has called "allochronic discourse."[8] The 'foreign' can appear in any number of roles here, but always in contrast to the culture in which it is pronounced. Either it is essentially undeveloped, or it stands at the very beginning of a developmental process. In

---

[6] Friedrich Schiller, "On Naïve and Sentimental Poetry," trans. Julias A. Elias, in *German Aesthetic and Literary Criticism: Winckelmann, Lessing, Hamann, Herder, Schiller, and Goethe*, ed. H.S. Nisbet (Cambridge: Cambridge University Press, 1985), 181. Italics modified to reflect German publication.

[7] Friedrich Schiller, "On Naïve and Sentimental Poetry," 181, 180.

[8] Johannes Fabian, *Time and the Other: How Anthropology Makes Its Object* (New York: Columbia University Press, 2014).

any case, its 'proper' place lies at the point of origin, and inasmuch as it persists in the present, it constitutes an anachronistic survival. Such a view readily translates into the opposition between nature and culture, whereby the foreign, located outside or at the beginning of history, is understood as nature and European society, which is thought to display history and progress, as culture.

Reflecting its durability, this figure of thought has varied with historical and scientific circumstances.[9] Medieval accounts of travel – undertaken in the hope of discovering a paradisiacal world far from Europe – clearly differ from what Schiller had in mind. The decisive difference is that from a medieval and biblical perspective, the origin did not represent the beginning of a historical development; rather, it was seen to lie *outside* of history (i.e., in paradise). The border between nature (that is, the "state of nature") and culture ("history") was drawn just as clearly. This is, for example, still true for Rousseau and his idea of the 'noble savage,' but not for Schiller.[10] According to Michel Foucault, around 1800 the "age of representation" passed into the "age of history,"[11] at which point the primal source ceased to be located outside the sphere of historical development. In Foucault's words, "In modern thought, such an origin is no longer conceivable."[12] The contemporary world and the birthplace it was supposed to have left behind were no longer opposites – instead they were connected, even continuous. This also holds for Schiller's reflections, as Sven Werkmeister has demonstrated.[13] In the foreign identity of "they are," Schiller finds his and his intended readers' own past: "what we were." The study of faraway peoples now furthers self-understanding. He declares, "A wise hand seems to have preserved these raw tribes for us down to our times, where we would be advanced enough in our own culture to make fruitful application of this discovery upon ourselves, and to restore out of this mirror the forgotten origin of our species."[14] Engagement with "raw tribes" then serves a reassuring function. Simultaneously,

---

9 For a summary and discussion of this figure's evolution, cf. Wolfgang Riedel, "Wandlungen und Symbole des Todestriebs. Benns Lyrik im Kontext eines metapsychologischen Gedankens," in *Sigmund Freud und das Wissen der Literatur*, ed. Peter-André Alt and Thomas Anz (Berlin: De Gruyter, 2008), 111–112.
10 See Sven Werkmeister, *Kulturen jenseits der Schrift*, 59.
11 Michel Foucault, *The Order of Things: An Archaeology of the Human Sciences* (New York: Vintage, 1994), 217.
12 Foucault, *The Order of Things*, 329.
13 Werkmeister, *Kulturen jenseits der Schrift*, 59–65.
14 Friedrich Schiller, "What Means, and for What Purpose Do We Study, Universal History?", in *Complete Works in Two Volumes*, ed. and trans. Charles J. Hempell, M.D. (Philadelphia: Kohler, 1861), 2: 348 (translation slightly modified). Also quoted in Werkmeister, *Kulturen jenseits der Schrift*, 60.

however, it destabilizes identity by confronting the unassimilably alien that lies at the foundation of the self.

According to "On Naïve and Sentimental Poetry," knowledge of what one was does not inspire alienation so much as melancholy. For Schiller, what one loves in nature is "quietly working life, the calm effects out from itself, existence under its own laws, the inner necessity, the eternal unity with itself."[15] This idealization bears clear traces of the paradisiacal tradition mentioned above. Following the triadic scheme of history, Schiller expresses the wish to ultimately return to the Golden Age.[16] That said, Schiller's inaugural address at the university in Jena presents a rather different picture of humankind's origins when he states, "how shaming and sad is the picture these people give us of our childhood!" Speaking of civilized man's supposed counterpart, Schiller declares, "His crude taste seeks joy in stupor, beauty in distortion, glory in exaggeration; even his virtue awakens horror in us, and what he calls his bliss can only arouse our disgust and pity. So were *we*."[17] The ambivalence of modern thought concerning human origins is in full evidence here. On the one hand, faraway peoples epitomize the other, that is, the opposite of one's own self-image. Schiller's and other contemporaries' self-understanding as mature, reasonable, self-disciplined, socialized, and cultivated contrasts with 'children of nature' who are portrayed as irrational, governed by emotion and drives, and potentially antisocial. In this manner, indigenous peoples are used as a screen for projecting everything that must be excluded from one's own sense of identity. On the other hand, these foreigners are supposed to represent modern society's point of departure. In this way, encounters with others who are meant to affirm one's own identity by revealing one's origins actually also unsettle this same sense of self.

The 'Primitive' as a Scientific Paradigm

The formula exemplified by Schiller's phrase, "They are what we were," ran through the cultural history of the nineteenth and early twentieth centuries. However, it assumed a new form in the context of the emergent human scien-

---

15 Schiller, "On Naïve and Sentimental Poetry," 181 (translation slightly modified).
16 The triadic model of history premises that humanity developed in three phases: a paradisical original condition, a state of alienation and disharmony between nature and society, and finally an anticipated reentry to paradise, which, however, is not a mere return but is thought of as a return to a higher level (e. g., through reflection by means of art), especially in theories of history around 1800.
17 Schiller, "What Means, and for What Purpose," 348.

ces.¹⁸ Around 1800, equating humanity's origins with indigenous peoples, children, and even animals amounted to no more than an analogy. Schiller recognizes as much and advocates caution: "The method of drawing conclusions by analogies is as powerful an aid in history, as everywhere else, but it must be [...] exercised with as much circumspection as judgment."¹⁹ But in the anthropological configuration of modernity and the new human sciences, it soon achieved the status of scientific fact. Ethnologists, developmental psychologists, and psychopathologists now believed they had found empirical evidence that the origins of humankind really were present in indigenous cultures, children, and the mentally ill. Indeed, the 'primitive' – understood as the presence of an origin – was the very *paradigm* that framed and organized their questions and answers.

Ethnologists and social anthropologists believed they had discovered that the thought and behavior of indigenous peoples corresponded to those of early humans. The supposed simplicity of these cultures was seen as evidence that they had not developed, thus had no history of their own, and were still living in a 'state of nature.' Developmental psychologists and biologists claimed the same of children, who, in the process of maturation, were thought to recapitulate the development of the species (in keeping with the biogenetic law established by the physician Ernst Haeckel). Finally, psychologists and psychiatrists exposed the 'primitive' in the mentally ill (especially schizophrenics), whom they considered to have regressed to an earlier stage of human development archived in the unconscious mind and specific organs of the body, such as the brain stem.

This pattern of reasoning secured the 'primitive' as a common point of reference among the different disciplines. Identical mental operations ('primitive thinking') were supposed to prevail across the board: scientists assumed that the mentally ill think like members of indigenous tribes, who think like children, who think like the mentally ill, and so on. In other words, the 'primitive' provided a platform for homogenizing objects of study. At the same time, it offered a paradigm for speculating about the origins and essence of humanity. Thus, Heinz Werner's *Einführung in die Entwicklungspsychologie* (1926; *Comparative Psychology of Mental Development*, 1940) identifies indigenous peoples, children, and the mentally ill as "primitive types" whose common features were supposed to shed light on the "general developmental laws of mental life."²⁰ These three groups were the epistemic objects where the search for first beginnings was car-

---

18 Foucault, *The Order of Things*, 330.
19 Schiller, "What Means, and for What Purpose," 352 (translation slightly modified).
20 Heinz Werner, *Einführung in die Entwicklungspsychologie* (Leipzig: Barth, 1926), 150, 23, 3. Tellingly, his 1919 study was devoted to the "origins of metaphor" (*Ursprünge der Metapher*).

ried out. And what began as an evolutionary enterprise quickly became an ontological one focused on the 'essence of man.'²¹ Since the 1910s, this ontological orientation was particularly pronounced in the study of art. A late, but well-known example is Martin Heidegger's 1935 lecture, "Ursprung des Kunstwerks" ("The Origin of the Work of Art," 1971), which understands "origin" to mean essence. From this perspective, the 'primitive' no longer represents a historical beginning so much as an a-historical core, on the basis of which general rules and tasks of art may be formulated.

The 'Primitive' in the Service of Cultural Critique

In 1917, the German sociologist Max Weber delivered his lecture, "Wissenschaft als Beruf" ("Science as a Vocation," 1946), to the *Freistudentischer Bund* (Union of Free Students) at the University of Munich. The speaker was addressing an audience he believed to be deeply disillusioned with modernity. Members of the rising generation, Weber observes, want "to return to their own nature and hence to nature as such" – "a life in communion with the divine" – and are hoping for "release from the rationalism and intellectualism of science."²² They are cultivating a cult of "experience" and "personality"²³ and yearning for a prophetic "leader"²⁴ in contempt of the "realm of artificial abstractions."²⁵ Instead of accepting the "destiny of our culture,"²⁶ they are fashioning "idols"²⁷ and looking for "professorial prophets"²⁸ to relieve them of thinking for themselves. Such "intellectualist romanticism of the irrational,"²⁹ Weber continues, amounts to "interpret[ing] human communities in religious, cosmic, or mystical terms," "furnish[ing] their souls, as it were, with antique objects," and "decorating [their] private chapel with pictures of the saints that they have picked up in

---

21 This shift also involves a change of historical models: "At the beginning of the twentieth century, the evolutionism proceeding by steps was increasingly replaced by synchronic approaches [...]. Now, reference to the primitive and the archaic could perform a critical function by opening the possibility for reversal [of perspective]" (Franke and Holert, *Neolithische Kindheit*, 319).
22 Max Weber, "Science as a Vocation," in *The Vocation Lectures: "Science as a Vocation, Politics as a Vocation,"* trans. Rodney Livingstone (Indianapolis: Hackett, 2004), 15.
23 Weber, "Science as a Vocation," 10.
24 Weber, "Science as a Vocation," 24.
25 Weber, "Science as a Vocation," 14.
26 Weber, "Science as a Vocation," 24.
27 Weber, "Science as a Vocation," 10.
28 Weber, "Science as a Vocation," 28.
29 Weber, "Science as a Vocation," 16.

all sorts of places," "creat[ing] a surrogate by collecting experiences of all kinds that they endow with the dignity of a mystical sanctity."³⁰

According to Weber's psychological sketch then, a comprehensive critique of civilization, progress, and rationality took hold among young people toward the end of the First World War. It was accompanied by a yearning for a different relationship to the world, which he describes in terms such as "nature," "life," "experience," "personality," "soul," "community," and "religion" – diction that also features prominently in artistic primitivism of the interwar period.³¹ Indeed, Weber himself refers to the 'primitive,' but not in his characterization of youth so much as his analysis of the state of society as a whole.

Weber criticizes the cowardice of the young generation, but he shares its disillusionment, expressing the conviction that the "process of intellectualization [...] at work in Western culture for thousands of years" – that is, "rationalization through science and a science-based technology" – leads to the wholesale "disenchantment of the world."³² In order to understand what "disenchantment" means in this context, it is important to note that he employs the term *ex negativo*, that is, by way of a counterpole predicated on the phantasm of the 'primitive.' By disenchantment, he means that "[u]nlike the savage for whom such

---

30 Weber, "Science as a Vocation," 30. The critical edition of this text notes oblique reference to the ideals of the youth movement surrounding Gustav Wyneken and particularly the "neo-mystical" publishing house Eugen Diederichs, which provided a forum for critics of modern rationalism (Weber, "Wissenschaft als Beruf," in *Gesamtausgabe*, ed. Horst Baier et al., vol. 17, section 1, *Schriften und Reden* [Tübingen: Mohr, 1992], 109n61 and 62).

31 Franke and Holert describe the interwar period as a time when "the loss of a unifying collectivity through the dividing processes of labor, scientification, and the 'liberal' individualization of capitalist modernity [...] led intellectuals and artists [...] to ideologically extreme and diverging scenarios of flight and evasion" (Franke and Holert, "Einführung," in *Neolithische Kindheit*, 10).

32 Weber, "Science as a Vocation," 30–31. Weber could "only grasp the rationalization process after he had discovered the process of disenchantment. Therefore, the question of when and where this happened leads to the heart of his sociology" (Friedrich H. Tenbruck, "Das Werk Max Webers," *Kölner Zeitschrift für Soziologie und Sozialpsychologie* 69, Suppl. 1 [2017]: 384) – and also leads, I would add, to the counter-image of "disenchantment" or the paradigm of the 'primitive' in which the latter takes shape. As Hartmut Lehmann has put matters, "As of now, nobody has been able to explain the origin of the term 'disenchantment of the world'" (*Die Entzauberung der Welt: Studien zu Themen von Max Weber* [Göttingen: Wallstein, 2009], 13). The answer, in my estimation, may be found by examining the discourse of the 'primitive.' Thus, Fuyuki Kurasava observes that "[t]he ethnological imagination is [...] essential to Weber's identification of the dynamics of rationalization that distinguish modern Euro-American societies from other sociohistorical formations" (*The Ethnological Imagination: A Cross-Cultural Critique of Modernity* [Minneapolis: University of Minnesota Press, 2004], 84).

forces existed, we need no longer have recourse to magic in order to control the spirits or pray to them. Instead, technology and calculation achieve our ends."[33] Weber's reflection is preceded by another direct comparison between "the savage" and "us": "The savage has an incomparably greater knowledge of his tools" and "knows how to obtain his daily food and what institutions enable him to do so." "We," on the other hand, know that we *could* arrive at the same understanding "if *only we wished* to."[34] Later in the lecture, Weber again speaks of a time "before the world had been divested of the magic of its gods and demons" and the "bygone days" when a "prophetic spirit [...] swept through great communities like a firestorm and welded them together."[35] Weber's diagnosis of the "disenchantment of the world" is predicated on its opposite, then: the idea of a *primitive worldview* under the spell of magical plenitude.[36]

The 'Primitive' as a Figure of Thought

The paradigm of the 'primitive' determining Weber's conception of modernity as its counterpole is characterized by the assumption of a distinct form of non-rational thinking and thus a fundamentally *other* relationship to the world. Therefore, the concept of the 'primitive' functions not only as a paradigm, but also as a pervasive "figure of thought"[37] in primitivist discourse in the early twentieth century. The adjective "primitive" (like the German *primitiv* and French *primitif*) comes from the Latin noun, *primitivus*, or "first of its kind."[38] Turned back into a

---

[33] Weber, "Science as a Vocation," 13.
[34] Weber, "Science as a Vocation," 12. Whenever two or more quotes in a paragraph derive from the same page number of the same source, they will be cited together with a concluding note. Thus, the citation in this note documents material from the previous two sentences.
[35] Weber, "Science as a Vocation," 30.
[36] That said – and in contrast to the thinkers discussed below – Weber himself did not think of such a worldview as a utopian ideal, but in terms of an irretrievable loss.
[37] "Figure of thought" is used here in the sense defined by Jutta Müller-Tamm, "Die Denkfigur als wissensgeschichtliche Kategorie," in Nicola Gess and Sandra Janssen, eds., *Wissens-Ordnungen: Zu einer historischen Epistemologie der Literatur* (Berlin: De Gruyter, 2014). In *Scenes from the Drama of European Literature* (Manchester: Manchester University Press, 1984), 11–78, Müller-Tamm draws on Erich Auerbach's "Figura" as well as on the essays collected in Gabriele Brandstetter and Sibylle Peters, eds., *De figura: Rhetorik – Bewegung – Gestalt* (Munich: Fink, 2002). See also George Lakoff and Mark Johnson, *Metaphors We Live By* (Chicago: University of Chicago Press, 1980), and Hans Blumenberg, *Paradigms for a Metaphorology*, trans. Robert Savage (Ithaca: Cornell University Press, 2010).
[38] "primitiv," in Friedrich Kluge, *Etymologisches Wörterbuch der deutschen Sprache* (Berlin: De Gruyter, 1995), 647.

noun in modern languages, the adjective yields the 'primitive,' i.e., a human being in the primal state, which, as mentioned above, was 'rediscovered' around 1900 in indigenous peoples, children, and the mentally ill. The result was a flexible anthropomorphic *figure* that organized scientific *thought* about the origins of humanity, culture, and nature. And the most important feature of this *figure* was that they were supposed to embody a different kind of *thinking* and mentality. According to early twentieth-century human scientists, this other frame of mind was defined by alogical relations and associations that determined how indigenous peoples, children, and the mentally ill perceived and understood the world. For the ethnologist Karl von den Steinen, this meant, for example, that members of the Bororo tribe in Brazil believed they were humans and parrots at the same time. Or for the psychiatrist Ernst Kretschmer, that schizophrenics thought distant magical powers control their lives. And for the developmental psychologist Jean Piaget, that children believed they could talk to animals, steer the path of the sun, or transform one object into another.[39] By turns, such thinking is characterized as 'magical,' 'mythical,' 'prelogical,' or 'mystical.' In 1925, for instance, the philosopher Ernst Cassirer offers a comprehensive philosophical assessment of what he calls "mythical thought."[40] For him, such thought does not differentiate between perception and imagination. Also, it entails a different sense of causality insofar as all things that seem similar or were once in contact with each other are believed to exercise necessary and reciprocal influence on each other.

Significantly, Cassirer and other scholars in the human sciences affirmed a direct connection between 'primitive thinking' and the essence of art and artistic creation. For instance, in his foundational study, *Primitive Culture,* the social anthropologist Edward B. Tylor presents the mental constitution of indigenous peoples as the key to poetry and literature:

---

[39] I will return to these examples in Chapters 2–4.
[40] Ernst Cassirer, *The Philosophy of Symbolic Forms,* vol. 2, *Mythical Thought,* trans. Ralph Manheim (New Haven, CT: Yale University Press, 1955). For another good example, see Richard Thurnwald's entry, "Primitives Denken," in Max Ebert, ed., *Reallexikon der Vorgeschichte* (Berlin: De Gruyter, 1927/1928), 10: 294–316. From a resolutely ethnological perspective, Thurnwald describes 'primitive thinking' as "adhering predominantly to a complex of phenomena without discrimination, [that is,] without having learned to distinguish between the reality of thought and that of its object" (296).

> In so far as myth [...] is the subject of poetry, and in so far as it is couched in languages whose characteristic is that wild and rambling metaphor which represents the habitual expression of savage thought, the mental condition of the lower races is the key to poetry.[41]

Similarly, psychiatrist Ernst Kretschmer advances the thesis that Expressionist art can be explained on the basis of how schizophrenics think – which is to say, on the basis of the thinking of early humans, which for him resurfaces in schizophrenia.[42] And in *Der Genius im Kinde* (The Genius in the Child, 1922), the pedagogical reformer and art historian Gustav Friedrich Hartlaub claims, "Only the poet and the artist preserve [...] this general imaginative potential of the child [...]. The 'artist' alone knows how to salvage, more or less, [what remains of] the immense inner life of childhood."[43] Here, art and origins condition and ground each other reciprocally.

These theses were also adopted by scholars of philosophy and art seeking to affirm the scientific nature of their enterprise of *Kunstwissenschaften* (art studies, but literally meaning science of art) and present a new set of aesthetic concepts. They strive to fill the causal void that 'genius' meant for earlier aesthetic theories by no longer linking artists' creativity to their talent, but to 'primitive thinking,' a capacity that – in contrast to the notion of 'genius' – presented no mysteries inasmuch as the human sciences had investigated it empirically and declared that it might still be found in the core essence of any and every human.[44] Similarly, scholars of art adopted theories of 'primitive language.' Thus, the literary historian Alfred Biese, in *Philosophie des Metaphorischen* (Philosophy of the Metaphorical, 1893), expresses the conviction that metaphor represents a survival of primeval language whose words are directly motivated by the objects they signify. Accordingly, poetry and the literary arts in general (*Dichtung* in German) afford privileged access to the world-in-itself.[45]

---

41 Edward Burnett Tylor, *Primitive Culture: Researches into the Development of Mythology, Philosophy, Religion, Art, and Custom* (London: Murray, 1871), 2: 404.
42 Ernst Kretschmer, *A Text-book of Medical Psychology* (London: Hogarth Press, 1952), 134–138.
43 Gustav Friedrich Hartlaub, *Der Genius im Kinde: Zeichnungen und Malversuche begabter Kinder* (Breslau: F. Hirt, 1922), 30.
44 And with that, as Leeb notes, art is declared a "generic feature" of the human essence (Leeb, *Die Kunst der Anderen*, 12).
45 Alfred Biese, *Die Philosophie des Metaphorischen In Grundlinien dargestellt* (Leipzig: L. Voss, 1893). On the fascination with 'primitive thinking' held by theories of language and metaphor, see Chapter 6; key points of reference are Wolfgang Riedel's essays, "Arara ist Bororo oder die metaphorische Synthesis," in *Anthropologie der Literatur. Poetogene Strukturen und ästhetisch-soziale Handlungsfelder*, ed. Rüdiger Zymner and Manfred Engel (Paderborn: Mentis, 2004); and "Archäologie des Geistes. Theorien des wilden Denken um 1900," in *Das schwierige*

In this way, recourse to the figure of the 'primitive' enabled scholars to find a new social function and justification for art, that is, as a way to reconnect with origins, gain acccess to a different kind of thinking, and explore its potentials. Artists did the same by presenting themselves as the spiritual kinsmen of indigenous peoples and other figurations of the 'primitive.' As the artist Paul Klee succinctly puts it,

> For these are primitive beginnings in art, such as one usually finds in ethnographic collections or at home in one's nursery. [...] Parallel phenomena are provided by the works of the mentally diseased; [...]. All this is to be taken very seriously, more seriously than all the public galleries, when it comes to reforming today's art. [This is how far we must go in order to not simply become antiquated.][46]

The 'Primitive' as Literary Utopia

The generation weary of civilization addressed by Weber gladly took up the figure of the 'primitive.' Or, to put it the other way around, the idealizing or utopian primitivism that flourished in the arts from the 1910s to the 1930s expressed the same yearning for (re)enchantment that he notes.[47] Indeed, the arts played a leading role in bringing about a primitivist perspective meant to counter the ills of civilization. This was due first to the dependence of the phantasm of the 'primitive' on aesthetic procedures to illustrate and animate it. Ben Etherington has recently stressed this point: "Owing to its inherently speculative nature, the wish for the primitive could only be realized in the kinesis of aesthetic mak-

---

neunzehnte Jahrhundert, ed. Jürgen Barkhoff, Gilbert Carr, and Roger Paulin (Tübingen: Niemeyer, 2000); as well as Werkmeister, *Kulturen jenseits der Schrift*, 197–247.
46 Paul Klee, *The Diaries of Paul Klee: 1898–1918*, ed. Felix Klee (Berkeley: University of California Press, 1964), 266. The bracketed sentence of this quote was translated directly from the German text because it was not included in the published translation.
47 Joyce S. Cheng points out the difference between "modernist primitivism of the prewar period" (in Cubism, for example) and the "dialectical and subversive primitivisms of the late 1920s and 1930s," which "did not view aesthetic claims as ends unto themselves so much as tools for critical thinking. The latter discerned the possibility, in contemporary criticism, to dismantle the foundations of Western, post-Enlightenment humanism, especially its legitimation of Reason." Now, "epistemological authority" was lent to "modes of experience" that traditionally had been viewed as the "Other of Reason: childhood, dreams, hallucinations, trance states, and madness" ("Primitivismen," in *Neolithische Kindheit. Kunst in einer falschen Gegenwart*, ed. Anselm Franke and Tom Holert [Zurich: diaphanes, 2018], 185). However, this turn – as I show below – had been made previously in the discourses of the human sciences and study of art, as well as in certain works of literature.

ing."[48] Second, as already noted, artistic activity was deemed a survival of 'primitive thinking.' Whether contemplating or producing art, one already seemed to stand closer to the 'primitive' than in other forms of activity. Finally, those inspired by primitivism considered it their mission to make the primitivistic utopia of a reenchanted world a reality by artistic creation. In a literary context, this meant *bringing about* 'primitive thinking,' its particular relationship to (figurative) language, and the transformation from a disenchanted to a reenchanted world, *in* and *as* literature – and by doing so to potentially reshape reality as well.

Insofar as these literary texts often projected *possible* worlds, thereby situating the 'primitive' in a space of open potential, it is correct to describe them as utopias, as Etherington does. Nevertheless, it is important to note that these literary utopias did not represent an exclusively affirmative stance. Rather, as I will demonstrate in later chapters, they were marked by a high degree of ambivalence, for example states of anxiety and self-loss, as portrayed by Robert Müller (see chapter 7). Furthermore, they head in different directions, in keeping with the different figurations of the 'primitive.' That is, they needed not straightforwardly imagine a different kind of society, but could just as well envisage aesthetic or epistemic utopias that simply concerned other ways of seeing the world. And while the primitivist longing for a magical state of the world often focused on indigenous cultures or a supposedly archaic way of life, it was equally characteristic of them to seek out the 'primitive' among children or to understand it as a hidden *other side* of oneself, as the following brief examples, not all of which can be treated in the book, intend to show. For instance, this other side was understood as an archaic stage of development, reactualized in states of frenzy or intoxication. This is the case in Alfred Kubin's dystopian novel, *Die andere Seite* (1909; *The Other Side,* 1969), where the god/dictator Patera gathers people with deviant behaviors in a "Dream Empire" in the Far East. They regress to an archaic level through collective states of intoxication and annex indigenous practices that open a different perspective onto the world.

Others claimed that modern Europeans had not grown out of, but only neglected their inborn capacity to think differently. By pushing it aside, it follows, they prevented further development of such thinking and forced it to *become* 'primitive' (i.e., undeveloped) in the first place. In this case, the primitivistic utopia involved reactivating this atrophied ability, cultivating it, and using it to improve European civilization. This is the case, for example, in Robert Musil's novelistic reflections on the so-called "other condition," which I deal with in great

---

48 Ben Etherington, *Literary Primitivism,* xiv.

detail in chapter 8. Closely related to this kind of primitivism is also the "magical Surrealist art."[49] The manifesto André Breton wrote for the movement (*Manifeste du Surréalisme*, 1924) levels a withering critique at dogmatic rationalism, which leaves no room for the imagination and other parts of the psyche. Breton and his associates set out to sound the depths of these stifled aspects of mental life, appealing to the well-known constellation of children, the insane, and archaic societies to foreground other possible ways of seeing the world. By fusing waking life and dream states as well as conscious and unconscious registers of meaning, they aimed to bring forth a higher plane of experience that would capture the "marvelous" essence of the world and make it visible.[50]

By contrast, Carl Einstein, an early proponent of primitivism in *Bebuquin oder die Dilettanten des Wunders* (1912; *Bebuquin, or the Dilettantes of the Miracle*, 2017) and the forerunner of Dada and Surrealism, came in his late work to reject in no uncertain terms the idealist notion that the world could be changed by means of the imagination.[51] Looking back in *Die Fabrikation der Fiktionen* (The Fabrication of Fiction, 1973 [but written 1933–1934]), he writes self-critically,

---

[49] André Breton, "First Manifesto of Surrealism," trans. Richard Seaver and Helen Lane, in Breton, *Manifestos of Surrealism* (Ann Arbor: University of Michigan Press, 1990), 29.

[50] Breton, "First Manifesto of Surrealism," 14–18. Somewhat unimaginatively, Breton's novel *Nadja* (1928) enlists the exoticizing cliché of a mysterious woman who opens the way for the male narrator to enter primitivist surreality through the transports of inspired madness. The author/narrator deems Nadja – who sees herself as the "mythological character" of "Melusina" (and calls Breton [or his alter-ego] "a god" or "the sun") – "one of those spirits of the air which certain magical practices momentarily permit us to entertain," and she becomes his muse (*Nadja*, trans. Richard Howard [New York: Grove, 1960], 106, 111). Through her mediation, the author/narrator experiences a fusion of dream and reality, a mode of thinking and perception guided by the power of imagination. As is typical for so many works since the age of Romanticism, these flights occur at the expense of the muse herself. (On this common fate of inspiring women, see Elisabeth Bronfen, *Over Her Dead Body: Death, Femininity and the Aesthetic* [Manchester: Manchester University Press, 1992]). At the end of the tale, the narrator abandons her when she is admitted to an institution. This turn may be read as a critique of the utopia of another kind of thinking or as a rebuke to the society that pathologizes such logic (which is Breton's intention). However, one can also see it as an admission of guilt on the part of the male author/narrator, who encourages his muse to join him in delusional journeys until they finally do her in. (Breton already anticipates and seeks to defend himself against charges of this kind. Nevertheless, as Karl Heinz Bohrer writes, critics early on faulted the author for "aesthetic, indeed factual, cannibalism (Roger Shattuck), … a reproach subsequently reformulated as inhumanity (Peter Bürger)" ("Nachwort," in André Breton, *Nadja*, trans. Bernd Schwibs [Frankfurt am Main: Suhrkamp, 2002], 153).

[51] Here, one sees clear proximity to the so-called Expressionism debate. Einstein's position is related to that of Lukács – whose prominent opponents included Ernst Bloch (discussed below).

Liberal nominalists, we thought to change the world with signs alone. Ambiguous pictures had made us forget Being, and we believed that a change in imagery would bring about the actual transformation of Being. We had sunk back to archaic magic and infantile fantasy. (The delusion that fiction should provide the starting-point and primary elements of Being. The gap between private, mythical notions and the collective reality at our disposal couldn't be bridged.)[52]

Einstein's critique evokes the primitivist utopia articulated in the works of Breton and Kubin: being able to change the world by simply thinking it or imagining it differently. And his words make something else plain, too: he accuses his younger self and his primitivist contemporaries of having succumbed to mythical thinking in their benighted quest for the 'primitive.' Instead he now casts his support for art serving the proletarian collective and ultimately for the necessity not of dreaming, but of political battle. Einstein says, in other words, that the assumption that reality can be changed through the power of imagination was *already* magical thinking. Thus, in these beliefs magical thinking *had already returned*, even though people still thought they had to seek it out.

At the same time, even in voicing this critique, Einstein still adheres to the primitivistic conviction that art has affinities with the 'primitive': "Something exceptionally conservative seems inherent to art, because vigorous art always takes us back to ancient strata."[53] In other words, for Einstein, surrealist "expectations of art" stem from "its magical-religious prehistory. Under modern conditions, its heightened claims are atavistic."[54] Thus, while Einstein accuses the surrealist movement of having succumbed to mythical thinking, his own critique is itself still located in the paradigm of the 'primitive.'

The Presence of the 'Primitive' in Disenchanted Modernity

With Einstein's remarks in mind, let us return to Weber's "Science as a Vocation." In a "disenchanted world," it seems, everything can be calculated and controlled. Weber writes that "mysterious, unpredictable forces"[55] no longer exist for modern

---

52 Carl Einstein, *Die Fabrikation der Fiktionen*, ed. Sibylle Penkert (Reinbek bei Hamburg: Rowohlt, 1973), 32.
53 Carl Einstein, "Berliner Vortrag über den Surrealismus," in Sibylle Penkert, ed., *Carl Einstein. Existenz und Ästhetik. Einführung mit einem Anhang unveröffentlichter Nachlaßtexte* (Wiesbaden: Steiner, 1970), 61; quoted in Katrin Sello, "Zur 'Fabrikation der Fiktionen,'" in Carl Einstein, *Die Fabrikation der Fiktionen*, ed. Sibylle Penkert (Reinbek bei Hamburg: Rowolt, 1973), 365.
54 Sello, "Zur 'Fabrikation der Fiktionen,'" 365.
55 Weber, "Science as a Vocation," 13.

men and women. Yet it is clear that Weber himself perceives his own times quite differently when he speaks of how "numerous gods of yore, divested of their magic and hence assuming the shape of impersonal forces, rise from their graves, strive for power over our lives, and resume their eternal struggle among themselves."[56] Weber made this declaration in the middle of the First World War, specifically with an eye toward the "conflict" about the "*value* of French and German culture,"[57] which in his estimation did not concern science, economy, or politics so much as an existential "destiny."[58] For "a millennium," Weber asserts, "reliance on the glorious pathos of the Christian ethic had blinded us" to this mythical reality, that is, to the fateful necessity of an overriding struggle commanding human existence, whether one attributes it to "impersonal forces" or "gods."[59] Weber's demand "to look the fate of the age full in the face"[60] then means both to confront the condition of transcendental homelessness and disorientation and to choose one god over another without a sure bearing.

> As long as life is left to itself and is understood on its own terms, it knows only that the conflict between these gods is never-ending. Or, in nonfigurative language, life is about the incompatibility of ultimate *possible* attitudes and hence the inability to ever resolve the conflicts between them. Hence the necessity of *deciding* between them.[61]

---

[56] Weber, "Science as a Vocation," 24. According to Friedrich H. Tenbruck, "this declaration," which "most sociologists [would] consider utter nonsense" shows that the late Weber was convinced that "we are taking leave from thousands of years of history" shaped by "an ethically unified way of life" backed by religion. "Modern man calls only by abstract names on the multiplicity of deities," who "engage in interminable battle on the social and political stage"; now, "decision is only possible through fanaticism," and "moderation takes the form of conformity or apathy. [This is] the point where Weber's sociology breaks down" (412–413). Hans-Peter Müller notes that here, "Weber relies on John Stuart Mill's formula of the polytheism of values and the eternal battle of the gods" (Hans Peter Müller, "Wissenschaft als Beruf," in *Max Weber Handbuch*, ed. Hans-Peter Müller and Steffen Sigmund [Stuttgart: Metzler, 2020], 263). Matthias Bormuth refers to this as Weber's "antique-Christian metaphorics" ("Max Weber im Lichte Nietzches," in Max Weber, *Wissenschaft als Beruf: Mit zeitgenössischen Resonanzen und einem Gespräch mit Dieter Henrich*, ed. Matthias Bormuth [Berlin: Matthes and Seitz, 2018], 26–27).
[57] The backdrop here includes the nationalistic and implicitly racist assumption, widespread in Germany after the First World War, of conflict between "civilization" (largely "French" in the nineteenth century and increasingly "English/North American" in the twentieth), which was thought to be "superficial" and "alienated," and German "culture," which counted as "deep" and "authentic"; cf. Thomas Mann's *Betrachtungen eines Unpolitischen* (1918; *Reflections of a Nonpolitical Man*, 1983) or Werner Sombart's *Händler und Helden* (1915; Merchants and Heroes).
[58] Weber, "Science as a Vocation," 23.
[59] Weber, "Science as a Vocation," 24.
[60] Weber, "Science as a Vocation," 24
[61] Weber, "Science as a Vocation," 27.

Thus, in Weber's own experience it is *not* predictability that characterizes disenchanted modernity. Instead, it is the feeling to be once again at the mercy of "impersonal forces," which, because they surpass human understanding, are experienced as ineluctable fatality. No possibility for controlling circumstances exists, only the bare necessity of opting for one of two mutually hostile poles of life; the result can only be tragic insofar as "serving [one] particular god [...] will *give offense to every other god.*"[62]

Weber thus diagnoses disenchantment, and at the same time – albeit without reflecting on the matter along dialectical lines – he enacts its turn into a mythically experienced world.[63] In this light, the rising generation's longing for transcendental sources of comfort and reassurance is entirely logical. Calls for prophecy and the balms of religious community speak volumes about how these young people saw their world: however "disenchanted" it may have seemed, the world had become mythical to them. Notably, from this it follows that the 'primitive' was simultaneously relevant on two registers: as both an *archaic* utopia and as an image of the *present*. The 'primitive' had not come and gone – and this is a crucial point – it was extant in the here and now as the *presence and reality of modernity itself*.

This presence of the 'primitive' was conceived in three different ways at the time. A model inspired by depth psychology saw it as the collective return of remnants from an unfinished past. A second, synchronistic, and anti-evolutionary model posited the co-presence of 'modern' and 'primitive' forces in human existence and therefore questioned the modern self-image.[64] Finally a dialectical

---

[62] Weber, "Science as a Vocation," 26.

[63] Hartmut Lehmann refers to this passage when looking for evidence that Weber's work "come[s] very close to the idea of re-enchantment" (*Die Entzauberung der Welt*, 14). Thus, according to the recollections of Karl Löwith, who had attended the lecture, Weber's appearance and his discourse exuded the "somber glow" of a "prophet" (qtd. in Matthias Bormuth, "Max Weber im Lichte Nietzsches," in *Wissenschaft als Beruf: Mit zeitgenössischen Resonanzen und einem Gespräch mit Dieter Henrich* [Berlin: Matthes und Seitz, 2018], 7). See also Bormuth's question of whether Weber himself had also given "catheter-prophecies" (19), as well as his numerous references to Weber's pathos-laden style (e.g., 27).

[64] Current scholarship in the fields of cultural and literary history argues along similar lines. Hartmut Böhme, for instance, claims that it is necessary to incorporate "magic" into the theory of modernity, that is, to not leave today's "fetishes, idols, and cultic forms" out of the equation (*Fetischismus und Kultur: Eine andere Theorie der Moderne* [Hamburg: Reinbek, 2006], 23). Iris Därmann concludes her study by calling for scholars to take not only European tradition into account, but also to "consult foreign philosophies" in equal measure (*Fremde Monde der Vernunft: Die ethnologische Provokation der Philosophie* [Paderborn: Fink, 2005], 725). Erhard Schüttpelz voices a plea for a perspective that connects to Latour's thesis that "We have never been modern": if "the moderns" have conceived "the foreignness of worlds outside

model viewed the presence of the 'primitive' as the mythological result of a misguided Enlightenment. Here, the 'primitive' no longer represents modernity's origin but its present, no longer constitutes modernization's opposite but its consequence. The 'primitive' is no longer seen as original but mediated, not as nature but culture, not as imago but reality. In this sense in 1934 Theodor W. Adorno writes of the "coincidence of the modern with the archaic" to Walter Benjamin:

> I have come to realize that just as the modern is the most ancient, so too the archaic itself is a function of the new: it is thus first produced historically as the archaic, and to that extent it is dialectical in character and not 'pre-historical,' but rather the exact opposite.[65]

For Weber, the presence of the 'primitive' had led to a resigned mythical worldview, in which people were caught between warring powers with only the possibility of making decisions that would ultimately prove tragic.[66] Einstein, the erstwhile celebrant of primitivism confronting the rise of fascism in Europe some ten years later, responded to the same circumstance by adopting a cooler attitude toward art, which he hoped would stand in the service of the proletarian collective and communist revolution. Still others, e.g., Robert Müller and Gottfried Benn (see chapter 7), used the ongoing vigor of the 'primitive' to affirm a barbarian status quo: the 'will to power' is simply the 'essence' of human existence and society.

Along these lines, but from the opposing political standpoint, Alfred Döblin indicts the "false primitivity" of National Socialism in his essay, "Prometheus und das Primitive" (1938, Prometheus and the Primitive). Cold-blooded pursuit of power, he argues, leads to "estrangement and brutalization" in social relations: "Here we have barbarism as the result of a denatured, Promethean im-

---

their own through a cosmological time-barrier, other cultures and societies" have done the same (Schüttpelz, *Die Moderne im Spiegel des Primitiven*, 410).

[65] Theodor Adorno to Walter Benjamin, 5 April 1934, in *The Complete Correspondence, 1928–1940*, ed. Henri Lonitz, trans. Nicholas Walker (Cambridge, MA: Harvard University Press, 2000), 38. Christopher Bracken in his reading of this quote stresses the inscription of the archaic (or magical) in modern forms of critique; thus, "Benjamin suggests that a properly 'magical' criticism does not decipher the meaning of the artwork. Instead it brings it back to life. Interpretation is therefore animation" (*Magical Criticism: The Recourse of Savage Philosophy*. [Chicago: The University of Chicago Press, 2007], 17). As will be shown in Chapter 9, the dialectical moment goes missing in this account.

[66] Cf. Wolfgang Mommsen: "The ultimate message of Max Weber's sociology was resignation; he offered no answer to the great ethical questions. The demagification of the world, the universal process of rationalization, which Weber described and fatefully affirmed, resulted ironically in the emergence of a new irrationalism" (*Max Weber and German Politics: 1890–1920*, trans. Michael S. Steinberg [Chicago: Chicago University Press, 1984], 66).

pulse."⁶⁷ Indeed, Döblin originally intended to call his piece "Das wahre und das falsche Primitive" (The True and the False Primitive).⁶⁸ The "false primitive," which is synonymous with "barbarism," emerges when man's excessive control over nature switches over into rule by violence; its corollary is an "illusory mysticism" (*Scheinmystik*) that glorifies the "absolute state."⁶⁹ This barbarism invokes "origins" only to legitimize its rule; its goal is appropriation. Döblin counters such a vision with a utopian appeal to the "true primitive." Such a different turn to "origins" would involve readiness to give one's self up. This move would even be required inasmuch as the primal state, as Döblin imagines it, occupies a space *before* individuation.⁷⁰ Encounter with the other – whether a foreign people or nature itself – is then a matter of participation, not acquisition.⁷¹

Whereas Döblin only hints at the need to preserve the critical and utopian impulse against the negative presence of the 'primitive,' the philosopher Ernst Bloch makes it a political demand. In essays written between 1929 and 1935, most of which are collected in *Erbschaft dieser Zeit* (1935; *Heritage of Our Times*, 1991), Bloch endeavors to think in dialectical terms, that is, to wrest a utopian 'primitive' from the archaic 'primitive,' thereby saving it from the fascists.⁷²

---

67 Alfred Döblin, "Prometheus und das Primitive (1938)," in *Schriften zur Politik und Gesellschaft* (Freiburg: Walter, 1972), 364.
68 Döblin, Note on "Prometheus und das Primitive," *Schriften zur Politik und Gesellschaft*, 508.
69 Döblin, "Prometheus und das Primitive," 365.
70 Döblin, "Prometheus und das Primitive," 349.
71 This view finds expression in the first book of the *Amazonas* trilogy, which Döblin wrote in the years immediately preceding the essay's publication. Here, the author crafts a mythical narrative of his own (on this, see Vera Hildenbrandt, *Europa in Alfred Döblins Amazonas-Trilogie: Diagnose eines kranken Kontinents* [Göttingen: Vandenhoeck & Ruprecht, 2011], 209–233), "transforming himself," as Jorge Luis Borges would write in a 1938 review, "into his own creatures" (Borges, "Die Fahrt ins Land ohne Tod, de Alfred Döblin," quoted in the afterword to Alfred Döblin, *Amazonas. Romantrilogie*, 234). The same dynamic is also at work in the events narrated. Thus, the first European soldiers to penetrate the jungle adopt the ways of its inhabitants to such a degree that their captain, worried they are losing their European identity, remarks that "it is their attachment to the brown-skinned people among whom the soldiers live, the life with the animals and on the water; and the priests are quite right: it's like the people are under the spell of the [natives]" (Döblin, *Amazonas*, 175. See Werkmeister, *Kulturen jenseits der Schrift*, 312).
72 Bloch was not alone in such an effort. Thus, only a few years later in France, members of the Collège de Sociologie undertook something similar: "The forms of a mythopoiesis constituting the world, as reconstructed in cultural-philosophical and ethnological research, were supposed to wrest the force of the collective from the hands of fascists" (Franke and Holert, "Einführung," in *Neolithische Kindheit*, 10).

In the essay "Zusammenfassender Übergang" (1935, "Summary Transition," 1990) Bloch declares, "Not all people exist in the same Now." A materialist and teleological view of history prompts him to locate certain social groups (e. g., peasants, lower-level office staff) in a temporal dimension that is out of sync with that of other groups: "depending on where someone stands physically, above all in terms of class, [there] he has his times."[73] Bloch stresses that *"real non-contemporaneity"*[74] stands at issue here – anachronistic modes of production and consciousness. For him, this anachronism is also responsible for the masses of clerks and other low-level employees turning to National Socialism, not to Communism, in the late Weimar Republic: "Impulses and reserves from pre-capitalist times and superstructures are then at work, [...] which a sinking class revives [...] in its consciousness."[75] Bloch describes such a return of what belongs to the distant past by way of paradigm of the 'primitive,' thereby participating in the allochronistic discourse of primitivism. For example, he notes that the "excess" of nationalism in his time calls to mind an "atavistic 'participation mystique,' of the attachment of the primitive man to the soil which contains the spirits of his ancestors." He also notes an "orgiastic hatred of reason, [...] in which – with a non-contemporaneity which becomes extraterritoriality in places – negro drums rumble and central Africa rises."[76]

For Bloch, then, the present of the interwar period had fallen under the spell of a form of the 'primitive' that displays all the negative traits warned about in evolutionist theories of the late nineteenth century: regression, irrationality, violence, and superstition. This primitivist superstructure, he argues, is unsuited to resolve the real contradictions of social life; it only obscures the "rift" "between the non-contemporaneous contradiction and capitalism."[77] Nevertheless, Bloch hopes to turn things around. For one, he does not wish to abandon the "'archaically' anticapitalist" forces to the fascists.[78] Secondly, he insists that the "not yet Past" – in other words, the 'primitive' – still harbors "subversive and utopian el-

---

73 Ernst Bloch, "Summary Transition: Non-Contemporaneity and Obligation to its Dialectic," in *Heritage of Our Times*, trans. Neville and Stephen Plaice (Berkeley: University of California Press, 1990), 97.
74 Bloch, "Summary Transition," 106.
75 Bloch, "Summary Transition," 105–106.
76 Bloch, "Summary Transition," 102. He also notes "non-contemporaneous" phenomena emerging from "even 'deeper' backwardness, namely from *barbarism*"; now, "needs and resources of olden times consequently break through [...] like magma through a thin crust," summoning forth the "darkest primitivization, of a totally non-contemporaneous, indeed disparate insanity," which might also be called "anachronistic degeneration" (107, 109).
77 Bloch, "Summary Transition,"110.
78 Bloch, "Summary Transition," 113.

ements,"⁷⁹ and that it is up to the coming "socialist revolution" to "recover[]" that "primitiveness concretely."⁸⁰

Thus, in Bloch's thinking, the 'primitive' bears not only a reactionary but also a progressive signature.⁸¹ He understands the 'primitive' as the product of a fertile imagination in a time of crisis, in which the yearning for the archaic is actually the desire for a better future.⁸² According to Bloch, the 'primitive' (as a figure for "the 'real' *new Adam*"⁸³) does not come from the past; though related to backward elements ("real non-contemporaneity") in society, it represents a task for the future.⁸⁴ However, in order to "recover" this 'primitive' potential from the unredeemed past, its utopian elements must first be freed from their "banishment" to false archaism.⁸⁵ Therefore its substance must not be grasped irrationally, but with the "new, more concrete rationalism" of socialism.⁸⁶ Whether or not the utopian content of the 'primitive' can be secured is thus ultimately a class question: "only a class with a future can use the 'distant fragrance of the *horizon*' and the 'images' which stand in it, and blast out the encapsulated element: namely the future significance of the images encapsulated into an undischarged past."⁸⁷

---

79 Bloch, "Summary Transition," 114. See also 115. For support, he invokes Karl Marx, according to whom the "'social childhood of humanity'" represents a "stimulus" that capitalism has not quieted (114).
80 Bloch, "Imago as Appearance from the 'Depths': Romanticism of Diluvium," in *Heritage of Our Times*, trans. Neville and Stephen Plaice (Berkeley: University of California Press, 1990), 317.
81 Its "emergence" occurs "in all times of genuine revolution" (Bloch, "Philosophies of Unrest, Process, Dionysus," in *Heritage of Our Times*, 316).
82 Bloch's essay, "Die Felstaube, das Neandertal und der wirkliche Mensch" (The Rock Dove, the Neanderthal, and the Real Human, 1929) in *Literarische Aufsätze. Werkausgabe* (Frankfurt am Main: Suhrkamp, 1965), makes this point particularly clear. Many contemporaries (e.g., Ludwig Klages, Carl Jung, Edgar Dacqué, and Gottfried Benn), he argues, want to go back "into primeval forests," which, however, have never really existed (462). In fact, Bloch stresses, cultivated individuals at all times and places have always seen their "negative image" in people living in a supposed state of nature or irrational Dionysian transports. Such images have never said anything about primordial reality or nascent humanity at all; instead, they bear witness to their own day, at times when the wish for a different future has arisen.
83 Bloch, "Die Felstaube, das Neandertal und der wirkliche Mensch," 468.
84 "[It is meaningless] to look for a *fact* [*ein Gegebenes*] when, from the beginning, it was a task [*ein Aufgegebenes*]" (Bloch, "Die Felstaube, das Neandertal und der wirkliche Mensch," 469).
85 The "utopia of the first 'beginning' seeks to escape from [...] mere 'primeval times'" (Bloch, "Final Form: Romantic Hook-Formation," in *Heritage of Our Times*, 150).
86 Bloch, "Loch Ness, die Seeschlange und Dacqués Urweltsage," in *Literarische Aufsätze*, 470.
87 Bloch, "Imago as Appearance," 308. Etherington (*Literary Primitivism*, xv, 8–9, 33) overlooks this matter when he invokes Bloch's "nonsynchronicity," failing to note his historical materialist

I have reported on Bloch's thought in such detail to demonstrate two important aspects of 1920s–1930s primitivist discourse: First, the 'primitive' of the present and the critical-utopian 'primitive' do not simply oppose each other as dichotomies; they are dialectically intertwined. Second, even this dialectical perspective retains the standing paradigm of the 'primitive' by assuming non-simultaneity between different populations or by thinking of a better future under the spell of the 'primitive.'

So far, I hope to have shown that the 'primitive' held an abiding fascination for European modernity as a narrative of origin, an instrument for critiquing civilization, a literary utopia, and a diagnosis of the present. While doing so, I have also reconceptualized the category of the 'primitive' as a scientific paradigm and a figure of thought. In what follows I will discuss the relationship between my literary and scholarly sources, thereby proposing a third reconceptualization of the category of the 'primitive' as a scientific *poème*.

Two Cultures: The 'Primitive' as *Poème*

In this study, I understand primitivism as a discourse that, on the one hand, was *produced by* certain texts, disciplines, and artistic practices and, on the other, itself *generated* the latter insofar as it shaped their questions, answers, scenarios, and blind spots. Even those texts from the late 1920s and 1930s that already take a critical look at the widespread fascination with the 'primitive' remain trapped in this very same paradigm.[88] In this book I therefore adopt the more removed perspective of discourse analysis. The task is not to substantiate, condemn, or rehabilitate the category of the 'primitive.' Rather, the study at hand seeks to understand the function of this phantasm for the thinking of modernity.

---

understanding of history. For Bloch, "remnants" are positive only insofar as they, however false or distorted, point to a future that is not *anti*- so much as *post*-capitalist.

88 This is also the case for later critics from the ranks of ethnology – for instance, Lévi-Strauss, who remains within the paradigm of the 'primitive' in spite of the critique he makes of it (cf. Francis L. K. Hsu, "Rethinking the Concept 'Primitive,'" *Current Anthropology* 5, no. 3 [1964]: 169–178; Adam Kuper, *The Invention of Primitive Society* [London: Routledge, 1988]; and Jacques Derrida, *Of Grammatology*, trans. Gayatri Chakravorty Spivak [Baltimore: Johns Hopkins University Press, 1997], 95–140). It also applies, as Victor Li has convincingly demonstrated, to the neo-primitivistic anti-primitivism of postmodern theories, which are supposed to have abandoned the concept of the 'primitive' but hold fast to the phantasm of the 'radical Other'; see Li, "Primitivism and Postcolonial Literature," in *The Cambridge History of Postcolonial Literature*, ed. Ato Quayson (Cambridge: Cambridge University Press, 2012), and *The Neo-Primitivist Turn: Critical Reflections on Alterity, Culture, and Modernity* (Toronto: University of Toronto Press, 2006).

When I say discourse analysis, this discourse includes literary as well as scientific sources – for both participate equally in the discourse of primitivism. Although I attend to both literary and scholarly texts under the single heading of the 'primitive,' it is important to note that by the turn of the century it was already standard practice to distinguish between the "two cultures" of natural scientists and literary intellectuals.[89] The literary and scientific sources treated in this book try to do justice to these two different horizons and expectations. As Nicolas Pethes has shown, different conceptions of knowledge prevailed in each "culture":[90] the "natural-scientific idea of empirically verifiable hypotheses" stood opposed to the "hermeneutic concept of historical understanding." Each realm also had its own methods and aims: empirical analysis and schematic representation with the goal of formulating general laws in the natural sciences versus describing the singularity of phenomena to account for individual expression and aesthetic autonomy in literature.[91]

In order to mark their distance from the *belles-lettres* and secure their status as indispensible disciplines, the young and inchoate fields of ethnology and psychology in particular, but also the study of language and the theory of art, sought to acquire the air of the 'hard sciences' wherever possible. To take just one example, Ernst Grosse's *Die Anfänge der Kunst* (1894; *The Beginnings of Art*, 1897) proposed to sound the laws governing the "nature and life of art" by empirical means starting "at the bottom" – that is, with the study of artifacts made by 'children of nature' (*Naturvölker*).[92] For the same reason, researchers in the fields of developmental psychology and psychopathology conducted laboratory experiments, collected data, and wrote up case histories in order to demonstrate their rigor. Conversely, literature of the period sought to affirm its integrity by distinguishing itself from scientific culture. Robert Musil, for instance, made a point of holding academic psychology at arm's length in order to plumb the depths of

---

[89] This turn of phrase was coined in 1959 by the English literary historian Charles Percy Snow in a lecture of the same name: "I believe the intellectual life of the whole of western society is increasingly being split into two polar groups" (*The Two Cultures* [Cambridge: Cambridge University Press, 1959], 4) consisting of literary intellectuals and natural scientists. Of course, this thesis prompted a counter-reaction – e. g., on the part of F.R. Leavis, who pointed out that his colleague was overlooking how scientific-technological culture also belongs to the humanist sphere in the West (Frank Raymond Leavis, *Two Cultures? The Significance of C.P. Snow* [London: Chatto & Windus, 1962]).
[90] On the prehistory of the debate, see Nicolas Pethes, "Literatur- und Wissenschaftsgeschichte. Ein Forschungsbericht," *Internationales Archiv für Sozialgeschichte der deutschen Literatur* 28, no. 1 (2003): 186–191.
[91] Pethes, "Literatur- und Wissenschaftsgeschichte," 182, 195.
[92] Ernst Grosse, *The Beginnings of Art* (New York: D. Appleton and Company, 1987), 18, 20.

the human soul and the emotions that rage within it in ways that refuse schematization.

Despite scholars' efforts to prove the scientific nature of their fields of endeavor, however, the constructed and imaginary quality of the 'primitive' is obvious. To use the term put into circulation by Gaston Bachelard, it stands at the heart of an epistemological "reverie," that is, a fiction of origination shaped by the affects, needs, and ideas of scientists that is largely devoid of scientific basis. Bachelard writes in *The Psychoanalysis of Fire*,

> We are going to study a problem that no one has managed to approach objectively, one in which the initial charm of the object is so strong that it still has the power to warp the minds of the clearest thinkers and to keep bringing them back to the poetic fold in which dreams replace thought and poems [*poèmes*] conceal theorems.[93]

In contrast to the *théorème* and "objective thought" in general, Bachelard defines the *poème* through its affinities with creative language, dreaming, and "[childlike] wonder."[94] What the *théorème* accomplishes in the field of objective science, the *poème* performs in the realm of a type of pre-scientific knowledge that is encumbered by affective and unconscious impulses. Bachelard also likens it to a "psychology of primitiveness."

> In order to construct a psychology of *primitiveness* it is sufficient, then, to consider an essentially new piece of scientific knowledge and to follow the reactions of non-scientific, ill-educated minds that are ignorant of the methods of effective scientific discovery.[95]

While "constrained by the thought of his times" to apply the scheme of the "primitive,"[96] Bachelard's view also subverts commonplace notions because it ascribes primitiveness to European scholars. In other words, in fashioning the *poème* of the 'primitive,' these scholars themselves exhibit, in Bachelard's terms, a "primitive" mindset.

Bachelard's talk of scientific reverie draws the human sciences into the orbit of literature and sheds light on their rhetorical features. As I will show, considerations of genre (e. g., case history, diary, travelogue, mythical tale) in scientific texts are particularly significant here, as are patterns of argument – for instance,

---

[93] Gaston Bachelard, *The Psychoanalysis of Fire*, trans. Alan C. M. Ross (Boston: Beacon, 1968), 2.
[94] Bachelard, *The Psychoanalysis of Fire*, 1.
[95] Bachelard, *The Psychoanalysis of Fire*, 25.
[96] Erich Hoerl, *Sacred Channels: The Archaic Illusion of Communication*, trans. Nils F. Schott (Amsterdam: Amsterdam University Press, 2018), 36.

analogy, appeals to origins, and levels of meaning extending from the literal to the figurative. Also, an analytical and argumentative approach is often combined with narrative stylization. Finally, due notice will be taken of tropes and figures of speech, which not only embellish scholarly works but also shape their conceptual content.

Characteristics like these affirm the truism that literary and scientific texts alike are rhetorically determined and that there are fewer categorical boundaries to be drawn between them than the "two cultures" tradition would like. But of course this also applies the other way around. The literary texts dealt with in this book react intensely and in different ways to the sciences of their time. Basically, one can distinguish four types of models conceptualizing the relationship of literature to scientific discourse:[97] models of influence, when scientific topics appear in literature; models of reflection, when literary texts criticize the sciences for their rationalistic and abstract conceptions; models of formal relationship, when procedures such as experiment are adapted by literature and modified for its use; and finally the model of co-evolution, which understands both literature and the sciences as two forms of representation that have grown from one discursive terrain. The present study is mostly based on the fourth model. Nevertheless, these sources will also be used to investigate precisely how the two forms of discourse differ and to what extent literature may play the role of a counter-discourse.

Literary Primitivism

Until fairly recently, scholars have not dedicated much discussion to literary primitivism. This is because studies of primitivism were informed not by the history of discourse so much as by a tradition of art history devoted to formalist modes of critique. At least since Robert Goldwater's *Primitivism in Modern Painting* (1938), and certainly since William Rubin's *"Primitivism" in 20th Century Art: Affinity of the Tribal and the Modern* (1984), discussions of primitivism had centered on how European artists took the artifacts of tribal societies in West Africa and Oceania as inspiration for their own creations.[98]

---

[97] As suggested by Nicolas Pethes in "Literatur- und Wissenschaftsgeschichte."
[98] William Rubin, ed., *"Primitivism" in 20th Century Art. Affinity of the Tribal and the Modern* (New York: Museum of Modern Art, 1984). Rubin's exhibition catalog was not exempt from criticism; see, for example, James Clifford, *The Predicament of Culture: Twentieth-Century Ethnography, Literature, and Art* (Cambridge MA: Harvard University Press, 1988). In the nineteenth century, "primitive art" still referred to European art of the late Middle Ages and early Renaissance.

Needless to say, it has proven difficult to detect this form of primitivism in literature. A linguistic barrier separated Europeans from the indigenous verbal arts of the above-mentioned tribal societies. Very few writers could understand the relevant languages, and those translations that were available failed to take the full range of meaning into account (to say nothing of stylistic peculiarities). As Erhard Schüttpelz has observed, the supposed "self-evidence" of visual art stood in contrast to "language itself" in the literary realm.[99] Attempts were made to rework oral traditions from Africa in French or German (e. g., Blaise Cendrars's *Anthologie Nègre* [1921; *The African Saga*, 1927] or Einstein's *Afrikanische Legenden* [African Legends, 1925]) and to imitate indigenous lyrical forms (Einstein's "Drei Negerlieder" [Three Negro Songs, 1916] or Tristan Tzara's "Negerlieder" [Negro Songs, 1916–1917]). Yet such efforts were small in number, and they hardly comprised a movement.

More recently, however, other notions have begun to establish themselves in art history that break from the formalist conception of primitivism advocated by Rubin. They have granted more room for developments in literature analogous to those in the visual arts by broadening the scope of inquiry in terms of the definition of primitivism and the diversity of artistic approaches to it available in the early twentieth century. Scholarly literature along these lines stresses that the label of 'primitive' at the time was applied not just to tribal artifacts from West Africa and Oceania, but also to other foreign cultural products – and, significantly, to the art of medieval (or popular) tradition, children, and the mentally ill.[100] In this light, the artists of Der Blaue Reiter (The Blue Rider) group, for example, merit attention because the exhibitions organized by Wassily Kandinsky and Franz Marc (as well as the almanac documenting them) included children's, medieval, and folk art to suggest how modernists might reorient their own creative activity. (The passage by Paul Klee quoted above is representative in this regard.)

The first exhibition of primitivistic works was organized in 1910 by Roger Fry (Grafton Galleries, London), featuring Gauguin, Picasso, and others. In this context, Fry engaged intensively with children's drawings, which thereby acquired a certain predominance in the aesthetics of the avant-garde. See Roger Fry, "Children's Drawings," *The Burlington Magazine for Connoisseurs* 30, no.171 (1917).
99 Schüttpelz, *Die Moderne im Spiegel des Primitiven*, 360.
100 Karla Bilang has aptly noted, "For [these] poets and sculptors, 'primitive' is a fluctuating term, which, in keeping with its etymology ('elemental,' 'original'), applies to a vast complex of non-classical, or non-professional, forms of expression such as those from early civilizations, archaic times, and ethnographic art, as well as to naïve painting, European popular tradition, and children's creations" (*Bild und Gegenbild: Das Ursprüngliche in der Kunst des 20. Jahrhunderts* [Stuttgart: Kohlhammer, 1990], 8).

Additionally, alternative approaches to primitivism in art history stress that attraction to 'primitive' art belonged to a larger trend of interest in 'primitive cultures' in general.[101] Therefore, not only those currents of European art that took certain objects as models but also those that drew inspiration from a supposedly 'primitive worldview' should be understood as primitivistic. As Colin Rhodes puts it,

> there is a large body of Primitivist art, [...] which bears no direct relationship to primitive art – its Primitivism lies in the artists' interest in the primitive mind and it is usually marked by attempts to gain access to what are considered to be more fundamental modes of thinking and seeing.[102]

Thus, in addition to the formalist, analytical perspective exemplified by Rubin (which focuses on Cubism), other currents, such as the perception of reality extolled by Dada and Surrealism, have recently come into view.[103]

This broader understanding of primitivism is fruitful for literary study. For one, including European cultural productions eliminates the language problem: folk tales, works by schizophrenics, children's stories, and medieval writings admit comprehension more readily and lend themselves to adaptation more easily. Secondly, shifting the perspective from artifacts to cultures (or, alternatively, to 'primitive' worldviews and ways of thinking) is especially significant for literature. This is not just because of the concomitant thesis that tropological language stands at the origin of such thought, but also because this approach allows literature to play to one of its strengths, its ability to carry out and reflect on 'primitive thinking' as constructed and realized by discourses of its time.

Aside from the alternative conceptions of primitivism proposed by art historians, the opening of literary studies to the *discursive history* of the 'primitive' has also brought literary texts into view as part of a broader 'Western primitivism.' From the perspective of discursive history, the question is no longer what distinguishes primitivism in the literary and visual arts, but what relationship literary primitivism entertains with wider discourse on the 'primitive.' Thus, in her pioneering study, *Gone Primitive* (1990), Marianna Torgovnick explores how literature participates in an all-encompassing "primitivist discourse" that, while

---

101 Cf. Cheng, "Primitivismen," who finds a change in objective from the modernist primitivism of the prewar period to primitivism as an instrument of critical thinking in the interwar period.
102 Colin Rhodes, *Primitivism and Modern Art* (London: Thames & Hudson, 1994), 7.
103 As a matter of course, recent studies, such as Susanne Leeb's *Die Kunst der Anderen*, make this perspective their starting point and incorporate the insights of discourse analysis and postcolonial scholarship (22–24).

"fundamental to the Western sense of self and Other," makes "the primitive" mean "whatever Euro-Americans want it to be."[104] In turn, scholars like Erhard Schüttpelz and Sven Werkmeister have foregrounded the media history of primitivism.[105] Thus, Schüttpelz's *Die Moderne im Spiegel des Primitiven* (Modernism in the Mirror of the Primitive, 2005) shows how from the very beginning discourse about the 'primitive' has had a literary cast. Modernist texts carry on an established genre tradition exemplified by Montesquieu's *Persian Letters*, which serves the "representation of foreign experiences"[106] and features the figure of the "primitive philosopher," who "affirms and translates a foreign claim to truth."[107] Finally, Etherington's *Literary Primitivism* (2017) stresses the *speculative* impulse underlying literary constructions of the 'primitive,' that is, the fact that it is preceded by primitivism: "If we assume that ideas [...] of the primitive prefigure primitivizing idealization, we have put the cart before the horse."[108] Etherington is interested less in general features of discourse about the 'primitive,' however, than in how literary texts attempt to realize the utopian project they promise.

Primitive Thinking (an earlier version of which appeared in German in 2013 under the title *Primitives Denken. Wilde, Kinder und Wahnsinnige in der literarischen Moderne*) takes a somewhat different angle by focusing on epistemological primitivism, how it was shaped by the human sciences of the early twentieth century, and its relevance for the theory and practice of the arts at that time. The paradigm of the 'primitive' informed the search for the beginnings and essence of humanity as well as scientific interest in fundamentally other ways of thinking. These enterprises were referenced in discourse on the arts, and literature and art in general came to be seen as a contemporary preserve of 'primitive thinking.' Along the same lines, aesthetic products were elevated to a source of knowledge on human origins and treated as the first step on the utopian path of embracing the 'primitive' and rejecting or reforming modern rationality. It is no wonder, then, that self-reflection was a key characteristic of primitivistic literature of the day, as it was concerned with the question of what literature aware of its affinity with the 'primitive' could or should achieve. This also meant reflecting on the dialectics of the 'primitive,' which involved not so much origins, essence, or utopia as much as the actual historical conditions of a society that

---

104 Marianna Torgovnick, *Gone Primitive*, 8, 9.
105 Sven Werkmeister's *Kulturen jenseits der Schrift* explores primitivism as a "reflection on and of the conditions of [literary production] in the age of technological media" (27).
106 Schüttpelz, *Die Moderne im Spiegel des Primitiven*, 390.
107 Schüttpelz, *Die Moderne im Spiegel des Primitiven*, 365.
108 Etherington, *Literary Primitivism*, 7.

was about to embrace fascism. Here, too, questions arose concerning art under the sign of the 'primitive': does it pave the way for the rise of fascism (e.g., in the sense of Döblin's and Bloch's criticism described above) or does it open a space for critical reflection? Thus, realizing 'primitive thinking' in and as literature also implied questioning it and potentially constituting a counter-discourse. This, not only insofar as these literary texts claim autonomy from the general discourse on the 'primitive,' but also because they develop modes of critically engaging with modernity's longing for the 'primitive.'

The book is divided into three parts: The first (chapters 2–4) treats scientific discourse on 'primitive thinking' in ethnology, developmental psychology, and psychopathology and addresses the literary aspects of such texts, namely their uses of analogies based on temporal schemes (survival, recapitulation, and regression), of core motifs (e.g., community, play, and delusion), and of a rhetoric of 'first beginnings.' The second part (chapters 5–6) examines theories on the origins of art, language, and metaphor that make use of the human sciences' theses on the affinity between 'primitive thinking' and artistic creation. On the basis of works by Robert Müller, Gottfried Benn, Robert Musil, and Walter Benjamin, the third part (chapters 7–9) examines how, in German literature of the 1910s to 1930s, indigenous cultures, children, and the mentally ill were treated as figures of 'primitive thought' and became the starting point for imperialist and proto-fascist deliria of progress, self-critical utopias of a different rationality, sentimental fantasies of regression, and dialectical transformations of 'primitive thinking,' each accompanied by renewals of literary language and form. As much as the four authors differ in their handling of the discourse of primitivism, Müller and Benn's reactionary engagement with 'primitive thinking' contrast starkly with efforts by Musil and Benjamin to combine the critique of instrumental reason with a critique of a cult of origins that attempts to make the 'primitive' its own. It is only against the background of such a criticism that a productive engagement with the discourse of 'primitive thinking' becomes possible.

Part One: **Figures of 'Primitive Thinking'**

Chapter 2
The Ethnological Paradigm of the 'Primitive'

European ethnology in the decades around 1900 was shaped by the paradigm of the 'primitive.' The 'primitive' did not simply replace the older concept of the 'savage,' nor did it merely refer to an object of study specific to ethnological discourse. Rather, the term distilled a perspective on indigenous cultures specific to colonialist modernity.[1] As Sven Werkmeister shows, at the vanishing point of this perspective was the search for the origin of (European) culture,[2] an endeavor that was a feature of what Michel Foucault calls the "age of history," which had replaced the "age of representation" around 1800.[3] Foucault finds that "in modern thought," looking for an origin situated outside of history, as previous thinking had done, "is no longer conceivable."[4] Rather, the modern awareness of one's own historicity was precisely what now made thinking about origins necessary. The present and the time of humanity's first beginnings were no longer regarded as opposed and separate epochs, with one belonging to history and the other located outside of it. Instead, the past and present now occupied points on a single continuous spectrum and were connected by one developmental process.

The human sciences, which include ethnology, emerged against this backdrop. According to Foucault, their object of study had come into being over time: "Man […] can be revealed only when bound to a previously existing historicity." His origins appear as both distant and near, foreign and familiar, inaccessible and well known. The 'modern European' is bound to this origin by means of

---

[1] On the prehistory and transformation of the topos around 1850, cf. Sebastian Kaufmann, *Ästhetik des "Wilden": Zur Verschränkung von Ethno-Anthropologie und ästhetischer Theorie 1750–1850. Mit einem Ausblick auf die Debatte über 'primitive' Kunst um 1900* (Basel: Schwabe, 2020), 647–653; Lucas Marco Gisi, "Die Genese des modernen Primitivismus als wissenschaftliche Methode," in *Literarischer Primitivismus*, ed. Nicola Gess (Berlin: De Gruyter, 2013); Bernd Weiler, *Die Ordnung des Fortschritts. Zum Aufstieg und Fall der Fortschrittsidee in der "jungen" Anthropologie* (Bielefeld: transcript, 2006); and, for a concise summary, Li, "Primitivism and Postcolonial Literature," 984.
[2] Werkmeister, *Kulturen jenseits der Schrift*, 57–70. Werkmeister refers to Foucault, but above all to Schiller, showing that modern notions about indigenous peoples had already come into effect around 1800. On the transformation of Schiller's dictum, "They are what we were," see Nicola Gess, "Sie sind, was wir waren. Literarische Reflexionen einer biologischen Träumerei von Schiller bis Benn," *Jahrbuch der deutschen Schillergesellschaft*, 56 (2012).
[3] Foucault, *The Order of Things*, 217.
[4] Foucault, *The Order of Things*, 329.

a historical development, yet removed from it by the vast temporal abyss of the "already begun": "It is always against a background of the already begun that man is able to reflect on what may serve for him as origin."⁵ Thus, the quest for origins proves both affirmative and unsettling inasmuch as it brings into view something foreign while simultaneously constituting the very basis of the self. "The original in man," Foucault writes,

> is that which articulates him from the very outset upon something other than himself; it is that which introduces into his experience contents and forms older than him, which he cannot master. [...] It links him to that which does not have the same time as himself; and it sets free in him everything that is not contemporaneous with him.⁶

From this conception of origins arises a perspective on indigenous peoples specific to colonial modernity: the *paradigm of the 'primitive,'* which underlies the emergence and consolidation of ethnology. Under this paradigm, indigenous peoples do not receive attention as exemplars of a prehistorical condition, as had been the case for the 'savages' of the eighteenth century. Nor are they investigated for their own sake, so that researchers might learn how those societies function (as would occur in later ethnology).

Instead, at the turn of the twentieth century, ethnologists looked to indigenous peoples in order to understand the origins of *their own* culture. In 1898, Leo Frobenius answered the question, "Where does our history begin?" by declaring:

> Those simple, exotic forms of culture represent documents of world history! What historians have bootlessly sought in ancient hieroglyphs and inscriptions, they are able to say. Taken as a whole, they tell the tale, wrapped in the wondrous language of images, of the origin of human culture.⁷

Examining foreign cultures thus served to promote understanding of the development of (European) culture. In this framework, indigenous peoples were, on the one hand, perceived and represented as epitomizing foreignness, the opposite of the image the ethnologists had of themselves and their own culture. The mature, rational, self-disciplined, sociable, and cultivated construct of the European was set against the irrational 'child of nature' dominated by feelings, drives, and potentially antisocial impulses. Indigenous cultures provided a

---

5 Foucault, *The Order of Things*, 330. Cf. the following quote from Tylor, which leaves open "whatever yet earlier state may in reality have lain behind it" (Tylor, *Primitive Culture*, 1: 19).
6 Foucault, *The Order of Things*, 331. Also quoted in Werkmeister, *Kulturen jenseits der Schrift*, 64.
7 Leo Frobenius, *Der Ursprung der Kultur* (Berlin: Bornträger, 1898), viii–ix.

screen onto which to project everything considered taboo or antithetical to the researcher's own culture.[8] As Fritz Kramer remarks:

> With a view to its "own" culture, nineteenth-century ethnography devised the "upside-down" world of foreigners. [...] As a representation of "alien" culture, it openly expressed the truth taboo in polite society. [...] Therefore I would like to call it *imaginary ethnography*.[9]

Likewise, in *The Invention of Primitive Society*, Adam Kuper stresses the mechanism of projection.

> In the end [...] it may be that something yet more fundamental than political and religious concerns informed the new wave of interest in human origins. In the second half of the nineteenth century, Europeans believed themselves to be witnessing a revolutionary transition in the type of their society. [...] Each conceived of the new world in contrast to "traditional society"; and behind this "traditional society" they discerned a primitive or primeval society. The anthropologists took this primitive society as their special subject, but in practice primitive society proved to be their own society (as they understood it) seen in a distorting mirror. For them modern society was defined above all by the territorial state, the monogamous family and private property. Primitive society therefore must have been nomadic, ordered by blood ties, sexually promiscuous and communist. [...] Primitive man was illogical and given to magic. [...] Modern man, however, had invented science. [...] They looked back in order to understand the nature of the present, on the assumption that modern society had evolved from its antithesis.[10]

On the other hand, the modern focus on origins also means that these antithetical others always already formed part of the researcher's own culture too. The process of historical development links the inhabitants of both worlds. In this light, it becomes difficult to know when and how to separate the 'savage' and 'civilized' realms of culture or to determine where one ends and the other begins.[11] Without intending to, ethnology turned into a "counter-science."[12]

---

8 On the basic scheme of projection in cultural theory around 1900, cf. Jutta Müller-Tamm, *Abstraktion als Einfühlung. Zur Denkfigur der Projektion in Psychophysiologie, Kulturtheorie, Ästhetik und Literatur der frühen Moderne* (Freiburg: Rombach, 2005). For a list of critiques of the "'savage' slot and [...] related manifestations," as well as remarks concerning "neo-primitivism as an anti-primitivist primitivism without primitives," see Li, *The Neo-Primitivist Turn*, viii–ix, and "Primitivism and Postcolonial Literature," 987–989.
9 Fritz Kramer, *Verkehrte Welten. Zur imaginären Ethnographie des 19. Jahrhunderts* (Frankfurt: Syndikat, 1977), 7–8.
10 Kuper, *The Invention of Primitive Society*, 4–5.
11 With reference to Wilhelm Wundt, Werkmeister also speaks of a discourse of "relative difference between one's own and the foreign" (*Kulturen jenseits der Schrift*, 67).

Around 1900, these issues consolidated into a new scientific term, the 'primitive.'[13] Producing the ambivalent consequences described above, early ethnology's 'primitives' were indigeneous peoples that European ethnology located at the origin of a general cultural evolution, while also seeking to retrace a universal course of human development. Often this happened by way of delineating different stages through which humankind must pass. Examples include the scheme outlined in Wilhelm Wundt's *Elemente der Völkerpsychologie* (1912; *Elements of Folk Psychology*, 1916), where the culture of "primitive man" gives way to the age of totemism, then to the age of heroes and gods, which leads finally to a state of (full) humanity. This developmental discourse never loses sight of the question of how to position the 'primitive' in relation to modernity – in other words, how to simultaneously liken the two while keeping them at a distance from each other. This is why early ethnology was profoundly shaped by the paradigm of the 'primitive.' It not only determined the field's emergence but also its basic assumptions, inquiries, and methods, as I would like to show in my examination of Edward Tylor's foundational work in the next section.

The Paradigm of the 'Primitive' in Tylor's *Primitive Culture*

In *Primitive Culture* (1871), Tylor sets out to examine two fundamental principles of human culture: the "uniform action of uniform causes" and "its various

---

[12] Foucault, *The Order of Things*, 381. Därmann agrees with this assessment but voices criticism inasmuch as the "privileged place that ethnology is accorded in the structure of our knowledge [...] proves to be [...] a self-conferred European privilege of cultural experience and representation by others" (*Fremde Monde der Vernunft*, 10). In agreement with Kramer, Michael Taussig, Schüttpelz, and others, she stresses the role of "foreign foreign experiences," i.e., "practices and forms of inversion of foreign cultural experiences and representations that shake the self-evident nature of European culture and science" (11).

[13] The entry "primitiv, der oder das Primitive" in the *Historisches Wörterbuch der Philosophie* notes that the term *primitive* appears in English ethnology by 1870, but not yet in a semantically fixed form. From the 1880s it appears in German ethnology (in the writings of Alfred Vierkandt, among others) but is still used interchangeably with "simple," "original," and "natural" (*Naturvolk*). By the early 1910s, attempts to fix the term's meaning are more frequent, for example in the works of Lévy-Bruhl (1910), Durkheim (1912), and Wundt (1912). On the surface, "the model of simple, small, archaic societies" ("primitiv, der bzw. das Primitive," in *Historisches Wörterbuch der Philosophie*, ed. Joachim Ritter and Karlfried Gründer. [Darmstadt: Wissenschaftliches Buchgesellschaft, 1971–2007], 7: 1318) applied to peoples "without [their] own written tradition and with 'little developed technology'" ("Primitive," in *Wörterbuch der Völkerkunde* [Berlin: Reimer, 1999], 295). However, ethnologists still in fact understood "'primitives' as petrified representatives of earlier stages of the history of the genus."

grades," which "may be regarded as stages of development or evolution, each the outcome of previous history, and about to do its proper part in shaping the history of the future." To this end, he focuses on the relationship between the "civilization of the lower tribes" and that of the "higher nations," in particular.[14] Thus, his study of "savage life" uses European 'high culture' as its reference point. Together these form two extremities of a scale measuring the levels of civilization:

> Civilization actually existing among mankind in different grades, we are enabled to estimate and compare it by positive examples. The educated world of Europe and America practically settles a standard by simply placing its own nations at one end of the social series and [...] arranging the rest of mankind between these limits according as they correspond more closely to savage or to cultured life.[15]

These two extremities also delineate the temporal span of historical development. Tylor presupposes that the "savage tribes" of his own day correlate with early humankind and names this correspondence the "primitive condition."

> By comparing the various stages of civilization among races known to history, with the aid of archaeological inference from the remains of pre-historic tribes, it seems possible to judge in a rough way of an early general condition of man, which from our point of view is to be regarded as a primitive condition, whatever yet earlier state may in reality have lain behind it. This hypothetical primitive condition corresponds in a considerable degree to that of modern savage tribes, who [...] have in common certain elements of civilization, which seem remains of an early state of the human race at large. If this hypothesis be true, then, [...] the main tendency of culture from primaeval up to modern times has been from savagery towards civilization.[16]

"Primitive culture" – the phrase that lends the book its title – refers to a hypothetical origin supposedly prevalent among contemporary indigenous peoples. Among them, so Tylor's thinking goes, this culture could be studied, and studied as the starting point of a cultural development that would ultimately culminate in the European achievements of the modern age.

Tylor pictures a process of evolution leading from one pole to the other. In doing so, he borrows methodological assumptions from the natural sciences and applies them to the analysis of culture and society.

---

14 Tylor, *Primitive Culture*, 1: 1.
15 Tylor, *Primitive Culture*, 1: 23.
16 Tylor, *Primitive Culture*, 1: 19.

> The ethnographer's business is to classify such details with a view to making out their distribution in geography and history, and the relations which exist among them. What this task is like, may be almost perfectly illustrated by comparing these details of culture with the species of plants and animals as studied by the naturalist. To the ethnographer, the bow and arrow is a species, the habit of flattening children's skulls is a species [...].[17]

He draws from a biological model not only in his understanding of distribution and classification, but also in his investigation of evolutionary lines:

> The consideration comes next how far the facts arranged in these groups are produced by evolution from one another. [...] Among ethnographers there is no [...] question as to the possibility of species of implements or habits or beliefs being developed one out of another.[18]

To validate this thesis, Tylor elaborates the concept of *survivals*, which provides one reason his study remains known to this day. Survivals represent "processes, customs, opinions, and so forth, which have been carried on by force of habit into a new state of society different from that in which they had their original home."[19] In the context of the life sciences, they amount to atavistic features of the cultural organism.[20]

For Tylor, survivals are proof that more advanced stages of culture evolved from older ones. Inasmuch as they defy being understood in the operative terms of newer developments, they challenge researchers to trace back to their first point of emergence, where they served as sensible cultural practices. As he writes:

> On the strength of these survivals, it becomes possible to declare that the civilization of the people they are observed among must have been derived from an earlier state, in which the proper home and meaning of these things are to be found; and thus collections of such facts are to be worked as mines of historical knowledge.[21]

---

[17] Tylor, *Primitive Culture*, 1: 7.
[18] Tylor, *Primitive Culture*, 1: 13.
[19] Tylor, *Primitive Culture*, 1: 15.
[20] Tylor does not assume that these remainders are inherited, but rather that they are handed down, which distinguishes his perspective from that of the recapitulation theorists discussed in the next chapter. At the same time, his employment of biological models contradicts this assumption.
[21] Tylor, *Primitive Culture*, 1: 64.

Survivals offer Tylor confirmation that cultures evolve and pass incrementally from the state of "savagery" to that of "civilization." He musters an array of outmoded practices and superstitious activities that, inasmuch as they have increasingly been abandoned, affirm the superiority of his contemporary culture over its alien past. For him, the incomprehensibility of these practices to and in the present attests to the advances, or distance at any rate, that had been gained in the interim.

Yet this stabilization of his own cultural and historical identity brings with it a reverse effect: survivals not only underscore a reassuring distance from the estranged past, but also prove its persistence in the present.[22] Tylor paints a picture of a culture saturated with rudiments of a past from which it has grown estranged: "there are thousands of cases of this kind which have become [...] landmarks in the course of culture."[23] Survivals of the "savage condition" abound in contemporary Europe, he holds, where the 'primitive' haunts the present: "In our midst," one still finds numerous "primaeval monuments of barbaric thought and life."[24] Basic achievements such as language and mathematics are said to derive from a time before time: "Language is one of those intellectual departments in which we have gone too little beyond the savage state."[25] Survivals of primordial mythology are evident in superstitions, works of the imagination, and instances of madness. Tylor pursues the "transmission, expansion, restriction, [and] modification" of the animistic beliefs of archaic people through to "our own modern thought."[26]

The ambivalent consequences of the above-mentioned reflection on origins are clearly expressed in Tylor's idea that civilized Europe harbors survivals of "primitive culture." The study of indigenous peoples from elsewhere, who supposedly remained in "savage conditions," reassured Tylor and his readers of the progressive development that their own culture had already undergone. Although survivals were still to be found, their puzzling nature appeared to confirm how much ground had been covered in the process. Yet for all that, their very existence indicated that those earlier stages of historical development did not belong to the past alone. Contemporary civilization did not simply 'evolve away' from its origins, and the latter were not over and done with. On one hand, this view affirms 'the human' as it was defined by the discourse of the

---

[22] Werkmeister points this out as well, but he does not discuss Tylor in detail (*Kulturen jenseits der Schrift*, 68–70).
[23] Tylor, *Primitive Culture*, 1: 64.
[24] Tylor, *Primitive Culture*, 1: 19.
[25] Tylor, *Primitive Culture*, 2: 404.
[26] Tylor, *Primitive Culture*, 1: 21.

nineteenth-century human sciences. This human figure identifies themselves through their history: They *are* who they *were*. On the other hand, survivals attest to a past that remains dialectically unresolved, that is, to patterns of thought and behavior running counter to how enlightened Europeans saw themselves. The latter did not wish to identify with that past, even though it underlay their own culture. Thus, "primitive culture," as conceived by Tylor, proves to be both of the past as well as doubly in the present, that is, found both in its complete state in indigenous cultures and in scattered survivals in Europe. It also proves both to be doubly alien: chronologically, inasmuch as "primitive culture" represents the origins of human history, and spatially, inasmuch as it prevails on other parts of the globe. It stabilizes and at the same time destabilizes conceptions of European identity that rest on notions like origin, history, and progress.

Ambivalence also shapes Tylor's own reaction to his discovery. Despite the omnipresence of survivals and their necessity to the recognition of cultural evolution, Tylor holds fast to the idea of progressive advancement, which involves an ultimate overcoming of the old by means of enlightenment and technology. Apropos of magical practices still observed in contemporary Europe, he describes survivals as an "unsatisfactory [...] fact"[27] of life. They pose the danger that advancement will turn into "degeneration,"[28] that is, that European culture will revert to an archaic stage of development. Put differently, survivals' potential to reverse the historical process is revealed in their tendency to bring about "revivals."

> Sometimes old thoughts and practices will burst out afresh, to the amazement of a world that thought them long since dead or dying; here survival passes into revival, as has lately happened in so remarkable a way in the history of modern spiritualism.[29]

At various points, Tylor also expresses unease about the contemporary phenomenon of spiritualism, which in his eyes resurrects "primitive culture."

> This shows modern spiritualism to be in great measure a direct revival from the regions of savage philosophy and peasant folklore. [...] The world is again swarming with intelligent and powerful disembodied spiritual beings, whose direct action on thought and matter is again confidently asserted as in those times and countries.[30]

---

27 Tylor, *Primitive Culture*, 1: 123.
28 Tylor, *Primitive Culture*, 1: 46.
29 Tylor, *Primitive Culture*, 1: 15.
30 Tylor, *Primitive Culture*, 1: 129.

Studying survivals thus serves the purpose of exposing them so they – and revival phenomena like spiritualism – may be eliminated altogether.

> It is a harsher [...] office of ethnography to expose the remains of crude old culture which have passed into harmful superstition, and to mark these out for destruction. [...] Thus, active at once in aiding progress and in removing hindrance, the science of culture is essentially a reformer's science.[31]

To summarize: as its developmental orientation follows the paradigm of the 'primitive' Tylor articulates, early ethnology distances indigenous peoples with one hand and with the other draws them close, affirming their essential kinship with European civilization.[32] The 'primitive' is the site where the cultural alien and archaic end and the researcher's own culture begins. The wish for clear oppositions and demarcations stands opposed to the suspicion that its fulfillment is impossible. Studying the history of one's own culture leads to a supposed point of origin, the familiarity of which unsettles the researcher's position instead of confirming it.

Analogy, Allochrony, and Survival

The paradigm of the 'primitive' establishes analogy as the foundational argumentation scheme for ethnological texts at the turn of the twentieth century. Exemplarily, Tylor declares in the citation above that the "correspond[ence]" between primeval cultures of humankind and those of present-day 'savages' provides the basis for studying the features of "primitive culture." In his classic work of critical anthropology, *Time and the Other*, Johannes Fabian critiques such "allochronic discourse" because it refuses indigenous peoples "coevalness" with the ethnologists who study them by excluding them from those researchers' physical ("synchronous") as well as typological ("contemporary") time.[33] This approach both constitutes and degrades the studied object by positing a temporal distance from the researcher. By "denial of coevalness," Fabian writes, "I mean a persistent and systematic tendency to place the referent(s) of anthropology in a Time other than the present of the producer of anthropological dis-

---

31 Tylor, *Primitive Culture*, 2: 410.
32 On this twofold strategy, cf. Michael C. Frank, "Überlebsel. Das Primitive in Anthropologie und Evolutionstheorie des 19. Jahrhunderts," in *Literarischer Primitivismus*, ed. Nicola Gess (Berlin: De Gruyter, 2013), 160.
33 Fabian, *Time and the Other*, 31, 37.

course."³⁴ The argumentative schema of analogy is fundamental to this procedure, which equates indigenous peoples in the present with people living in a primal state. Once both terms of the analogy are established, indigenous culture is rendered obsolete – in contrast to the investigator's own, which is supposed to stand at a more advanced point of evolution. The analogy produces a putative identity: indigenous culture is *in actuality* 'primitive'; *in actuality* it belongs to another time.

What warrants such an approach? How is it possible for human beings who are now alive to belong to a wholly different age? To make this case, early ethnology had to adopt the assumption that some cultures have withdrawn from the progress of history and remain stuck in time. Accordingly, in *Elements of Folk Psychology*, Wilhelm Wundt answers the question of "Who is primitive man?" by pointing elsewhere on the globe: "there are other parts of the earth which, in all probability, really harbour men who are primitive."³⁵ Wundt affirms this thesis by pointing out that their cultures seem to be very simple, and that in order to understand them, there would be no need to return to any earlier conditions of humanity. No significant "mental development" should be assumed to have taken place among them.³⁶ Wundt therefore locates such peoples at the initial stage of their own cultures, and of civilization in general. He – like others of his day – equates indigenous culture with prehistoric culture. Contemporary peoples are denied their own history; both physically and typologically, they are said to belong to another time, to the first beginnings of humanity.

Thus, ethnology's analogical scheme of argumention is closely linked to the temporal models of allochrony and ahistoricity. Their counterpart is the model of survival, in which, instead of the present being relocated to the past, the past is (re)discovered in the present. Accordingly, in *Primitive Culture*, Tylor declares that contemporary "savages" and their ways of life represent the "remains" of archaic culture. Collectively, "savages" count as survivals – leftovers from another time, which stands still and does not evolve. They function as an ever present primal state. In this way, the 'primitive' is revealed to be a temporal category, or in Fabian's words, "*Primitive* being essentially a temporal concept, is a category, not an object, of Western thought."³⁷

---

34 Fabian, *Time and the Other*, 31.
35 Wilhelm Wundt, *Elements of Folk Psychology: Outlines of a Psychological History of the Development of Mankind*, trans. Edward Leroy Schaub (London: Allen & Unwin, 1916), 18.
36 Wundt, *Elements of Folk Psychology*, 20.
37 Fabian, *Time and the Other*, 18. Emphasis in original. This quotation is also featured on the cover of Schüttpelz, *Die Moderne im Spiegel des Primitiven*.

Whatever is designated as 'primitive' is catapulted into the past and stripped of its capacity for development. In works by Tylor and of his evolution-minded contemporaries, this judgment is negative, and the 'primitive' represents all that the researchers' own culture is seen to have evolved away from. But as we will see, the term may have positive connotations as well that serve to critique the observer's own society; in this light, the 'primitive' stands for a utopia achieved in another time and place.[38]

The 'Primitive' as a Figure of Thought in Early Ethnology

The category of the 'primitive' in early ethnological discourse operates quite literally as a figure of thought. First of all, the 'primitive' assumes the form of a concrete *figure* – the indigenous person – through which it can be *thought about*. Secondly, this person's supposedly other way of *thinking* is one of the most important characteristics attributed to them by early ethnologists: The 'primitive' functions as a figure for a way of thinking labeled either magical, mythic, mythological, mystic, or prelogical. This focus is not surprising, given that the 'primitive' was constructed from the outset as a platform for modern European self-reflection, which included reflection on the conditions of their own knowledge. As Sven Werkmeister aptly observes, ethnological discourse centered on "the question about the historical conditions for the laws of thinking itself. [...] At the beginning of the twentieth century, the *primitive* took the stage [...] more and more [...] as an *epistemological figure*."[39] Early ethnological studies thus devoted a great deal of attention to the allegedly other ways of thinking performed by indigenous peoples, as well as the worldview such thinking gives rise to, whether they qualified it in positive or negative terms. Broadly speaking, these studies may be mapped out along lines drawn by Edward E. Evans-Pritchard, who distinguishes between intellectualist, emotionalist, and sociological theories of "primitive religion."[40] These schools of thought prevailed in England, Germany, and France, more or less succeeding one another around the turn of the century.

---

38 Cf. the short summary in Franke and Holert, eds., *Neolithische Kindheit*, 319; Torgovnick, *Gone Primitive*, 8–10: "The primitive does what we ask it to do" (9); as well as Armin Geertz, "'Can We Move Beyond Primitivism?' On Recovering the Indigenes of Indigenous Religions in the Academic Study of Religion," in *Beyond Primitivism: Indigenous Religious Traditions and Modernity*, ed. Jacob K. Olupona (New York: Routledge, 2004), 52–53.
39 Werkmeister, *Kulturen jenseits der Schrift*, 77. Emphasis in original.
40 In his *Theories of Primitive Religion* (Oxford: At the Clarendon Press, 1965), a lecture already written in part in 1934.

Intellectualist Theories

To describe the "mental state"[41] of past and present "savage tribes," Tylor enlists the concept of animism. Animism, according to Tylor, is the "deep-lying doctrine of Spiritual Beings, which embodies the very essence of Spiritualistic as opposed to Materialistic philosophy."[42] Two basic principles guide his considerations. Their experience of dreams, sickness, and death leads "primitives" to conclude "logically"[43] (in keeping with their "low[er]" level of intellectual development) that souls can exist detached from physical bodies and that there is a realm of spirits extending up to the level of gods.[44] Tylor devotes a significant portion of the second volume of his study to cases illustrating this claim. In the process, he neglects a thesis advanced earlier in the work derived from the associationist psychology of the time, according to which animism "belongs to that great doctrine of analogy, from which we have gained so much of our apprehension of the world around us."[45] In contrast to modern society, indigenous cultures consider analogical relations to be matters of actual fact: "They could see the flame licking its yet undevoured prey with tongues of fire."[46] 'Primitive thinking' for Tylor thus operates by means of analogies that are deemed to be reality. Tylor concludes that the people he studies are involuntarily transferring their own thoughts, feelings, and actions onto objects that belong to the external world, which accounts for their belief in spirits and ghosts. As we will see, the same holds for ethnologists.

Tylor did not explore the further ramifications of this thesis, but James Frazer took it up in *The Golden Bough* (1890). To describe the worldview of "savages," he develops the concept of "sympathetic magic." For Frazer, belief in magic is based on two principles of thought, the laws of similarity and of contact. As in Tylor's theory of analogy, things that are similar in kind or once stood in some form of contact are not merely associated with one another, but instead through a sequence of associations are thought to be related or even identical and to entertain a causal relationship with each other.[47] For Frazer,

---

[41] Tylor, *Primitive Culture*, 1: 256.
[42] Tylor, *Primitive Culture*, 1: 384.
[43] Tylor, *Primitive Culture*, 1: 423.
[44] Tylor, *Primitive Culture*, 1: 385.
[45] Tylor, *Primitive Culture*, 1: 268–269.
[46] Tylor, *Primitive Culture*, 1: 269.
[47] James George Frazer, *The Golden Bough: A Study in Magic and Religion* (New York: Penguin, 1996), 13.

this "secret sympathy"[48] between objects underlies magic in its theory and practice, and he emphasizes that the people he studies take the laws of similarity and contact to be laws of nature. Frazer sees no need to explain magic by means of a belief in spirits or ghosts. On the contrary, it is precisely the lack of such belief that characterizes magic, which "assumes that in nature one event follows another necessarily and invariably without the intervention of any spiritual or personal agency."[49] He recognizes in the absence of this belief a certain kinship to contemporary natural science. The key difference, in his estimation, is that "sympathetic magic" is based on "laws of nature" that are false because they do not admit empirical verification. Instead of observing phenomena precisely, indigenous peoples rely on "an extension, by false analogy, of the order in which ideas present themselves to our minds."[50]

Like Tylor, Frazer derives two basic principles of thought from the principle of association, which he links to a naive confusion between reality and ideality:[51] indigenous individuals mistake mental connections for actual ones. But while affirming differences between them and contemporary Europeans, Frazer points to mental operations that they share: Both attempt to explain their world using the same functions of the "human mind."[52] And both worldviews exhibit "logical consistency,"[53] in the broad sense that they are formed through rules. The sole difference between them is the ability – or inability – to distinguish between abstract ideas and empirically verifiable reality. This perspective enables Frazer to embed magical thinking in a theory of development and progressive history informed by evolution. Frazer understands the indigenous individual as an undeveloped predecessor to the contemporary European and devalues the former's worldview as "fatal[ly] flaw[ed]."[54] Accordingly, remnants of those beliefs, which endure as superstition, represent a "standing menace,"[55] in contrast to the "germ"[56] of progress auguring enlightenment and science.

---

48 Frazer, *The Golden Bough*, 14.
49 Frazer, *The Golden Bough*, 58.
50 Frazer, *The Golden Bough*, 854.
51 Frazer: *The Golden Bough*, 14.
52 Frazer, *The Golden Bough*, 59.
53 Frazer, *The Golden Bough*, 314.
54 Frazer, *The Golden Bough*, 59.
55 Frazer, *The Golden Bough*, 67.
56 Frazer, *The Golden Bough*, 12.

Emotionalist Theories

Tylor and Frazer premise that indigenous peoples' mental habits are based on the nature and interrelationship of the phenomena most meaningful to their existence. They enlist the psychology of association to affirm that 'primitive thinking' treats associative connections as though they were objective, real relationships. This produces magical (for Frazer) or religious (for Tylor) ideas. While scholars from neighboring fields hailed their theories, Tylor and Frazer also encountered opposition from other schools of ethnology. Critics challenged the assertion that 'primitive thinking' is animated by protoscientific epistemological interests and argued instead that it derives from affect.

For example, in *Elements of Folk Psychology*, Wilhelm Wundt underscores the difference between the disciplines of individual psychology and folk psychology (or *Völkerpsychologie*, of which he was a leading authority). For one, he contends, the study of individual psychic life does not provide insight into the history of the human spirit, nor does it grasp the central role of "community life"[57] (which Wundt does not systematically investigate either). Second, Wundt distinguishes between folk psychology and ethnology: the former concerns the "mental development" (*geistige Entwicklung*) of peoples studied, over and above their other characteristics.[58] Consequently, his work displays the same limiting evolutionary assumptions that the early ethnological projects had. Wundt sets out to retrace the progress of the human spirit from "primitive conditions" by way of an "almost continuous series of intermediate steps to the more developed and higher civilizations."[59] He defines the first level – that of "primitive man" – by the latter's habit of associative, intuitive thought bearing on both sensory and supersensory matters. But in contrast to his English counterparts, Wundt derives the supersensory mental operations of indigenous peoples from their affective experiences. More specifically, these mental operations are involuntary, affective projections onto objects in the surrounding world.

The outcome, what he calls "mythological thinking,"[60] operates within the confines of emotion, following the path laid down by the affective projections

---

57 Wundt, *Elements of Folk Psychology*, 6.
58 Wundt, *Elements of Folk Psychology*, 5.
59 Wundt, *Elements of Folk Psychology*, 4.
60 Wundt, *Elements of Folk Psychology*, 91. Before Wundt, Alfred Vierkandt had already spoken of the "mythological way of thinking" in his seminal work, *Naturvölker und Kulturvölker* (Leipzig: Duncker & Humblot, 1896). Such thought differs from scientific reasoning because it rests on different premises; for instance, it demonstrates a "lack of the idea of universal regularity" and relies on "belief in the influences of [...] spiritual beings" (252), exemplifying the "adherence of the

just noted. Like Tylor before him, Wundt traces such thinking (which he also describes as "belief in magic and demons") back to the experiences of death and illness.[61] That said, intellectual engagement with such experiences is less important in his eyes than fear, the affect occasioned by sudden change. Fear is involuntarily expressed in the notion of the demon, the maleficent force embodied by the dead or triggering disease, which only a magician might counteract.[62] Wundt explicitly turns against the assumptions of the English ethnologists when he discounts efforts to explain existentially significant phenomena: "it is not intelligence nor reflection as to the origin and interconnection of phenomena that gives rise to mythological thinking, but emotion."[63]

Karl Theodor Preuss, in *Die geistige Kultur der Naturvölker* (The Spiritual Culture of Primitive Peoples, 1914), likewise addresses the "magical thinking"[64] characteristic of "humanity long ago [and] its representatives today,"[65] that is, "peoples living in a state of nature."[66] Like Wundt, he is convinced that emotional excitation, not intellectual curiosity, shapes such thinking: existential experiences such as death, illness, combat, or hunting prompt spontaneous actions that yield items of reflection only after the fact, in the form of magical practices.[67] Preuss offers as an example the mimetic representations of desired objects. Unlike Frazer, Preuss considers them not a willfull operation of "magical analogy," but rather a spontaneous expression of desire that is only later formalized into magical ritual.[68] In contrast to Wundt, who argues for the projection of affect, Preuss takes up a thesis first articulated by Robert R. Marett: that magic serves a cathartic function. Acts of magic discharge emotional energy.

Sociological Theories

Preuss was influenced by the theories of Emile Durkheim and his pupils. At the turn of the century, the latter offered a radically new, sociological approach to

---

consciousness to the sensually given," a lack of understanding for the abstract, and the preponderance of associative connections.
61 Wundt, *Elements of Folk Psychology*, 81.
62 Wundt, *Elements of Folk Psychology*, 82, 92–93.
63 Wundt, *Elements of Folk Psychology*, 92–93.
64 Preuss, *Die geistige Kultur der Naturvölker* (Leipzig: Teubner, 1914), 8.
65 Preuss, *Die geistige Kultur der Naturvölker*, 1.
66 Cf. Christoph Gardian, *Sprachvisionen. Poetik und Mediologie der inneren Bilder bei Robert Müller und Gottfried Benn* (Zurich: Chronos, 2014), 103–113.
67 Preuss, *Die geistige Kultur der Naturvölker*, 20–21.
68 Preuss, *Die geistige Kultur der Naturvölker*, 30.

'primitive thinking' and elaborated a genealogy of logical operations on its basis. English and German ethnologists had already recognized that examining 'primitive thinking' might afford insights into the origins of their own powers of cognition. But the Durkheim school would be the first to make the relativizing consequences of these comparisons obvious.[69]

De quelques formes primitives de la classification (1902; *Primitive Classification*, 1963), by Durkheim and his nephew Marcel Mauss, advances the provocative thesis that "the faculties of definition, deduction, and induction" are not "immediately given in the constitution of the individual understanding." Instead, they have emerged historically and developed within the social collective: "In these methods of scientific thought [there are] veritable social institutions whose origin sociology alone can retrace and explain." They illustrate this claim by tracing the social origins of the "classificatory function"[70] – that is, the "rudimentary"[71] modes of mental organization to be found in the "least evolved societies."[72]

Further studies by Durkheim and Mauss broaden the scope of inquiry. In 1902, in collaboration with Henri Hubert, Mauss also published *Esquisse d'une théorie générale de la magie* (1902; *A General Theory of Magic*, 1972). In its sociological approach, this book finds the key to magical beliefs and practices in collective representations completely foreign to "adult European understanding."[73] The notion of "magical potential" is said to animate an overall "milieu" of magic, whose elements obey rules other than those that govern the "world of the senses."[74] As in *Primitive Classification*, the authors distance themselves from English ethnologists who derive their concept of magic from an intellectualist psychology of the individual. At the same time, Hubert and Mauss argue against German scholars who appeal to *individual* psychology based on affect[75] and posit instead that the idea of magical potentiality is evident when one focuses on the

---

69 On Durkheim and Mauss's project, see, for example, Vincent Crapanzano, "The Moment of Prestidigitation. Magic, Illusion, and Mana in the Thought of Emile Durkheim and Marcel Mauss," in *Prehistories of the Future. The Primitivist Project and the Culture of Modernism*, ed. Elazar Barkan and Ronald Bush (Stanford: Stanford University Press, 1995). In *The Mind of Primitive Man* (New York, 1961), Franz Boas also took an explicitly relativist position; because of its focus on Europe, this work is not discussed here.
70 Emile Durkheim and Marcel Mauss, *Primitive Classification*, trans. Rodney Needham (London: Cohen & West, 1963), 2.
71 Durkheim and Mauss, *Primitive Classification*, 4.
72 Durkheim and Mauss, *Primitive Classification*, 3.
73 Marcel Mauss, *A General Theory of Magic*, trans. Robert Brain (London: Routledge, 1972), 107.
74 Mauss, *A General Theory of Magic*, 107.
75 Mauss, *A General Theory of Magic*, 107–108.

"psychology of man as a community."[76] Objects only acquire magical potential through the attributions of the collective. Because all of its members share a particular need, the medium they collectively identify for bringing about the fulfillment of that wish is really accorded that capacity in the performative act. "The whole society" entertains "the false images of its dream," and "public opinion" achieves "the synthesis between cause and effect."[77]

There is "nothing intellectual or experimental" about magical potentiality (which according to Mauss and Hubert finds expression in the term *mana*). It involves only "the feeling of society's existence and society's prejudices." In other words, the authors do not understand the power of magic in the sense of an individual delusive projection – as Frazer does in terms of ideas and Wundt in terms of affect. Nor is magic a matter of catharsis, as Preuss contends. Instead, Mauss and Hubert view magic as the performative power of the collective to devise the means needed to satisfy their desires and, by doing so, to fulfill them at the same time. By explaining the phenomenon, the scholars seek to strip away its apparent "absurdit[y]" and reveal its inner logic. Inasmuch as "magical potentiality" is granted credence, they observe, the magical act appears altogether rational, that is, as a calculated harnessing by means of *mana*, through which objects are lent their magical powers and the desired outcome is achieved.[78]

In *Les formes élémentaires de la vie religieuse* (1912; *The Elementary Forms of Religious Life*, 1915), Durkheim adopts the same perspective and stresses the *actual* effectiveness of rites. Members of the collective

> take away with them a feeling of well-being [...] How could this sort of well-being fail to give them a feeling that the rite has succeeded, that it has been what it set out to be, and that it has attained the ends at which it was aimed? [...] The moral efficacy of the rite, which is real, leads to the belief in its physical efficacy, which is imaginary.[79]

This collective event accounts for the rise of totemism, which Durkheim considers the most elementary form of religion. In the collective experience of both psychic and physical "violent super-excitation" offered by the ritual,[80] participants are overwhelmed by a force greater than that of the individual: "very intense social life [...] does a sort of violence to the organism, as well as to the individual

---

76 Mauss, *A General Theory of Magic*, 108.
77 Mauss, *A General Theory of Magic*, 126.
78 Mauss, *A General Theory of Magic*, 126.
79 Emile Durkheim, *The Elementary Forms of the Religious Life*, trans. Joseph Ward Swain (London: Allen & Unwin, 1915), 359.
80 Durkheim, *The Elementary Forms of the Religious Life*, 216.

consciousness."[81] The "sacred" refers to this force, which is conceived as an independently existing substance (e.g., *mana*) that takes possession of people and things and can transfer from one of them to the other. When performing sacred rites, society is actually honoring itself, that is, its own transcendence and authority over its members. The "profane," in contrast, is the mundane lifeworld, unaffected by this force. "Above the real world where his profane life passes [man] has placed another which, in one sense, does not exist except in thought, but to which he attributes a higher sort of dignity than to the first."[82] By clearly distinguishing between the sacred and profane, Durkheim answers the question that he had left open in *Primitive Classification* regarding the fundamental bifurcation underlying all differences operative in the world. Attending to the ritual scene also reveals that all categories possess an affective charge. The force that engulfs the collective performing its rites amounts to an "avalanche" of passions.[83]

Durkheim also revisits in-depth the relationship between primeval and logical ways of thinking, which he had first addressed in his study of classification. *The Elementary Forms of Religious Life* once again takes up the role that collective rituals play in the history of human thought. Religion represents the matrix in which the faculty of judgment first takes shape; its terms likewise represent products of the collective.[84] Hereby, Durkheim offers an answer to the old question of whether the categories of understanding are given a priori or constructed by the individual. In his eyes, they are constructed, but by the collective, and as such they possess a given, necessary, and ineluctable quality for the individuals constituting the group.[85] Durkheim avoids the charge of relativism through his sociological method of deriving categories of thought. Indeed, he views their social origin as the best guarantee of their objectivity and naturalness: they have stood the test of generations. As he puts it, "If [the category] were not founded in the nature of things, it would have encountered in the facts a resistance over which it could never have triumphed."[86] Because society itself forms "a part of nature" and is, in fact, its "highest representation,"[87] it plays a decisive role in shaping "human nature."[88] Durkheim sees thinking, religion, and humanity as

---

81 Durkheim, *The Elementary Forms of the Religious Life*, 227.
82 Durkheim, *The Elementary Forms of the Religious Life*, 422.
83 Durkheim, *The Elementary Forms of the Religious Life*, 216.
84 Durkheim, *The Elementary Forms of the Religious Life*, 36–42.
85 Durkheim, *The Elementary Forms of the Religious Life*, 42–47.
86 Durkheim, *The Elementary Forms of the Religious Life*, 2.
87 Durkheim, *The Elementary Forms of the Religious Life*, 18.
88 Durkheim, *The Elementary Forms of the Religious Life*, 53.

deriving from social processes and postulates society as a new transcendental subject.

In addition to adopting a more philosophical mode of engagement, the theories of the French school differ from those of England and Germany in their sociological orientation, which informs their tracing of a genealogy of human thought. By deeming rational thinking to be historically formed, French ethnologists reject the narrative of its progressively widening distance from 'primitive thinking.' Continuity prevails, they emphasize: *all* intellectual operations are socially contingent. In this framework, "magical thinking" and "logic" overlap rather than stand opposed. The former proves to be more logical and the latter more magical than previously theorized. In contrast to English ethnology, the French school does not contend that a deluded projection of ideality into reality takes place in 'primitive thinking.' Instead, the starting point is the very real power that the collective holds over its members, which sets up categories that subsequently operate a priori. German theories on 'primitive thinking' adopt a similar approach inasmuch as they acknowledge the very real power of affect, but their focus on individual psychology necessarily leaves key aspects of affect unexplored. The sociological perspective eliminates this methodological shortcoming and offers a plausible answer to the question of how a sustained belief in magic is possible in the first place.

Lucien Lévy-Bruhl's Notion of Participatory Thinking

Durkheim's broader circle included Lucien Lévy-Bruhl, whose work from the 1910s and 1920s presented the most powerful and widely influential theory on 'primitive thinking' at the time. In seven books written over some thirty years, Lévy-Bruhl set out to understand the mental structures of "primitive"[89] peoples. Like Durkheim and his pupils, he critiqued the theories of the English ethnologists on two basic points, faulting them for positing "the identity of a 'human mind,' which from the logical point of view is always exactly the same at all times and in all places,"[90] and for making "the mental processes of the *individual human mind*"[91] their point of departure. Because of these assumptions, he argued, ethnologists had constructed a "native" whose thinking differs from

---

[89] The first book speaks of "sociétés inférieures" (*Les fonctions mentales dans les sociétés inférieures*, 1910); the second opts for "mentalité primitive" (*La mentalité primitive*, 1922).
[90] Lévy-Bruhl, *How Natives Think*, trans. Lilian A. Clare (New York: Washington Square Press, 1966), 8.
[91] Lévy-Bruhl, *How Natives Think*, 13. Emphasis in original.

the contemporary European's only in its defectiveness and immaturity. Consequently, the individual ethnologist would only need to ask himself how he, if he were a member of an indigenous people, would have arrived at the ideas of 'primitives'.

For Lévy-Bruhl, the hypothesis of animism suitably describes 'primitive thinking,' yet, because of its anachronistic presuppositions, it is incapable of explaining it. Lévy-Bruhl therefore follows Durkheim in stressing that 'primitive' ideas and concepts represent "social phenomena"[92] – not products of individual reasoning so much as the results of collective activity, which impose themselves on individuals as an "article of faith"[93] in passing from one generation to the next. Also, he posits a stark difference between indigenous and contemporary European societies along with the two "types of mentality"[94] supposedly developed in them.

For Lévy-Bruhl, emotional and motoric factors play a big role in the "primitive mentality's" collective representations brought about and renewed by ritualized threshold experiences. Thus, in the context of these representations, a given object is not only processed through cognition, but it also simultaneously triggers particular feelings and actions. On this basis, the object accrues a potency, which "is always real for the primitive and forms an integral part of his representation." Lévy-Bruhl is less interested in the origin of this potency than in analyzing its ontological state: how it manifests itself in the compulsion to certain actions and emotions, how it appears to have always already existed in the collective representation, and how it is handed down to members of the collective and reinforced through rituals. For want of better terminology, he calls such power "mystical," by which he means the "belief in forces [...] which, though imperceptible to sense, are nevertheless real."[95] These forces are also evident in the influence one thing can exercise on another according to 'primitive thinking,' as they spin elaborate nets of relations that determine how the world is perceived.

Lévy-Bruhl emphasizes, however, that this mystical dimension does not create a second reality. Instead, the members of what he calls "lower societies" perceive only *one* reality, which is itself *both* mystical and objective.

> The superstitious man, and frequently also the religious man, among us, believes in a twofold order of reality. [...] But the primitive's mentality does not recognize two distinct worlds

---

92 Lévy-Bruhl, *How Natives Think*, 13.
93 Lévy-Bruhl, *How Natives Think*, 15.
94 Lévy-Bruhl, *How Natives Think*, 18; cf. 118.
95 Lévy-Bruhl, *How Natives Think*, 25.

in contact with each other, and more or less interpenetrating. To him there is but one. Every reality, like every influence, is mystic, and consequently every perception is also mystic.[96]

Accordingly, the mystical thought of 'primitives' is not based on mental associations. Association presupposes a dissociation, for example, between human beings, animals, and things. The mystical view does not acknowledge such discontinuity.[97] In this reality, every member of the Bororo tribe, to take up the often-quoted examples by Karl von den Steinen and Aby Warburg, is not just a human being, but also a parrot *in reality*; a snake *is* a lightning bolt *as well as* an animal.[98]

For Lévy-Bruhl, the 'primitive' grasp of reality depends on how mystical thought shapes the mental processing of sensory stimuli. He does not attribute a different kind of sense perception to 'primitives' so much as another form of understanding.

> Primitives see with eyes like ours, but they do not perceive with the same minds. We might almost say that their perceptions are made up of a nucleus surrounded by a layer [...] of representations which are social in their origin.[99]

Notably, "representations are connected" differently in this understanding.[100] They obey the "law of participation," which Lévy-Bruhl glosses as follows:

> I should be inclined to say that in the collective representations of primitive mentality, objects, beings, phenomena can be, though in a way incomprehensible to us, both themselves and something other than themselves. In a fashion which is no less incomprehensible, they give forth and they receive mystic powers, virtues, qualities, influences, which make themselves felt outside, without ceasing to remain where they are.[101]

At the center of this law is its vexing acceptance of difference and identity at the same time, thus disregarding the notion of logical contradiction.[102] Lévy-Bruhl therefore calls this quality of 'primitive thinking' "prelogical" (rather

---

96 Lévy-Bruhl, *How Natives Think*, 54.
97 Lévy-Bruhl, *How Natives Think*, 31.
98 Karl von den Steinen, *Unter den Naturvölkern Zentral-Brasiliens. Reiseschilderungen und Ergebnisse der zweiten Schingú-Expedition 1887–1888* (Berlin: Dietrich Reimer, 1894), 352. Aby Warburg, *Das Schlangenritual. Ein Reisebericht* [1923] (Berlin: Wagenbach, 1988).
99 Lévy-Bruhl, *How Natives Think*, 31.
100 Lévy-Bruhl, *How Natives Think*, 54.
101 Lévy-Bruhl, *How Natives Think*, 61.
102 Riedel, "Arara ist Bororo," 222.

than anti- or alogical).¹⁰³ Prelogical thinking proceeds more synthetically than analytically. The mental connections between its representations do not rely on prior analyses (as consolidated in concepts, for example), but are always already supplied with the representations: "The syntheses [...] are nearly always both undecomposed and undecomposable."¹⁰⁴ That is also why this type of thinking is extraordinarily enduring: it refuses modification by experience, contradiction, or other forms of disproof. Rather, analysis is often replaced by memory, which plays a great role in Lévy-Bruhl's view of the "mentalité primitive." He acknowledges that indigenous peoples possess concepts, but these concepts obey the law of participation. Because they concentrate on mystical relations, the European observer is at pains to grasp them. The concepts follow the path laid by "preconnections" that are always already given by collective representations.¹⁰⁵ Thus, "primitive" concepts are immersed in an "atmosphere of mystic possibilities,"¹⁰⁶ where they summon forth feelings of a universal, reciprocal, and amorphous action and reaction of all things and beings – a dynamic back-and-forth in which the human being is included. The "strange" quality of these operative categories is most evident in the notion of *mana*, which refers to the deeper unity of the one and the many, the individual and the species, and the widest diversity and shared identity of all.¹⁰⁷

Lévy-Bruhl's first book is obviously indebted to Durkheim and his school, but it also diverges from them at key points. The author is not interested in a genealogy of thought so much as an analysis of the actual state of 'primitive thinking.' Two theses drive such analysis: First, different kinds of society possess different "types of mentality."¹⁰⁸ Therefore, no direct path leads from 'primitive thinking' to that of contemporary Europeans. Time and again, Lévy-Bruhl stresses the otherness of such thinking, for example by situating it beyond logic or by clearly differentiating between collective representations and logical concepts.¹⁰⁹ Secondly, however, he recognizes to a greater degree than his predecessors the impossibility of recovering points of origin. No matter how far back one goes, "we shall never find any minds which are not socialized, if we may put it thus, not already concerned with an infinite number of collective representations which have been transmitted by tradition, the origin of which is lost in obscur-

---

103 Lévy-Bruhl, *How Natives Think*, 63.
104 Lévy-Bruhl, *How Natives Think*, 91.
105 Lévy-Bruhl, *How Natives Think*, 93.
106 Lévy-Bruhl, *How Natives Think*, 108.
107 Lévy-Bruhl, *How Natives Think*, 109.
108 Lévy-Bruhl, *How Natives Think*, 18.
109 Lévy-Bruhl, *How Natives Think*, 54.

ity."¹¹⁰ As a result, he does not stress the constructed quality of collective representations so much as their givenness, the way they "impose [...] claims on [...] individuals"¹¹¹ and shape each perception before it has even occurred.

For Lévy-Bruhl, the individual and collective do not have initial sensory experiences that are then overlaid with collective representations by means of tradition or collective events. The act of perception itself is conditioned equally by both sensory data and pregiven collective representations that always carry along embedded connections to other representations.¹¹² Again and again, as Erich Hoerl has shown, Lévy-Bruhl stresses the constitutive role of precedent.¹¹³ For example, the "syntheses" of magical thinking cannot be seen as its products so much as original or "fresh"¹¹⁴ events "always bound up with preperceptions, preconceptions, preconnections, and we might almost say with prejudgments." The process may be described as "a priori participation."¹¹⁵ The mystical dimension that gets transmitted through a network of beings, objects, emotions, and actions is not first brought forth by thinking. Instead, this network is always already there, decisively determining perception. It would be pointless, according to Lévy-Bruhl's reasoning, to investigate the "logical process" of "the primitive's mind" that are supposedly responsible for their peculiar interpretation of the events in their world. "This mentality," he affirms, "never perceives the phenomenon as distinct from the interpretation"; both occur at the same time. If anything, he simply reverses the question to inquire how "the phenomenon became by degrees detached from the complex in which it first found itself enveloped" and came to be understood in logical terms.¹¹⁶ Thus already in his first book, Lévy-Bruhl hints that his investigations will shift from the study of the mental structures of 'primitives' to an anthropological account.¹¹⁷

This project is most evident in the *Carnets* (1949; *Notebooks*, 1975), Lévy-Bruhl's final work. Almost all the author's observations (from January 1938 to February 1939) revolve around the puzzle of participation, or more specifically around the suspicion that this enigma concerns a phenomenon hardly graspable in European thinking and discourse because it is so utterly foreign to it: "Is there

---

110 Lévy-Bruhl, *How Natives Think*, 13–14.
111 Lévy-Bruhl, *How Natives Think*, 3.
112 Lévy-Bruhl, *How Natives Think*, 30–31, 9–10.
113 Cf. Hoerl, *Sacred Channels*, 212–222, on which the following relies.
114 Lévy-Bruhl, *How Natives Think*, 90.
115 Lévy-Bruhl, *How Natives Think*, 91.
116 Lévy-Bruhl, *How Natives Think*, 32.
117 This is Hoerl's thesis in his chapter on Lévy-Bruhl (*Sacred Channels*, 205–225), which draws mostly from *The Notebooks*.

a difference between the participation thus expressed in our language and what really exists in the consciousness of primitive man, and if so, what is it?" The answer given in the same notebook entry is that participation does not involve two separate ideas (e.g., a corpse and a ghost) so much as it always precedes and predetermines these representations:

> It is before them, or at least simultaneous with them. What is given *in the first place* is participation. [...] For the primitive man it is this duality-unity which is – not thought – but felt first, and it is then, if he reflects, that he recognizes a participation between the ghost on the one hand, and the corpse on the other.[118]

Here, Lévy-Bruhl gets to the heart of a matter that he had already touched upon in his first book. He had noted that 'primitive thinking' obeys the law of participation above all: everything is conceived according to the principle of affective and motoric participation in the other. Moreover, he sees participation not as an outcome of thinking so much as setting the course of perception in advance. Therefore, for 'primitives,' there is no stage prior to participation. At long last, in the *Notebooks*, Lévy-Bruhl asserts that not just their thinking but also their very "be[ing]" is determined by participation.

> They are what they are by virtue of participations: the member of the human group through participation in the group and in the ancestors; the animal or plant through participation with the archetype of the species, etc. ... If participation were not established, already real, the individuals would not exist.

Therefore, one should not ask how participation arises between beings and objects. Instead, the question is how beings and objects can possibly be released from this participation, which is always already in place. In the *Notebooks*, Lévy-Bruhl holds that such a separation of elements is inconceivable to the "mentalité primitive" because participation constitutes the basis of their being: "For the primitive mentality *to be is to participate*."[119]

Given Lévy-Bruhl's early debt to Durkheim, it is not surprising that he explains the relationship between being and participation in sociological terms: "Since the answer is not to be found in a particular form of mental activity (law, principle, general scheme, etc.), it is accordingly necessary for us to turn

---

118 Lucien Lévy-Bruhl, *The Notebooks on Primitive Mentality*, trans. Peter Rivière (New York: Harper & Row Publishers, 1975), 2. Emphasis in original.
119 Lévy-Bruhl, *Notebooks*, 18. Ellipses and emphasis in original. See Hoerl, *Sacred Channels*, 214.

to the content of the feelings of participation (between the individual and the other members of this group [...]).”[120] Despite himself, Durkheim time and again presupposes the anteriority of the individual to the collective. For instance, he discusses the need to develop means of communicating thoughts (i.e., language) because individuals must reach agreements with one another to build a society. In contrast, Lévy-Bruhl attempts the ultimately impossible project of abandoning the anteriority of the individual because this temporal scenario doesn't correspond to "the mentality of primitives,"[121] which is bound to the collective and oriented on participation. He also differs from other ethnologists of his generation in that he does not understand 'primitive thinking' as an early phase of logical thinking. He faults his precursors for wanting "to refer their mental activity to an inferior variety of our own."[122] Affirming that indigenous peoples outside of Europe think in a way that represents a radical alternative to familiar logic, Lévy-Bruhl emphasizes the fundamental otherness of such a mindset.

In this perspective, 'primitive thinking' receives a positive valorization and relevance for modern society. Unlike English ethnologists, who fear the residual traces of magic, Lévy-Bruhl affirms that the insights of logic never deliver complete satisfaction because one only ever comes to know objects "imperfect[ly]" and "external[ly]."[123] Participative thinking, on the other hand, provides an "intimate [...] communion between entities": "All idea of duality is effaced."[124] The "need of participation" – which he considers "more imperious and more intense [...] than the thirst for knowledge" – may also be observed in "people like ourselves." Indeed, the soul "aspires to something deeper than mere knowledge, which shall encompass and perfect it."[125] Instead of looking for evolutionary or genealogical development from 'primitive' to logical thought, Lévy-Bruhl declares that modern Europeans have a need for this other form of thinking.

All the same, Lévy-Bruhl's attitude to 'primitive thinking' remains ambivalent. Even though he turns away from the evolutionary scheme, his work abounds in formulations such as "not yet" and comparative and superlative wordings such as "lower" and "higher," which all imply a positive course of development. Like Tylor and Frazer, Lévy-Bruhl describes the participative thinking

---

120 Lévy-Bruhl, *Notebooks*, 92.
121 Lévy-Bruhl, *How Natives Think*, 87.
122 Lévy-Bruhl, *How Natives Think*, 61.
123 Lévy-Bruhl, *How Natives Think*, 344.
124 Lévy-Bruhl, *How Natives Think*, 344–345.
125 Lévy-Bruhl, *How Natives Think*, 346.

that arises in European societies as residual "mystic elements,"[126] which he opposes to the desideratum of the "rational unity of the thinking being."[127] The relationship between these two ways of thinking is characterized at times as dualistic – which attributes participative thinking to the 'primitives' and logical thinking to contemporary Europeans – or as differentiated – which finds both ways of thinking among both groups in more or less pronounced forms. Regarding the latter, he writes for instance, "Our mental activity is both rational and irrational. The prelogical and the mystic are co-existent with the logical."[128] Lévy-Bruhl bases his work on the antinomy of the two types of thought, but he has difficulty keeping them apart from one another.

In all the studies at issue so far, the 'primitive' surfaces as an epistemological figure, i.e., representing another way of thinking. Engagement with the thought of others is also a mean of reflecting on one's own thought processes. However, the role assigned to 'primitive thinking' varies quite a bit in this context. English ethnologists explain the idiosyncratic conceptual worlds of 'primitives' by pointing to their contemplation of existential experiences and their associational mental operations. Though these operations are thought to be universal, they supposedly lead to false judgments (mistaking the ideal for real) when carried out by the indigenous mind. Here, 'primitive thinking' appears as the evolutionary forerunner of scientific reasoning, which is vital to modern Europeans' self-understanding. By contrast, German ethnologists shine the spotlight on affect to search for the origins of the 'primitive mindset' in affective projections or cathartic gestures. They coin the concepts of "mythological" (Wundt) or "magical" (Preuss) thought, which they define not according to particular operations so much as to their specific content (e.g., demons born of affective projections or actions retroactively deemed magical). Here, too, 'primitive thinking' represents the point of origin of contemporary logic. On the one hand, the former is never too far away from the latter because their fundamental operations are considered so similar. On the other hand, they stand worlds apart because the former is thought to channel affect in a manner foreign to modern science. French ethnologists depart from the focus on individual psychology by their English and German counterparts in favor of social psychology, but they share an attachment to the great significance of affect. For them, the group dynamic sets the course for magical thinking, which they explain by turning to the collective and the performative power of its rituals, in which individuals experience a

---

126 Lévy-Bruhl, *How Natives Think*, 342.
127 Lévy-Bruhl, *How Natives Think*, 346.
128 Lévy-Bruhl, *How Natives Think*, 347.

force much stronger than themselves and become emotionally and physically overwhelmed. In contrast to the model proposed by English and German theorists, the sociological view holds that magic in fact works because rites have an undeniable effect on participants, whose experiences confirm their belief in its efficacy.

Whereas English and German ethnologists follow an evolutionary paradigm and examine 'primitive thinking' as a less developed precursor to rational thought, the French pursue a genealogical project. Tracing the provenance of the categories of understanding, they arrive at the peculiar conceptual worlds of primeval societies. The sacred serves a special role as a type of primal category whereby the collective worships itself, or rather its power over its members. But what holds for the sacred also applies to all other categories that emerge in this manner: they appear to the individual as given by nature, not because they are a priori, but because the collective has created and passed them down. Thus, the genealogical project, which recognizes even the principles of rational thought as products of society and history, also relativizes the types of thinking it attributes to Europeans – even though Durkheim, who understands society as nature and temporal duration as an index of objectivity, would wish otherwise.

Lévy-Bruhl draws the conclusion that magical thinking is an autonomous alternative to rational thinking. Although he adopts the social-psychological orientation of Mauss, Hubert, and Durkheim at first, he ultimately distances himself from them in that he does not analyze participation as the result of a collective event. Instead, he increasingly deems it to be a feature always already inherent to 'primitive being': something that does not determine thinking so much as feeling, or more precisely, a fundamental disposition, which in turn precedes and shapes all thought. In a sense, Lévy-Bruhl's level of analysis falls behind that of social psychology inasmuch as the constructed nature of participation goes amiss in his work. In his view, in the beginning there was participation, not the collective. This perspective leads to a further point of difference and not just with his French colleagues. Lévy-Bruhl holds that magical thinking operates by way of an a priori synthetic scheme of collective representations. Syntheses are not formed, but always already given. What's more, they do not involve hierarchy so much as posit identity. Thus, 'primitive thinking' does not associate a human being and a parrot with one another or subordinate one to the other. Instead, a human being *is* always already also a parrot.

Lévy-Bruhl breaks with the ethnological tradition of a progressive history that arrives at the logical thinking of the European subject. As we have seen, his reflections and judgments occasionally reveal traces of this tradition, but his overall theory insists that 'primitive thinking' represents an autonomous alternative to rationality. At some points, Lévy-Bruhl conceives this alternative in terms of an historical discontinuity, where a discrete period of 'primitive thinking' is succeeded by another of 'logical thinking.' At other points, the two intellectual patterns occur simultaneously, with 'primitive thinking' representing both an alternative to modern Europeans and a possible means by which they might cure themselves of alienation. This feature of Lévy-Bruhl's theory is the source of his extraordinary popularity from the 1910s to the 1930s among artists and writers, who turned to his writings in their development of artistic primitivism.

Time and again, the works I have been discussing also trace an arc from 'primitive thinking' to art, that is, to the production and reception of indigenous art as well as to European artists. This is due to the assumption by early ethnologists that 'primitive thinking' has survived in the contemporary creation of art: the artist is a survival of the 'primitive' and the creation of art is a survival of 'primitive thinking.' Tylor, for instance, considers the mental procedures of indigenous cultures to provide insight into the literary arts: "In so far as myth [...] is the subject of poetry, and in so far as it is couched in language whose characteristic is that wild and rambling metaphor which represents the habitual expression of savage thought, the mental condition of the lower races is the key to poetry."[129] The various accounts of 'primitive thinking' entail similarly varied conceptions of art. Depending on the ethnological school in question, art is said to conjure up a naïve and intuitive mode of explaining the world, to express elementary human affect, or to establish a collective identity. As for the artists who wish to return to 'primitive thinking,' the path may involve childlike inquiry, immediate and unconditional forms of expression, or – as per the French theories – rites through which the collective first constitutes itself.[130] In chapters 5 and 6, I will return to this role of 'primitive thinking' in studies of the arts.

The Ethnological *Poème* of the 'Primitive'

Later ethnological research moved away from the paradigm of the 'primitive' and exposed it as erroneous. The same judgment applies to the assertion of 'primitive thinking.' In 1959, Godfrey Lienhardt declared that "no one who studies savage

---

129 Tylor, *Primitive Culture*, 2: 404.
130 Authors belonging to the Collège de Sociologie are also very relevant in this regard. For discussion from a sociological perspective, see Stephan Moebius, *Die Zauberlehrlinge. Soziologiegeschichte des Collège de Sociologie (1937–1939)* (Konstanz: UVK, 2006). From the perspective of literary history, see Irene Albers and Stephan Moebius, "Nachwort," in Denis Hollier, *Das Collège de Sociologie, 1937–1939* (Frankfurt am Main: Suhrkamp, 2012).

societies would say, today, that there are modes of thought which are confined to primitive peoples."[131] Ten years earlier, in *Les structures élémentaires de la parenté* (1949; *The Elementary Structures of Kinship*, 1969), Claude Lévi-Strauss had already discounted the "so-called archaism of primitive thought" as an "illusion," rejecting the idea that indigenous peoples and children think in the same way.[132] Instead, he shows that the notion of childlike thinking serves as a "sort of common denominator for all thoughts and all cultures."[133] Thus, any given culture will appear childlike to another. In 1962, Lévi-Strauss's *Totémisme aujourd'hui* (*Totemism*, 1963) also exposed ethnology's own totemistic illusion: "totemism is [...] the projection outside our own universe [...] of mental attitudes incompatible with the exigency of a discontinuity between man and nature which Christian thought has held to be essential."[134] The same can be said of the construction of 'primitive thinking,' of which totemism is understood to be an expression. In *La pensée sauvage* (1962; *The Savage Mind*, 1966), Lévi-Strauss refutes the notions that 'primitive thinking' is undeveloped and prelogical. Instead,

> there are two distinct modes of scientific thought. These are certainly not a function of different stages of development of the human mind but rather of two strategic levels at which nature is accessible to scientific enquiry: one roughly adapted to that of perception and the imagination; the other at a remove from it.[135]

In other words, what Lévi-Strauss calls the "savage mind" proves to be as developed and complex in structure as the mind of "modern science."[136]

Needless to say, it could be argued that even the critics of the paradigm of 'primitive thinking' had blind spots – for instance, if they still posited two fundamentally different ways of thinking or continued to unreflectively speak of the 'primitive' without contextualizing the term in its discursive history. Scholars such as Francis L. K. Hsu, Adam Kuper, and Johannes Fabian have drawn attention to the latent persistence of this paradigm in ethnological research.[137] My pre-

---

131 Godfrey Lienhardt, "Modes of Thought," in Evans-Pritchard, Edward E., Raymond Firtz, John Layard et al., *The Institutions of Primitive Society. A Series of Broadcast Talks* (Oxford: Blackwell, 1959), 95.
132 Claude Lévi-Strauss, *The Elementary Structures of Kinship*, trans. James Harle Bell and John Richard von Sturmer (Boston: Beacon, 1969), 95.
133 Lévi-Strauss, *The Elementary Structures of Kinship*, 94.
134 Lévi-Strauss, *Totemism*, trans. Rodney Needham (Boston: Beacon, 1963), 3.
135 Lévi-Strauss, *The Savage Mind* (Chicago: University of Chicago Press, 1966), 15.
136 Lévi-Strauss, *The Savage Mind*, 13.
137 Hsu, "Rethinking the Concept 'Primitive'"; Fabian, *Time and the Other*; Kuper, *The Invention of Primitive Society*. On how critics of ethnocentrism tend to re-exoticize the peoples they

sent purpose, however, is to show that by critically attending to the theorems informing ethnology's earliest stages, it is possible to uncover what the French philosopher and historian of science Gaston Bachelard would call their "poetic" character. The same holds for the figure of the 'primitive' and for the idea of a 'primitive thinking.' Following Bachelard's lead, we can understand these terms as *poèmes*, that is, as the basic motifs structuring the "poetry" of early ethnological efforts. As I remarked in the introduction, Bachelard contends that the "poetry of science" typically takes form in the context of an initial encounter or "first contact."[138] The emerging field of ethnology not only thematized "first contacts," but indeed owed its substance to what happened when representatives of the nascent field encountered indigenous peoples in what was, in fact, a profoundly colonialist setting.

When Bachelard speaks of the "poetry of science," "scientific reveries," and their "poèmes," he himself brings the concept of the 'primitive' into play: "Reverie […] always operates as it would in primitive minds."[139] Given Bachelard's psychoanalytic orientation, the statement is not surprising. Freud likewise started from the assumption that mental states and modes of expression that belong to the "earliest and most obscure periods of the beginnings of the human race" are reactivated in dreams.[140] (Chapter 4 will explore this matter in detail.) But Bachelard's discussion of the 'primitive poetry of science' is especially interesting because in the scenario of "first contact" he attributes primitivity to the scientific observer, not the culture under observation.[141] In fact, when devising their *poème* of the 'primitive,' ethnologists themselves think in a manner that they describe as 'primitive,' namely taking analogies as proof of identity. Beginning with Tylor, a key feature in the ethnological construction of 'primitive thinking' has been the supposed reliance by indigenous peoples on analogies they consider to be actual matters of fact. At the same time, however, ethnology itself is based on an unwitting analogical operation. Because affinities are said to exist between how indigenous peoples think and the ways that the first (hypo-

---

study, see Derrida on Lévi-Strauss (*Of Grammatology*, 95–140). Cf. Rolf Parr, "Exotik, Kultur, Struktur. Tangenten dreier Perspektiven bei Claude Levi-Strauss," *kultuRRevolution. Zeitschrift für angewandte Diskurstheorie* 32–33 (1995). In turn, Därmann critiques "thoroughgoing efforts to shield Derrida against perspectives from foreign cultures and Native American materials" (*Fremde Monde*, 18).

138 Bachelard, *The Psychoanalysis of Fire*, 1.
139 Bachelard, *The Psychoanalysis of Fire*, 4.
140 Sigmund Freud, *The Interpretation of Dreams*, trans. James Strachey (New York: Basic Books, 2010), 550.
141 Bachelard, *The Psychoanalysis of Fire*, 1–4.

thetical) human communities must have, the former count as descendants of the latter and the two groups are ultimately treated as identical, i.e., as 'primitives.' The resulting construction of the anthropomorphic figure of the 'primitive' is informed by substantialization and animistic thinking alike – which many ethnologists would prefer to see only in indigenous others. Also, the historical development that ethnologists are tracking is given substance as well as life by means of this figure, i.e., the source of the modern self now receives a face and living presence.

Bachelard's "psychoanalysis of reason" examines how unconscious impulses, affects, and representations trigger the production of scientific reveries.[142] Factors include the "need to *possess*,"[143] as well as an animistic belief in "*living matter*."[144] Regarding early ethnologists, knowledge that a primal source still exists and is available for study in real life might be said to satisfy both needs. Bachelard likewise deems a belief in full and unmediated contact – direct sensory experience leading to unambiguous conclusions – to be reverie (rather than thought). This judgment would certainly apply to early ethnologists' study of indigenous peoples, when in reality 'first contact' often occured through multiple intermediaries. Finally, Bachelard stresses the key role of libido, which is expressed as a scientific "will to power."[145] This is evident in researchers' aversion to critically review their own scientific results, for example in the early days of ethnology, when ethnologists often relied on others' reports and did not conduct any fieldwork of their own to verify or disprove their claims. It is also found in the distancing and deprecating gestures that draw a fundamental line of separation between European and non-European cultures. The colonialist framework to which ethnology owes its very existence implies from the outset a will to dominate what is foreign.

Thus, Bachelard's picture of how unconscious motivations, affects, and representations may shape scientific poetry can very well be verified by early ethnology. However, this picture needs to be completed by attending to the basic patterns followed by ethnological poetry. For Bachelard's observations on *poèmes* imply that they play the same role in scholarly reveries that theorems play in science. In this sense *poèmes* would be understood as the structuring pattern of scientific "poetry." What does this circumstance mean in the context of early ethnological writings? To what extent does the 'primitive' represent such a *poème*?

---

[142] Gaston Bachelard, *The Formation of the Scientific Mind* (Manchester: Clinamen, 2002), 29.
[143] Bachelard, *The Formation of the Scientific Mind*, 173. Emphasis in the original
[144] Bachelard, *The Formation of the Scientific Mind*, 159. Emphasis in the original.
[145] Bachelard, *The Formation of the Scientific Mind*, 207.

What aesthetic properties inhere in the 'primitive' and in texts that enlist this category to define – and shape – their object?

Other historians of science have also suggested that aesthetic factors, a certain proximity to literature or visual art, contribute to the success of scientific works. In *The Structure of Scientific Revolutions*, for example, Thomas Kuhn points out that the acceptance of a new model also depends on its consistency, the inner coherence and unity it displays. He compares the shift from one paradigm to another with a "change in visual gestalt: the marks on paper that were first seen as a bird are now seen as an antelope, or vice versa."[146] Ludwig Fleck stresses consistency, too. Instead of employing metaphors from the realm of literature or the visual arts, he turns to music when he speaks of the "harmony" that closed systems exhibit.[147] By the same token, Fleck enlists the concept of *Stimmung*, a German word that, while difficult to translate, may be understood as a combination of atmosphere and mood. For him, *Stimmung* does not refer to an already constituted "thought style" (the structural equivalent to what Kuhn calls paradigm) but produces it in the first place: "Like any style, the thought style also consists of a certain mood and of the performance by which it is realized. [...] Whole eras will then be ruled by this thought constraint [...] until a different mood creates a different thought style and a different valuation."[148] More than any other academic field treated in this book, ethnology has confronted its past in the manner that these historical epistemologists demand. In particular, it has done so in response to the debate inaugurated by postcolonial studies and displayed in the edited volume *Writing Culture*, to which I will return below.

In *Orientalism*, which may be considered the charter of postcolonial studies, Edward Said writes:

> The phenomenon [...] as I study it here deals principally, not with a correspondence between Orientalism and Orient, but with the internal consistency of Orientalism and its ideas about the Orient [...] despite or beyond any correspondence [...] with a "real" Orient.[149]

Leaving aside Said's essentialism, much of what he says applies to the subject at hand. My task here is to examine discourse in the human sciences about 'prim-

---

[146] Thomas S. Kuhn, *The Structure of Scientific Revolutions* (Chicago: University of Chicago Press, 1970), 85.
[147] Ludwik Fleck, *Genesis and Development of a Scientific Fact* (Chicago: University of Chicago Press, 1979), 38.
[148] Fleck, *Genesis and Development of a Scientific Fact*, 99.
[149] Edward W. Said, *Orientalism* (New York: Vintage, 1994), 5.

itive thinking' and the ways it is reflected in literature. But my present discussion does not concern the adequacy of those studies, nor do I seek to engage with the real cultures of indigenous peoples. The 'primitive humanity' discoursed upon in these texts does not exist any more than the 'primitive thinking' it is supposed to exemplify. In keeping with Said's observations, this does not mean that mere fantasy stands at issue, however. The works treated here concern distorted representations of specific human beings and cultures, and such misrepresentation holds consequences. Even if the discourse on the 'primitive' is literary or academic, it has implications in terms of power politics. Said's Foucauldian perspective on orientalism applies here, too: it is "a Western style for dominating, restructuring, and having authority over [non-Westerners]."[150] Thus, at some point, "Orientalism" commandeers the "Orient";[151] that is, in many respects it becomes what orientalism misrepresents it to be. While the reasons for this are too varied and complex to be discussed here, the 'primitive' occasions a similar dynamic. This state of affairs is evident, for example, when art historians question the authenticity of so-called 'primitive artifacts.' From a critical perspective, 'primitivism' (which represents a Western or, at any rate, a non-indigenous bearing) is what produces 'primitive' art in the first place by inducing foreign peoples to fashion objects for European travelers that match the latter's preconceived notions. In extreme instances – as ethnographers (motivated by their own fantasies) searching for 'virgin' cultures have often observed – the process has led to a wholesale restructuring of native ways of life. Primitivism is therefore no longer just "here" but also "there." It is the primary agent, and the 'primitive' is its aftereffect.

That said, Said does not offer terribly inspiring individual readings.[152] Although his thesis is convincing, it is frustrating to find the readings of individual texts reproducing the same model of orientalism over and over. Said tends to suppress differences between texts and to overlook deviations and points of ambiguity in individual works to confirm that orientalism is inescapable. In his eyes, Europeans cannot occupy a position outside this discourse of power, which always already has conditioned their viewpoint and made them its exponents.[153] Fortunately, as Oliver Lubrich has shown, other approaches to postcolonial studies discover alternative models of representing alterity in the texts

---

150 Said, *Orientalism*, 3.
151 Said, *Orientalism*, 96.
152 Oliver Lubrich, "Welche Rolle spielt der literarische Text im postkolonialen Diskurs?" *Archiv für das Studium der neueren Sprachen und Literaturen* 1 (2005): 18.
153 Said, *Orientalism*, 11.

they study, a recognition that facilitates more nuanced readings of that literature.¹⁵⁴ Stephen Greenblatt presents an approach that, by proceeding typologically, facilitates comparisons between texts in terms of how they encounter the other. His *Magnificent Possessions* takes an emotional reaction to foreignness as its point of departure: "Wonder – thrilling, potentially dangerous, momentarily immobilizing, charged at once with desire, ignorance, and fear – is the quintessential human response to [...] a 'first encounter.'"¹⁵⁵ Using the travel accounts of Mandeville and Columbus (among others), Greenblatt points out that the transition from this emotion to the attempt at description can give rise to two contrary attitudes.

> One path leads to [...] discursive strategies to articulations of the hidden links between the radically opposed ways of being and hence to some form of acceptance of the other in the self and the self in the other. The movement is from radical alterity – you have nothing in common with the other – to a self-recognition that is also a mode of self-estrangement: you *are* the other and the other is you. The alternative path leads to [...] discursive strategies [...], that is, to articulations of the radical differences that make renaming, transformation, and appropriation possible. The movement here must pass through identification to complete estrangement: for a moment you see yourself confounded with the other, but then you make the other become an alien object, a thing, that you can destroy or incorporate at will.¹⁵⁶

The first perspective is described as metaphorical, since it is based on the perception of similarity that does not vanish into absolute difference or absolute sameness. The second perspective is considered metonymic, insofar as what is alien comes to be subsumed under the self and its possessions. Whereas he clearly identifies the latter mindset as colonialist, the former "abstains from taking possession."¹⁵⁷ For Greenblatt, such "disinterest" amounts to an aesthetic relationship to the foreign. While Greenblatt's method of reading, because of its typological orientation, may not do justice to all works, it is admirably suited to identifying the spectrum of otherness that diverse texts represent. In contrast to Said, Greenblatt brings out how colonial discourse also harbors countervailing tendencies, which posit a simultaneous self-alienation and familiarization with the other rather than a domestication of it. Particularly suggestive is his proposal that these tendencies are tied to an aesthetic attitude of the narrator to the foreigner he represents.

---

154 Lubrich, "Welche Rolle spielt der literarische Text im postkolonialen Diskurs?" 21–22.
155 Stephen Greenblatt, *Marvelous Possessions* (Oxford: Oxford University Press, 1991), 20.
156 Greenblatt, *Marvelous Possessions*, 135. Emphasis in original.
157 Greenblatt, *Marvelous Possessions*, 24.

Contemporary ethnologists have also risen to the challenge of Said's thesis. Their discipline is founded on the premise that indigenous cultures may be described impartially, and they have developed methods for countering the obstacles that stand in the way of fulfilling this task. Unlike post-colonialist readings of the representation of otherness in documents that, for the most part, predate the emergence of ethnology, their focus concerns the methods and stylistic devices employed in that process of representation. Seeking to expose the constructed nature of ethnological authorship and authority, James Clifford has identified four kinds of authorship:

The oldest model establishes itself by means of "experience" through testimony. Ever since the time of Bronislaw Malinowski, "the 'man on the spot' [...] and the [...] anthropologist in the metropole"[158] have constituted two complementary sides of ethnology – an arrangement that takes care of problems attending the earlier division of labor (in particular, the unreliability of sources). In this framework, the ideal field researcher serves as a neutral recorder of the foreign world, serving as a blank page where an objective image of the foreign takes shape. As Clifford notes, this method is subject to criticism inasmuch as the observer views other cultures in light of his own and thereby taints the record.

The second approach seeks to remedy such bias by enlisting interpretation as a means of authentication. This orientation is exemplified by the approach Clifford Geertz developed in light of Paul Ricoeur's discussion of hermeneutics. In "Thick Description," Geertz uses the example of winking, which can hold a broad range of meanings, to illustrate the difference between a given physical action and the cultural code framing it – or, more precisely, the interplay between them, which is what constitutes a meaningful gesture in the first place.

> Doing ethnography is like trying to read (in the sense of "construct a reading of") a manuscript – foreign, faded, full of ellipses, incoherences, suspicious emendations, and tendentious commentaries, but written not in conventionalized graphs of sound but in transient examples of shaped behavior.[159]

He emphasizes hereby that ethnographers' interpretations belong to the "second and third order"; only a member of the culture under observation is in the position to offer "first order ones." Thus, the manuscript appears to be unreadable to the ethnographer at first. The whole process makes evident that all interpreta-

---

158 James Clifford, "On Ethnographic Authority," in *The Predicament of Culture: Twentieth-century Ethnography, Literature, and Art* (Cambridge, MA: Harvard University Press, 1988), 26.
159 Clifford Geertz, "Thick Description. Towards an Interpretative Theory of Culture," in *The Interpretation of Cultures* (New York: Basic Books, 2000), 9.

tions (including "first order ones") "are [...] fictions; fictions, in the sense that they are 'something made,' 'something fashioned' – the original meaning of *fictio* – not that they are false, unfactual, or merely 'as if' thought experiments."[160] Clifford, however, critiques the interpretive model for its reliance on writing, which means detaching phenomena from their performative context: "The actuality of discursive situations and individual interlocutors is filtered out."[161]

Ethnology has responded to this deficit by adopting a third approach involving methods of authentication based on dialogue and even polyphony. The dialogue model operates by way of exchanges between the researcher and members of the foreign culture, often in an interview framework. The polyphony model aspires to an even greater restriction of ethnographic authority by aiming for a collage of information from diverse and native sources. Interestingly, compilations made by researchers such as Franz Boas and Malinowski in the early twentieth century already exemplify this approach: "In these works the ethnographic genre has not coalesced around the modern interpretational monograph closely identified with a personal fieldwork experience. [...] These older assemblages include much that is actually or all but written by informants."[162] To be sure, these texts are also under the control of ethnographers, who record, translate, and put in writing what informants tell them with greater and lesser accuracy. Clifford notes, however, the example of Malinowski, who published material he recognized he did not understand. For his own part, Clifford would like ethnographic texts to take the fourth approach by occupying an "arena of diversity," which he defines in reference to works of literature and literary theory[163] – for instance, Mikhail Bakhtin's concept of the "polyphonic novel," or the multivocality at work in Charles Dickens's novels.[164]

By enlisting literature as a model for, if not a component of, ethnography, Clifford is continuing a long tradition in ethnology. As should be clear by now, the 'primitive' represents a transitional figure in ethnological discourse. It stands at the border between the foreign and the familiar and between nature and culture, and it facilitates the constant renegotiations of that border. The transitional nature of the 'primitive' is also involved in ethnology's understandings of itself, especially in its early phases. As Sven Werkmeister has shown, the discipline swings between a philological orientation and one rooted in the nat-

---

160 Geertz, "Thick Description," 15.
161 Clifford, "On Ethnographic Authority," 40.
162 Clifford, "On Ethnographic Authority," 45.
163 Clifford, "On Ethnographic Authority," 46.
164 Clifford, "On Ethnographic Authority," 46–47.

ural sciences.¹⁶⁵ In some instances, it even exhibits an oscillation between a scientific and a literary orientation – though it often fails to acknowledge this circumstance. A look at two celebrated examples will make as much plain.¹⁶⁶

I have already remarked that early ethnological texts feature an analogical scheme of argument. Other fundamental rhetorical features include the topoi of origin, beginning, and evolution – to say nothing of the topos of the 'primitive' itself. Equally, it is important to note preferred choices of genre that inform ethnological works, including the beginnings of celebrated studies such as Frazer's *Golden Bough* and Malinowski's *Argonauts of the Western Pacific* (1922).¹⁶⁷ Frazer does not begin his book with an exposition of his theory of "sympathetic magic" (which is reserved for the third chapter). Instead, he invokes a work of visual art, the painting by William Turner that lends the study its name. The ekphrasis that follows leaves it open as to whether Frazer's description is based on the actual landscape or its depiction on the canvas – whether we are in "a realm 'transfigured' by the 'imagination'" or material reality. What is more, the author's own words reenact the process ascribed to the artist: "The scene, suffused with the golden glow of imagination in which the divine mind of Turner steeped and transfigured even the fairest natural landscape, is a dream-like vision of the little woodland lake of Nemi."¹⁶⁸ Frazer then proceeds to evoke features of the landscape that we would ascribe to his own 'transfiguring imagination.'

> No one who has seen that calm water, lapped in a green hollow of the Alban hills, can ever forget it. The two characteristic Italian villages which slumber on its banks, and the equally Italian palace whose terraced gardens descend steeply to the lake, hardly break the stillness and even the solitariness of the scene. Diana herself might still linger by this lonely shore, still haunt these woodlands wild.¹⁶⁹

These suggestive words do not describe the scenery so much as immerse the reader in it, for one is enjoined to envision the goddess Diana coursing through

---

165 Werkmeister, *Kulturen jenseits der Schrift*, 70–77.
166 Further examples for early ethnology that operates in a literary mode can be found in the writings of Leo Frobenius.
167 On Frazer's proximity to literature, cf. Stanley Edgar Hyman, *The Tangled Bank: Darwin, Marx, Frazer and Freud as Imaginative Writers* (New York: Atheneum, 1962), 187–292; and Christopher Herbert, "Frazer, Einstein, and Free Play," in *Prehistories of the Future*, ed. Barkan and Bush, who credits the author with a "modernist style of thought" and posits affinities with "early modernist writers" such as D. H. Lawrence (134).
168 Frazer, *The Golden Bough*, 1.
169 Frazer, *The Golden Bough*, 2.

the forest – that is, to enter a picture Frazer himself has drawn. The account of a "strange and recurring tragedy" follows:

> In this sacred grove there grew a certain tree round which at any time of the day, and probably far into the night, a grim figure might be seen to prowl. In his hand he carried a drawn sword, and he kept peering warily about him as if at every instant he expected to be set upon by an enemy. He was a priest and a murderer; and the man for whom he looked was sooner or later to murder him and hold the priesthood in his stead.[170]

Introducing an unknown man, the "grim figure" of a "murderer," and gruesome customs that are enigmatic because they remain unexplained serves to heighten suspense. The next paragraph in this narrative sequence again appeals to the reader's fantasy.

> We picture to ourselves the scene as it may have been witnessed by a belated wayfarer on one of those wild autumn nights when the dead leaves are falling thick, and the winds seem to sing the dirge of the dying year. It is a sombre picture, set to melancholy music – the background of forest showing black and jagged against a lowering and stormy sky.[171]

Now, description of the landscape resumes, but in a markedly different tone. The language abounds in metaphors and calls on a synaesthetic mode of perception because imaginary music complements the visual scenery. Frazer's study does not begin like a scholarly or scientific work, then, but much as a novel would – a book full of suspense falling somewhere between thriller, mystery, and crime fiction. The author adopts the role of a detective on the hunt for "a fairly probable explanation of the priesthood of Nemi."[172] The literary cast of the opening pages clearly serve to elicit interest on the part of the reader. In equal measure, it reveals the affinity between ethnology (as Frazer practices it), philology, literary technique, and the formation of fiction. On the pages that follow, the author-detective presents himself as cannily interpreting an array of myths and legends, "stories told to account for Diana's worship" that he bluntly qualifies as "unhistorical."[173]

The opening of Malinowski's *Argonauts of the Western Pacific*, which is equally famous, follows another literary strategy.[174] The organization is tripar-

---

170 Frazer, *The Golden Bough*, 1.
171 Frazer, *The Golden Bough*, 2.
172 Frazer, *The Golden Bough*, 3.
173 Frazer, *The Golden Bough*, 6.
174 Already in his preface to *Argonauts*, Frazer draws attention to the literary qualities of Malinowski's descriptions, comparing the artistry of his character sketches to Shakespeare's (Fraz-

tite: First comes a description of the population of the South Sea islanders, written in the style of an encyclopedia entry. Next, Malinowski offers a methodological reflection that focuses on the relationship between, "on the one hand, [...] direct observation, [...] native statements and interpretations, and on the other, the inferences of the author, based on his common sense and psychological insight."[175] In equal measure, he considers the relationship between field research and the (subsequent) tallying of results. The third component represents the item of interest for my purposes in this chapter: Malinowski means to provide a "brief outline of an Ethnographer's tribulations as lived through by myself."[176] Over and over, the autobiographical narration asks that readers use their imagination to picture themselves in the author's shoes.[177]

> Imagine yourself suddenly set down surrounded by your gear, alone on a tropical beach close to a native village while the launch or dinghy which has brought you sails away out of sight. [...] Imagine further that you are a beginner, without previous experience, with nothing to guide you and no one to help you. [...] This exactly describes my first initiation into field work on the south coast of New Guinea. I well remember the long visits I paid to the villages during the first weeks; the feeling of hopelessness and despair after many obstinate but futile attempts had entirely failed to bring me into real touch with the natives, or supply me with any material. I had periods of despondency, when I buried myself in the reading of novels [...]. Imagine yourself then, making your first entry into the village.[178]

---

er, "Preface," in Malinowski, *Argonauts of the Western Pacific*, x). For a thorough discussion see Harry C. Payne, "Malinowski's Style," *Proceedings of the American Philosophical Society* 125.6 (1981). Payne demonstrates, among other things, that Malinowski's writing is marked by three features reflecting his adoption of "native rhetoric" (424). Cf. also Clifford Geertz's reflections on "I-Witnessing" in Malinowski, especially his *Diary* (contemporaneous with *Argonauts*) ("I-Witnessing. Malinowski's Children," in *Works and Lives: The Anthropologist as Author* [Stanford: Stanford UP, 1988]). For discussion of Malinowski's relationship to Joseph Conrad's "Heart of Darkness" (especially in the *Diary*), see Clifford, "On Ethnographic Self-Fashioning: Conrad and Malinowski," in *The Predicament of Culture, Twentieth-Century Ethnography, Literature and Art* (Cambridge, MA: Harvard University Press, 1988).

175 Bronislaw Malinowski, *Argonauts of the Western Pacific: An Account of Native Enterprise and Adventure in the Archipelagoes of Melenesian New Guinea* (London: Routledge & Kegan Paul, 1932), 3.
176 Malinowski, *Argonauts of the Western Pacific*, 4.
177 For Frazer's influence on Malinowski, cf. George W. Stocking, Jr., "'Cultural Darwinism' and 'Philosophical Idealism,'" in *Race, Culture and Evolution: Essays in the History of Anthropology* (New York: Free Press, 1968), 53; Stocking goes on to show how *Argonauts* may be read as a "euhemerist myth" (56).
178 Malinowski, *Argonauts of the Western Pacific*, 4.

Autobiography also features in the following section, into which the author interlaces descriptions of the *"proper conditions for ethnographic work."*

> Soon after I had established myself in Omarakana [...], I began to take part [...] in the village life [...]. I would get out from under my mosquito net, to find around me the village life beginning to stir [...]. As I went on my morning walk through the village, I could see intimate details of family life, of toilet, cooking, taking of meals.[179]

In sum, Malinowski's introduction is constantly switching between analysis, description, and argument, on the one hand, and narrative autobiography, on the other. While this style serves to pique interest, even more importantly, it underscores the first-hand experience that affirms the writer's authority. As noted above, Clifford identifies this strategy as the earliest of four ways of establishing the veracity of ethnological claims.

Alternation between these two styles, which could be associated with the rhetorical level of *dispositio*, also occurs in other ethnological texts, particularly at points when the discussion concerns the customs of native peoples or the field researcher's experience gathering data for analysis. Often, such narrative passages, set apart from the rest of the text, are in fact quotations; the author himself has not performed any investigations on site and must rely on the stories of others. This is the case for Lévy-Bruhl. A great number of particularly impressive passages of this type are featured in chapter 8 of his first book, for example, which examines relations between the living and the dead. Thus, the story is told of a young girl who married her betrothed's ghost. Lévy-Bruhl then interprets the tale to demonstrate that "primitives" have "mystic" ideas about life and death that cannot be grasped with "our" concepts.[180] These narrative inlays admit comparison with case histories in clinical psychology, to which I will return in chapters 3, 4, and 8.[181] Such passages purport to be based on empirical facts – and even when myths or legends stand at issue, these are understood as empirical data documenting a collective's worldview. Also, these narrative inlays claim to report the specific beliefs and practices of one particular culture, which

---

179 Malinowski, *Argonauts of the Western Pacific*, 7. Emphasis in original.
180 Lévy-Bruhl, *How Natives Think*, 273.
181 Needless to say, numerous differences also exist, for instance, the absence of a fixed narrative scheme (e.g., the figures of doctor and patient or the arc from symptom to crisis to resolution); nor is it a matter of incidents that are subsequently brought into chronological and causal order. See Chapters 3 and 4 especially for fuller discussions.

are at the same time supposed to open a broader anthropological horizon – Lévy-Bruhl, for example, speaks of how 'primitives' view the dead in general.

The rhetorical and literary aspects of these (and other) ethnological texts point to their fictionality in a twofold sense: First, in that Geertz demonstrates how the ethnologist's 'reading' of a culture yields a fabrication – a fiction in the sense of the Latin *fingere*. Second, in the more radical meaning of fictionality that Clifford outlines in his introduction to *Writing Culture:* "the fact that [ethnography] is always caught up in the invention, not the representation, of cultures."[182] In contrast to Geertz, Clifford's concept of fiction explicitly incorporates *inventio*, the rhetorical canon for devising "things not actually real."[183] At the same time, another shade of meaning is at play in the closely related *invenire*, or discovery. He does not claim that ethnographic texts present a mere concoction, but rather a "true fiction," which occupies a space somewhere between invention and discovery. The researcher confronts what Clifford considers a moral demand to be cognizant of this unavoidable fact and to bear it in mind when writing: "Ethnographic truths are thus inherently *partial* – committed and incomplete. [...] A rigorous sense of partiality can be a source of representational tact."[184] One option for handling this situation is to go on the offense and ennoble ethnography's rhetorical and literary features as desired methods. Examples of such an approach include Tzvetan Todorov's *The Conquest of America*, which proposes "to narrate a history" on the model of the novel. With an eye to the three unities of classical drama, Todorov has the authors of the texts discussed speak both in monologue and in concert, to bring forth a polyphony of voices. That said, Todorov also takes pains to avoid presenting complete inventions. His aim is to provide an "exemplary story,"[185] that is, "one that will be as true as possible."[186]

Frazer and Malinowski do not share this goal and offer no reflections on the necessary fictionality of their studies, but the literary traits of their texts reveal it

---

182 Clifford, "Introduction: Partial Truths," in *Writing Culture: The Poetics and Politics of Ethnography*, ed. James Clifford and George E. Marcus (Berkeley: University of California Press, 1986), 2.
183 Clifford, "Introduction," 6.
184 Clifford, "Introduction," 7.
185 Tzvetan Todorov, *The Conquest of America: The Conquest of the Other*, trans. Richard Howard (Norman, OK: University of Oklahoma Press, 1999), 4.
186 This might be the only possible outcome of the dilemma identified by Kuper (*The Reinvention of Primitive Society: Transformation of a Myth* [London: Routledge, 2005], 201–224) and Li, among others, that recent postcolonial and "native" positions get entangled in an "anti-primitivist primitivism without primitives" (*The Neo-Primitivist Turn*, ix), which is essentialist instead of openly utopian (that is, fictional).

nevertheless. Above, I pointed out that early ethnological works often affirmed the proximity of 'primitive thinking' to artistic creation. The partial literariness these texts exhibit introduces yet another dimension where ethnology proves its relevance for literature. If ethnology is always already (also) literature, the opposite holds as well. Literature itself can claim to be ethnology – or, at any rate, an "imaginary ethnography"[187] that recognizes from its inception that 'primitive thinking' amounts to a fiction of the author's own culture, and that the author's own culture represents the actual focus of attention and conundrum to be explained.[188]

---

[187] Gabriele Schwab, *Imaginary Ethnographies. Literature, Culture, and Subjectivity* (New York: Columbia University Press, 2012).

[188] A particularly interesting author in this context is the writer *and* ethnologist Michel Leiris. For extensive discussion, cf. Irene Albers, *Der diskrete Charme der Anthropologie. Michel Leiris' ethnologische Poetik* (Konstanz: UVK, 2018), especially 25–46; see also Marie-Denise Shelton, "Primitive Self: Colonial Impulses in Michel Leiris's *L'Afrique fantôme*," in *Prehistories of the Future*, ed. Barkan and Bush; and Marjorie Perloff, "Tolerance and Taboo," in *Prehistories of the Future*, ed. Barkan and Bush.

Chapter 3
The Child as 'Primitive'

Like turn-of-the-century ethnology, developmental psychology of the same period was shaped by the paradigm of the 'primitive' and followed the principle of analogy.[1] In contrast to early ethnology, however, it equated 'prehistoric man' and the contemporary child (instead of indigenous peoples) under the heading of the 'primitive.'[2] Consequently, the analogy required a different temporal model whereby the desired relation is not inscribed by allochrony and arrested development, but rather by recapitulation of a past developmental process in the present. Whereas the ethnological concept of survival is best understood in idealist terms (in the sense of a transmission of older cultural properties), the model of recapitulation is exclusively materialist and must be understood in biological terms. Drawing on popularized evolutionism, the new field of child psychology presumed that children's thought is systematically programmed to steadily and progressively mature into adult thinking. This course of development was thought to recapitulate a phylogenetic cultural development all the way from 'primitive' to 'civilized' thinking.

---

[1] Developmental psychology emerged in the late nineteenth century in tandem with "folk psychology" (*Völkerpsychologie*) and animal psychology; at the time, it was largely referred to as "child psychology." See Georg Eckardt, Wolfgang G. Bringmann, and Lothar Sprung, eds., *Contributions to a History of Developmental Psychology: International William T. Preyer Symposium* (Berlin, New York, Amsterdam: Mouton, 1985), Part I.

[2] One of the few monographs on the child against the background of the broader historical discourse on primitivism is George Boas' book, *The Cult of Childhood* (London: Warburg, 1966). Boas sees the cult of the child/childhood as a substitute for the cult of the 'primitive' (or, more precisely, the "noble savage") after contradicting experiences had rendered the latter impossible (8–9). See also Wittmann, *Bedeutungsvolle Kritzeleien*, which deals with the recapitulationist concept of a "Neolithic childhood" in historical treatments of children's drawings (187–243). Yet she emphasizes that after 1910 the significance of the theory of recapitulation for embryology and anatomy quickly waned and that growing criticisms of its adaptation to other fields were mounted by cultural historians as well (230–237). See also Elisabeth Wesseling, ed., *The Child Savage, 1890–2010. From Comics to Games* (Farnham: Ashgate, 2016), which discusses the "child-savage analogy" as a "root metaphor" of "modern Western culture" (5); the first part of the volume examines "how the child-savage analogy was fleshed out by children's media during the heyday of imperialism" (14).

Recapitulating Phylogeny

The most important point of reference here is Ernst Haeckel. His writings – *Generelle Morphologie der Organismen* (General Morphology of the Organisms, 1866), *Natürliche Schöpfungsgeschichte* (1868; *The History of Creation*, 1873), and *Anthropogenie* (1874; *The Evolution of Man*, 1876) – had established the "biogenetic law" whereby the individual life recapitulates the life of the species.

> Ontogeny is a brief and rapid recapitulation of Phylogeny, dependent on the physiological functions of Heredity (reproduction) and Adaptation (nutrition). The individual organism reproduces in the rapid and short course of its own evolution the most important of the changes in form through which its ancestors, according to laws of Heredity and Adaptation, have passed in the slow and long course of their palaeontological evolution.[3]

For Haeckel ontogeny and phylogeny are not just similar processes; rather, the latter represents the mechanical cause of the former.[4] This sets his perspective apart from those of the natural philosophers before him who affirmed that similarities exist between ontogeny and phylogeny due to the grand-scale unity of nature, not to causal relations.[5] By contrast, Haeckel's thesis of a mechanical cause is based on two assumptions: First, he follows Jean-Baptiste Lamarck's conviction that acquired traits can be passed down by inheritance. (The textbook example runs like this: giraffes needed to stretch their necks to gather leaves from trees. Then their offspring were born with longer necks. The parents had transmitted this actively acquired trait to their young.) Second, Haeckel does not claim that ontogeny repeats every stage of phylogeny. Some stages get skipped, and only the most important ones are retained. Otherwise, ontogeny would extend to impossible lengths over the course of a human's development. As it is, its duration is the same from individual to individual and generation to generation.[6]

Haeckel's biogenetic law originally only concerned embryonic development. But as his theory quickly spread and became popularized, child psychologists applied it to infants and toddlers as well. One of the earliest works in child psy-

---

[3] Ernst Haeckel, *The Evolution of Man: A Popular Exposition on the Principal Points of Ontogeny and Phylogeny* (New York: D. Appleton, 1897), 1–2. On the "pervasive influence" of this "law" on "criminal anthropology, racism, child development, primary education, and Freudian psychoanalysis," see Stephen Jay Gould, *Ontogeny and Phylogeny* (Cambridge, MA: Harvard University Press, 1977), 115–166.
[4] Haeckel, *The Evolution of Man*, 4.
[5] Cf. Gould, *Ontogeny and Phylogeny*, 33–46, 76–84.
[6] On these two basic assumptions, cf. Gould, *Ontogeny and Phylogeny*, 80–84.

chology – indeed, the book that has come to stand as its founding document[7] – is William T. Preyer's *Die Seele des Kindes. Beobachtungen über die geistige Entwicklung des Menschen in seinen ersten Lebensjahren* (1882; *The Mind of the Child: Observations Concerning the Mental Development of the Human Being in the First Years of Life*, 1888). Here, Haeckel's authority is invoked when Preyer explains, among other things, that

> what we know [...] of the most ancient languages shows so great an agreement in regard to [...] the language of children [...] that we may say the human race [...] has behind it a course of development [...] similar to that which every normal child goes through in learning to speak.[8]

A decade later, James Sully, the founder of child psychology in England, went even further by making the ontogenetic recapitulation of phylogeny the guiding principle of his influential *Studies of Childhood* (1895). Indeed, in his book, analogies between the child and prehistoric humanity are omnipresent,[9] a relation programmatically articulated in the introduction.

> [The] evolutionary point of view enables the psychologist to connect the unfolding of an infant's mind [...] with the mental history of the race. [...] [The] first years of a child, with their imperfect verbal expression, their crude fanciful ideas, their seizures by rage and terror, their absorption in the present moment, acquire a new and antiquarian interest.[10]

---

7 "Wilhelm Preyer first put the psychology of early childhood on to a scientific basis" (William Stern, *Psychology of Early Childhood: Up to the Sixth Year of Age*, trans. Anna Barwell [New York: Henry Holt, 1924], 12). That said, Preyer was hardly "the first to tackle the issue of children's mental development. Indeed, the biographical and educational records published in English before Preyer provide a wealth of information on the subject" (John C. Cavanaugh, "Cognitive Developmental Psychology before Preyer: Biographical and Educational Records," in *Contributions to a History of Developmental Psychology*, 206). On "why Preyer's monograph [...] became the 'initial chapter' [...] of modern child psychology" (178), cf. Georg Eckardt, "Preyer's Road to Child Psychology," and Jaeger, "Origins of Child Psychology: William Preyer," in the same volume as above.
8 William Preyer, *Mental Development in the Child*, trans. H. W. Brown (New York: Appleton, 1894), 160.
9 E. g., James Sully, *Studies of Childhood* (London: Longmans Green, 1896), 9, 28, 61, 82, 91–94. On Sully, especially in the context of the Child Study Movement, cf. Sally Shuttleworth, "Child Study in the 1890s," in *The Mind of the Child. Child Development in Literature, Science and Medicine, 1840–1900* (Oxford: Oxford University Press, 2010), 267–289.
10 Sully, *Studies of Childhood*, 8.

The same view prevailed in the United States. In 1904, Sully's American counterpart G. Stanley Hall held the thesis that the child at play repeats both the biological and the cultural evolution of humankind.

> I regard play as the motor habits and spirits of the past of the race, persisting in the present [...]. The best index and guide to the [...] activities of adults in past ages is found in the instinctive, untaught, and non-imitative plays of children [...]. In play every mood and movement is instinct with heredity. Thus we rehearse the activities of our ancestors, back we know not how far [...]. It is reminiscent [...] of our line of descent, and each is the key to the other.[11]

On this basis, Hall concludes that "the child is vastly more ancient than the man. [...] Adulthood is comparatively a novel structure built upon very ancient foundations."[12] From this conviction his followers drew some daring conclusions. For example, in Switzerland, the child psychologist Pierre Bovet asserts the pedagogical value of Haeckel's notion of recapitulation as follows:

> Many of the child's instincts and likings, which were formerly a dead weight on his teacher's hands, take on a positive interest, as soon as the latter ceases to regard them as individual and passing whims, and accustoms himself to look on them as the living prolongation of the great forces which have fashioned mankind for thousands of years.[13]

Accordingly, Bovet contends that combat skills develop along phylogenetic lines. Young children, he claims, do not fight or show aggression until about the age of three – which corresponds to the peaceful and paradisiacal life imagined to have been in existence at the beginning of human history. After that, skills such as scratching, kicking, hitting, and the use of weapons supposedly develop in phylogenetic order.[14] Likewise, Karl Groos, in Germany, wrote in his introduction to *Das Seelenleben des Kindes* (The Inner Life of Children, 1904):

> We can also [...] hope that through our study we will be able to uncover the many connecting threads between the growth of the individual soul and the first beginnings of the human species. [...] [Child psychology] should [...] have the vocation to fathom the myster-

---

[11] G. Stanley Hall, *Adolescence: Its Psychology and Its Relations to Physiology, Anthropology, Sociology, Sex, Crime, Religion, and Education* (New York: Appleton, 1904), 1: 202.
[12] Hall, quoted in Gould, *Ontogeny and Phylogeny*, 141.
[13] Pierre Bovet, *The Fighting Instinct*, trans. J.Y.Y. Greig (New York: Dodd, Mead and Company, 1923), 150.
[14] Bovet, *The Fighting Instinct*, 152–154. Gould adduces the same examples from Bovet and Hall (*Ontogeny and Phylogeny*, 140–141).

ies of the spiritual development of humanity; conversely, what we know about the development of the species should shed a bright light on many phenomena of childhood life.[15]

Later studies in the field of developmental psychology were just as much shaped by the conviction that phylogeny repeats itself in ontogeny and that therefore the child is to be understood as a contemporaneous 'primitive.' In 1914, William Stern deemed it self-evident that children exhibit "psychic life" that is "primitive."[16] Right at the beginning of *Die geistige Entwicklung des Kindes* (1918; *The Mental Development of the Child*, 1924), Karl Bühler refers to how child psychology may greatly inform research on the "history of the species,"[17] where the "science of prehistory," he writes, may reap "its best source of information."[18] Throughout the work, he also refers to indigenous peoples – for instance, when treating the "animalistic phase" of the child in light of Leo Frobenius's discussion of "personifications [...] in the fairy tales of half-civilized North African tribes."[19] And chapters dedicated to children's art (a topic of great interest to child psychology ever since Sully's publication[20]) explore "ethnological parallels."[21] The figure of the 'child-primitive' remains in force in studies from the 1920s. In *Einführung in die Entwicklungspsychologie* (Introduction to Developmental Psychology, 1926), Heinz Werner cites structural similarities between forms of thinking shared by children and "children of nature," who he understands as two different manifestations of the same "primitive type" and its "primordial thought-processes."[22] Jean Piaget's studies of child psychology from the

---

15 Karl Groos, *Das Seelenleben des Kindes* (Berlin: Reuther & Reichard, 1904), 10.
16 Stern, *Psychology of Early Childhood*, 36.
17 Karl Bühler, *The Mental Development of the Child: A Summary of Modern Psychological Theory*, trans. Oscar Oeser (London: Routledge, 2002), 1.
18 Bühler, *The Mental Development of the Child*, 2.
19 Bühler, *Die geistige Entwicklung des Kindes* (Jena: Fischer, 1924), 139. Though this volume is available to the English reader in translation, it is abridged; therefore this passage and others where the German edition is cited have been translated by Erik Butler.
20 Cf. Barbara Wittmann, "Johnny-Head-in-the-Air in America: Aby Warburg's Experiment with Children's Drawings," in *New Perspectives in Iconology: Visual Studies and Anthropology*, ed. Barbara Baert, Ann-Sophie Lehmann, and Jenke Van den Akkerveken (Brussels: AspEditions, 2012), 120–142, and *Bedeutungsvolle Kritzeleien*, 161–171, 187–241.
21 Bühler, *Die geistige Entwicklung des Kindes*, 291. Strikingly, however, Bühler draws far fewer parallels between children and 'primitives' than he does between infants and animals. In general, he is inspired by "animal psychology," which emerged around the same time as child psychology. Accordingly, he speaks of early mental development as "the *humanization* of the child" (*The Mental Development of the Child*, 1).
22 Heinz Werner, *Einführung in die Entwicklungspsychologie*, 131, 140.

same period (and still relevant today) also repeatedly point out such similarities and use ethnologically inflected terms (e. g., participation) to describe the child's worldview.[23]

These examples from the early decades of developmental psychology demonstrate that the 'child-primitive' – constructed on the basis of the theory of recapitulation – represents a fundamental paradigm in the early stages of the field. In this case the 'primitive' also takes the form of a contemporary 'prehistoric human'; however, this time they are contemporary not in the figure of indigenous peoples, but in the figure of the child, who recapitulates phylogenetic development.

Othering: The 'Bad' Child

Conceiving of the child in the paradigm of the 'primitive' alienated the former. Suddenly – and especially against the backdrop of psychoanalysis, which considers the child to be driven by impulses – the child emerged as a wild, strange, and even threatening being inhabiting a world barely accessible to adult minds.[24] Examples include Sully's remarks on children's "crude fanciful ideas" and "seizures by rage and terror," as well as the parallels Bühler draws between them and animals (in his eyes, the child only becomes a human being over time).

For Freud, children are not just subject to the same urges as adults; they live out these urges directly because their inhibitions have not yet developed. Not only do children, according to Freud, have an infantile sexuality of their own, but this manifests itself as paraphilia, i.e., sexual ideas, needs, or activities associated with one's personal suffering or that of one's victims (sadomasochism). Whereas Freud's early theory of drives can only grasp this destructive form of sexuality as perversion (the child is "polymorphously perverse"), in the context

---

23 Jean Piaget, *Judgment and Reasoning in the Child* (London: Routledge, 1999), 255–256, and *The Child's Conception of the World*, trans. Joan and Andrew Tomlinson (Lanham, MD: Littlefield Adams 1989), 133.
24 On the history of the 'wild child,' cf. Nicolas Pethes, *Zöglinge der Natur: Der literarische Menschenversuch des 18. Jahrhunderts* (Göttingen: Wallstein, 2007), 62–122; and Dieter Richter, *Das fremde Kind: Zur Entstehung der Kindheitsbilder des bürgerlichen Zeitalters* (Frankfurt am Main: Fischer, 1987), 139–174, which draws parallels to ethnological discourse; and Reinhard Kuhn, *Corruption in Paradise: The Child in Western Literature* (Hanover, NH: University Press of New England, 1982). On the motif of the 'brute,' or 'insane child,' against the backdrop of recapitulation and degeneration theory in the English-speaking world, see also Shuttleworth, *The Mind of the Child*, 181–206.

of his later theory it can be understood as an expression of *thanatos,* the death drive. The latter's effects are also more evident in the child than in the adult: children act out destructive desires, whether in the form of aggression directed at the self or others, more openly than adults do. Thus, in *Jenseits des Lustprinzips* (1920; *Beyond the Pleasure Principle*, 1922), where Freud develops the concept of the death drive, he speaks of the destructive game of a small child that repeatedly hurls objects away from himself (*fort-da*). In Freud's estimation, the child uses the game to reproduce the painful absence of his mother; the "gain in pleasure" (*Lustgewinn*) lies in the child's taking control of the situation (by playing the active role) on the one hand, and on the other hand in his taking revenge on the mother herself. Incidentally, William Stern had already seen a striving for control at work in the "destructive games" of children in *Psychologie der frühen Kindheit* (1914; *Psychology of Early Childhood*, 1924). For him the child's pleasure lies in "being the cause, which [...] can never be exhibited in a more elemental form than in destruction."[25]

Groos even devotes two entire chapters to children's destructive activities in *Die Spiele der Menschen* (1899; *The Play of Man,* 1901). These first of all represent part of an analytical game, which dissects things and living beings in order to understand their structure. Yet they also express a "destructive impulse" that is particularly evident in fighting. His examples of such "wild destructiveness" include the tendency of infants to "tear paper, pull the heads off of flowers, rummage in boxes, and the like."[26] The child, in the use of such analytical-destructive acts against other living creatures, is likened to the 'child of nature': "since the child, like the savage, has not our clear perception of the difference between what is living and the lifeless, he will pull to pieces a beetle, a fly, or a bird with the same serenity which accompanies his demolition of a flower."[27] Groos adduces particularly drastic examples in the subchapter on "the destructive impulse." For him, a destructive act is characterized as game like whenever it is "continued simply for the sake of its intoxicating effects."[28] Like Freud and Stern, he explains it in terms of gaining power. For instance:

> An eight-year-old girl with an angelic face secretly put some pins in her little brother's food, and calmly awaited the catastrophe, which fortunately was averted. [...] A girl twelve years

---

25 Stern, *Psychology of Early Childhood,* 311.
26 Groos, *The Play of Man*, trans. Elizabeth A. Baldwin (New York: D. Appleton, 1913), 98.
27 Groos, *The Play of Man*, 98.
28 Groos, *The Play of Man,* 217.

old pushed a child of three, with whom she was playing, into a pile of paving stones for no other reason than that she might have the opportunity to tickle him cruelly.[29]

The fact that both cases concern girls heightens Groos's transformation of the "loveable"[30] child into a cruel one; for the pedagogical and psychological literature of the day usually credited girls with being less inclined to violence than boys.

Groos also sees affinity between the destructive child and the criminal adult inasmuch as he identifies a play-impulse at work in their misdeeds.

> Among criminals murders may sometimes result from following this impulse. Some time ago three peasants were tried for the murder, with incredible cruelty, of a servant. They were father, son, and mother. After the old man had throttled his victim he said to his accomplices, "Now he is dead enough." But the woman, to make sure, dealt a hard blow on the poor fellow's head. "Now I think he has had enough, this fine rabbit that we have caught." Here the bounds between play and earnest are hard to place, but probably belong at the point where the prearranged plan is no longer the leading thought, it having given place to mad delight in inflicting injury.[31]

Hall offers a biogenetic explanation for this affinity between the child and criminal, describing children's destructive actions as a phase that recapitulates phylogeny: "The child revels in savagery." However, he argues that children should not be denied their inclinations because they need to repeat this primal state in order to mature into civilized adults. As a kind of "catharsis,"[32] this stage needs to be lived out. Otherwise, he warns, development will either stop at this level[33] or "wild destructiveness" will return later: "Rudimentary organs of the soul now suppressed, perverted or delayed, [will] crop out in menacing forms later in adulthood."[34] After all, Hall contends, "criminals are much like overgrown children."[35] For him, the "child torturer"[36] and the "torturer" recapitulate the behavior of "primitive man."

---

29 Groos, *The Play of Man*, 219–220. Groos takes both examples from Friedrich Scholz's *Die Charakterfehler des Kindes: Eine Erziehungslehre für Haus und Schule* (Leipzig: E. H. Mayer, 1891), 148–149.
30 Groos, *Das Seelenleben des Kindes*, 2.
31 Groos, *The Play of Man*, 220. Groos takes this case from Scipio Sighele's *Psychologie des Auflaufs und der Massenverbrechen* (Dresden: Reissner, 1897), 13–14.
32 Hall, *Adolescence*, x.
33 E.g., Hall, *Adolescence*, 338.
34 Hall, *Adolescence*, x.
35 Hall, *Adolescence*, 338.
36 Hall, *Adolescence*, 359.

> The savage is a good father, perhaps husband and tribesman, with a kindly nature, but all his virtues are expended on those nearest him, and for all others he has suspicion, enmity, and bitter hostility. In the torturer the boundary between these two sentiments is disturbed. [...] He places the neighbor in the same position as the alien and enemy, whom he would capture and torture.[37]

Indeed, the turn of the century witnessed a spate of works of criminal anthropology based on the notion of a biogenetic law of crime. Examples include the criminologist Erich Wulffen's *Psychologie des Verbrechens* (Psychology of Crime, 1908), *Gauner- und Verbrechertypen* (Types of Crooks and Criminals, 1910), a handbook on sexual criminals (*Sexualverbrecher*, [Sexual Criminals, 1910]), and the 500-page book *Das Kind. Sein Wesen und seine Entartung* (The Child: His Nature and Degeneration, 1913). The author introduces the latter study by highlighting his professional interest in "the criminal soul and the origins of crime." "The task," he begins, "was to eavesdrop on emergent crime in the child's soul and determine its direct relationship to instincts, drives, and inclinations that the human being brings forth from nature's womb."[38] In Wulffen's estimation, "pedagogical doctrine and criminal psychology" confront the same "cardinal problem": "How do we learn to do good and avoid evil?" In this light, pedagogy actually represents for him a domain at the margins of criminal psychology. After all, "most young people go through a sort of half-criminal phase"; the task for educators, then, is to redirect "antisocial instincts and drives."[39]

Hall and Wulffen's reflections on the criminal nature of the child owe a great deal to the theories of Cesare Lombroso, whose influence persisted well into the twentieth century. The premise of his *L'uomo delinquente* (1876; *Criminal Man*, 1911) is that lawbreakers have remained, both physically and psychically, at an early stage of human development. Accordingly, Lombroso likens their behavior to that of children, whose natural antisociality and violence he describes in detail.[40]

> This fact, that the germs of moral insanity and criminality are found normally in mankind in the first stages of existence, in the same way as forms considered monstrous when exhibited by adults, frequently exist in the foetus, is such a simple and common phenomen-

---

37 Hall, *Adolescence*, 360.
38 Erich Wulffen, *Das Kind. Sein Wesen und seine Entartung* (Berlin: Langenscheidt, 1913), xix.
39 Wulffen, *Das Kind*, xxi.
40 In so doing, he refers to Paul Moreau, *De l'homicide commis par les enfants* (Paris: Asselin, 1862), among others.

on. The child [...] represents what is known to alienists as a morally insane being and to criminologists as a born criminal.[41]

For him, children and criminals are equally prone to anger, vengeance, jealousy and envy, lying, cruelty, sloth, imitating others' actions without foresight, and any number of other vices, all of which culminate in criminal activity.[42] He supports his thesis with numerous case histories to show children who he views already as "criminals." In contrast to Freud, Groos, and Stern, he freely evaluates children's actions in moral terms, which he explains as the result of "evil impulses,"[43] the intensity of their passions resembling those of "savages."[44] At the same time, Lombroso distinguishes between children who are "wicked" due only to their age from those who have inherited "perverse instincts." In the latter case, nothing can stop the child from becoming a criminal in adulthood: "the best and most careful education, moral and intellectual, is powerless to effect an improvement on the morally insane."[45]

For Lombroso, then, crime is the outgrowth of inherited moral atavism and of the "degeneration" of civilized European adults to an earlier onto- and phylogenetic stage of development. Moral atavism is attended by physical abnormalities that the author uses to identify born 'degenerates' and that he interprets as the result of "arrested development."[46] Thus, "the true criminal type is characterized by jug ears, low forehead, plagiocephaly or protuberances on the sides of the skull, large jaw, facial asymmetry and fuzz on the forehead," and delinquent children exhibit anomalies in "a proportion equal to that of adult criminals."[47]

Lombroso tries to support his thesis by examining the role of crime among "savages" and even animals. His discussion of "Moral Insanity and Crime among Children" is preceded by chapters entitled "Crime and Prostitution among Savages" and "Crime and Inferior Organisms." "Here" – among indigenous peoples

---

[41] Cesare Lombroso and Gina Lombroso-Ferrero, *Criminal Man According to the Classification of Cesare Lombroso* (New York: G. P. Putnam's Sons, 1911), 130. Because different translations are based on different editions of Lombroso's Italian original, which were themselves substantially revised, my citations of this source refer to different translations, differentiated in subsequent footnotes by co-authorship and date.

[42] Lombroso and Lombroso-Ferrero, *Criminal Man* (1911), 130–140.

[43] Lombroso and Lombroso-Ferrero, *Criminal Man* (1911), 206.

[44] Lombroso and Lombroso-Ferrero, *Criminal Man* (1911), 130, 131, 135, 136.

[45] Lombroso and Lombroso-Ferrero, *Criminal Man* (1911), 143.

[46] Lombroso, *Criminal Man*, trans. Mary Gibson and Nicole Hahn Rafter (Durham, NC: Duke University Press, 2006), 222.

[47] Lombroso, *Criminal Man* (2006), 195.

and animals, that is – "crime is not the exception but almost a general rule."⁴⁸ Inasmuch as he views crime as the exception in modern-day Europe, it represents a relapse to early stages of evolution when criminal behavior is supposed to have been the norm. Lombroso's work thus gives the "wicked" nature of children a biological basis and attributes it to their recapitulation of primal humankind's amorality. Similarly, criminals are criminals because they have remained at the level of the 'child-primitive': "The concept of atavism helps us to understand why punishment is ineffective against born criminality."⁴⁹

This barbarization and criminalization of children marked a serious departure from the Romantic ideal of the child, whose traces, though still perceptible around 1900, were now contradicted and put into question by a new figure. The appearance of the 'bad child'⁵⁰ is especially pronounced in contemporary literary works that carry out a characteristic reinterpretation, or rather re-evaluation, in which they recognize a creative potential in children's destructive activities. Consider, for instance, Walter Benjamin's remarks about the child as a "dehumanized being" or Robert Musil's fascination with children's cruelty (both of which I will return to in chapters 9 and 8, respectively). Another example is Joachim Ringelnatz's game manual *Geheimes Kinder-Spiel-Buch* (1924; *The Secret-Games-for-Children Book*, 1989),⁵¹ which affirms the supposedly amoral and violent tendencies of children by inviting his little readers to stamp a fish to death and then perform experiments on it,⁵² toy with homemade bombs,⁵³

---

48 Lombroso, *Criminal Man* (2006), 175.
49 Lombroso, *Criminal Man* (2006), 338. Lombroso remained influential well into the twentieth century. In March 1928, for example, a child psychology conference was held on his concept of the "born criminal." "In the discussion, two opposing positions emerged: a biological-psychiatric viewpoint [...] advocated by Karl Birnbaum, Hans Walter Gruhle and Johannes Lange, among others, who considered disposition to be of decisive importance, and that of Krames and von der Leyens, who insisted on the inseparable combination of milieu and disposition" (Wolfgang Rose, Petra Fuchs, and Thomas Beddies, *Diagnose "Psychopathie": Die urbane Moderne und das schwierige Kind. Berlin 1918–1933* [Vienna: Böhlau, 2016], 260).
50 Of course, the idea of the 'bad child' existed before this point, but until the end of the nineteenth century it was usually still based on religion and determined, directly or indirectly, by the doctrine of Original Sin. This changed around 1900, even if some aspects of moral-religious discourse undoubtedly continued. On the further history of the 'bad child,' see Nicola Gess, "Böse Kinder. Zu einer literarischen und psychologischen Figur um 1900 (Lombroso, Wulffen) 1950 (Golding, March) und 2000 (Hustvedt, Shriver)," in *Kindheit und Literatur. Konzepte – Poetik – Wissen*, ed. Davide Giuriato, Philipp Hubmann, and Mareike Schildmann (Freiburg i. Br.: Rombach, 2018).
51 As indicated, an English translation of Ringelnatz's book exists (trans. Andrew Lee), but it could not be obtained for reference.
52 Joachim Ringelnatz, *Geheimes Kinder-Spiel-Buch* (Potsdam: Kiepenheuer, 1924), 19.

spit at each other,⁵⁴ and produce clumps made of urine and excrement and then throw them onto the ceiling.⁵⁵ To summarize, the Romantic focus on the 'good child,' as the embodiment of innocence and naïveté, had given way to interest in the 'bad child,' in whom, according to the theory of recapitulation, the 'primitive' was present. In other words, the premise that children's development recapitulates phylogeny served to other them. As Barbara Wittmann has observed, a peculiar "hybridity" emerged whereby children were treated as something "between paleontological fossil and historical document, myth and history, and nature and culture."⁵⁶ And since, according to the 'biogenetic law', the "savage"⁵⁷ is fated to return, developmental psychologists such as Groos deemed it necessary to take appropriate measures in education and upbringing to ensure that children's transition to the final stage of phylogeny/ontogeny would proceed successfully, yielding rational and moral adults – along the very same lines as the 'civilizing mission' of colonial projects.⁵⁸ Without these appropriate steps, moral atavism – biologically predetermined stasis at the level of 'savagery' – would condemn children to a life of perversion, antisocial activity, and criminality.

The reversal of this othering of the child, however, was the nostrification of the 'primitive.' When embodied as the European child, the 'primitive' was incorporated even more powerfully into the modern self than it had been by ethnology. Even more than indigenous peoples, children brought the stakes of the 'primitive' close to home. The 'primitive' now represented not so much a survival of European culture's ancient origins as what every single civilized adult had once been themselves – primal conditions are to a certain extent permanently present in every childhood and thus an integral part of each life story and mem-

---

53 Ringelnatz, *Geheimes Kinder-Spiel-Buch*, 10.
54 Ringelnatz, *Geheimes Kinder-Spiel-Buch*, 15.
55 Ringelnatz, *Geheimes Kinder-Spiel-Buch*, 5.
56 Wittmann, *Bedeutungsvolle Kritzeleien*, 192.
57 Groos, *The Play of Man*, 98.
58 Perceptively, in his critique of contemporary pedagogical doctrine Walter Benjamin speaks of "colonial pedagogy" ("Kolonialpädagogik," in *Gesammelte Schriften*, ed. Rolf Tiedemann and Hermann Schweppenhäuser [Frankfurt am Main: Suhrkamp, 1991], 3: 272–274); see also Gould regarding the dubious ambivalence of this position's roots in biological determinism: "On the one hand, recapitulation is cited in the name of greater individual freedom and liberation from ancient constraints – mold education to the child's nature, for he is repeating his ancestry and it must be so; do not impose adult criteria for discipline and morality upon a savage child. On the other hand, it is used to deny freedom by consigning certain individuals to biological inferiority – criminals and 'lower' races" (*Ontogeny and Phylogeny*, 164–165; also quoted in Wittmann, *Bedeutungsvolle Kritzeleien*, 240).

ory. The othering of the child by means of the 'primitive' corresponds to the nostrification of the 'primitive' by means of the child.[59]

Moreover, through its (con)figuration as a child, the 'primitive' is removed from culture. Instead of viewing customs and thought in their cultural and social contexts, developmental psychology focused on individuals as though they were independent of the collective.[60] The consequences were twofold: For one, it meant that the thought and conduct of individuals were considered innate qualities to be evaluated in universal terms – that is, they were neither dictated by culture or society, nor the result of personal development. In other words, the relativism that the Durkheim school or Lévy-Bruhl adopted when examining 'primitive thinking' is nowhere in evidence in the discourse of developmental psychology. Instead the 'child-primitive's' conduct is embedded into a quasi automatic course of development. In contrast to ethnology, this theoretical framework of developmental psychology did not look for the motivations and purposes behind child development so much as predispositions and tendencies – in keeping with its underlying biological materialism. While contemporary ethnology lent more attention to how primal substance is handed down on the level of ideas and practices, the analogies devised by developmental psychology relied on biological foundations. From this perspective, phylogeny repeats itself in children due to a biogenetic law, not due to cultural institutions (e. g., language and customs). By the same token, the 'child-primitive' isn't seen as a survival frozen in a permanent state of arrested development. Instead it is understood as a recapitulation of that earlier time in an ontogenetic course of development determined by phylogenesis.

Second, refraining from the cultural-historical and sociological perspective prompted developmental psychologists to speculate about the disruptive, norm-violating potential of 'child-primitives.' Positing a biologically determined

---

59 Ruth Murphy points out the tensions within this construction, which the child simultaneously assigns to the other and – in keeping with the imperative of development (that is, more or less automatically) – to itself: "The child is both a colonized Other, allied with animals, savages and primitives against the power of the civilized adult, and a proto-colonialist who will soon assume the imperial power of the white adult over the 'lesser' animals and 'lower' races" ("Kipling's Just So Stories: The Recapitulative Child and Evolutionary Progress," in Wesseling, *The Child Savage*, 44).

60 An exception is Lev Semenovich Vygotsky (who I discuss in Chapter 9), a representative of the field who was open to socio-psychological factors, and therefore important for authors like Benjamin. His *Myshlenie y rech* (1934; *Thinking and Speech*, 1962) adopts a historical and sociological perspective that, like the approach taken by French ethnologists, yields a genealogy of the way both children and adults think. In this light, mental activities are legible as products of culture or, more precisely, of culture mediated by language.

course of development opened the prospect of a phylo- and ontogenic phase distinguished by *shaping* the world, not simply adapting to it. The importance of this creative activity becomes particularly clear in developmental psychologists' engagement with children's ways of thinking as it expresses itself through play. For them, child's play does not represent collective thinking – or thought shaped by the collective – and, as such, does not fuel reflection on the constitution of contemporary society, as it was approached, for example, by the Collège de Sociologie. Instead, at issue is the thinking of individuals who, by means of sovereign play, release themselves from prescribed norms and their corresponding worldviews. For developmental psychologists, the play-based thinking of 'child-primitives' represents a platform reflecting the possibility of dealing creatively with the world, a place where researchers can speculate on the essence of creativity and art production (see chapter 5). In this context, one question stands front and center: does the child at play perceive the game to be real or a mere illusion? For children's play can only be understood as a creative handling of their environment if they can differentiate play from reality and exercise sovereignty through play, an ability that the child deceived by illusion completely lacks.

The Question of Conscious Deception

How Children Think

Most studies of child psychology from the late nineteenth and early twentieth centuries offer normative descriptions of child development with the goal of communicating age-appropriate milestones against the backdrop of an established and defined sequence of developmental levels and phases. As a rule, these are based on endogenous theories that understand maturation to be genetically predetermined. It follows that thinking was held to result not from external (environmental, social, and/or cultural) sources or to be something children themselves work out; instead, it was viewed simply as a matter of inherited biological programming.

According to Karl Groos in *Das Seelenleben des Kindes*, the early stage of intellectual development is characterized by the tendency of thought to wander along lines of vaguely intuited association. Concepts are formed only with difficulty. He holds that prone to illusion and combining heterogeneous elements illogically, children have trouble grasping concepts and are highly suggestible. As much is evident in their fondness for inventing stories, up to the point of devising an "explanatory mythos"[61] for the world at large. William Stern notes similar qualities in *Psychology of Early Childhood*, but he does much more than his contemporary to situate them in the context of intellectual development. In other words, Stern does not describe the actual state of children's thinking, but focuses instead on its steady maturation, which begins in his estimation with the discovery of the "meaning of speech and the will to achieve it."[62] Characteristics of the early stages of this journey are the child's development of individual representations and concepts shaped by affect,[63] affective self-expression,[64] substantialization,[65] a lacking consciousness of relationality and of one's own mental processes,[66] surprise, and wonder.[67]

Karl Bühler, in *The Mental Development of the Child*, lists similar characteristics to those named by Stern, but he stresses a new feature: mastery of the principle of analogy is a key step in the systematic development of the non-thinking infant into a thinking child with nascent judgment abilities.[68] Analogical thinking translates into the child's belief that all beings and objects exist to serve human beings.[69] Bühler calls this "most primitive" form of worldview "purely teleological and egocentric – or, at any rate, anthropocentric."[70] Bühler emphasizes that children, unlike poets, do not give life to inanimate matter so much as they assume that everything is alive, since they do not know otherwise. Corresponding with the anthropocentric judgment above, there is a phase of object perception during which things are apprehended in such a way that they are enlivened by empathy (*Einfühlung*). Invoking Théodule-Armand Ribot, Bühler speaks of the child's "animistic phase" and compares it to the use of personifications in the tales of North African tribes.[71]

This is only one of many examples of how children's thinking was explained with traits attributed to indigenous peoples. That said – and in contrast to ethnological discourse – developmental psychologists rarely give a thorough answer to the question of what motivates children to think as they do. Sully is a bit of an exception when he makes the same assumptions typical of ethnologists in

---

61 Groos, *Das Seelenleben des Kindes*, 136.
62 Stern, *Psychology of Early Childhood*, 162.
63 Stern, *Psychology of Early Childhood*, 162–164, 171–173.
64 Stern, *Psychology of Early Childhood*, 165.
65 Stern, *Psychology of Early Childhood*, 172.
66 Stern, *Psychology of Early Childhood*, 380–383.
67 Stern, *Psychology of Early Childhood*, 383–397.
68 Bühler, *The Mental Development of the Child*, 133.
69 Bühler, *The Mental Development of the Child*, 140.
70 Bühler, *The Mental Development of the Child*, 155.
71 Bühler, *Die geistige Entwicklung des Kindes*, 139.

tracing how the "primal wonderment" at the "confusion of novelties" triggers intellectual development, which in turn gives rise to the "impulse to comprehend things, to reduce the confusing multiplicity to order and system."[72] In other words, Sully understands children's mental life in the same way that Frazer views that of indigenous peoples, as a kind of pre-scientific operation; accordingly, he refers to children as "young investigator[s]" or "little philosophers."[73]

At the same time, he takes a cue from German ethnology – which, as we have seen in chapter 2, focuses on emotion – and indicates that intense affect provides the impetus for cerebral activity. Though intellectualistic in orientation, Sully's thesis is that children will only begin thinking about a concrete item if they absolutely want it because of an existential need.[74] Thus, the scholarly debate about the primacy of intellect or affect is present in his work, but it does not play much of a role overall. The same ambivalence holds for the majority of studies by the developmental psychologists who followed him: a certain leaning toward the intellectualist position is often in evidence, but no significant discussion is pursued.

This is the case, first, because the endogenous orientation of developmental theories excludes external factors, and, second, because their background in individual psychology discounts research oriented in genealogy and sociology alike. Also, the authors shifted the debates over the relative significance of affect and intellect to a dispute over the origins of language. In the corresponding chapters of their studies, these points are controversially and exhaustively discussed. By turns, language is thought to develop in response to the urge to classify, in order to communicate or release affective experience, as the outcome of social interaction or as an always already given (by means of transmission) and learned cultural property. Chapter 6 will discuss these debates at length.

Deception – or Not?

A further and more significant difference between the theories of developmental psychology and those of ethnology lies in the former's thematization of the relationship between 'primitive thinking' and the question of (self-)deception, which is barely addressed by ethnologists. Preyer had already described the child's first

---

[72] Sully, *Studies of Childhood*, 70.
[73] Sully, *Studies of Childhood*, 79.
[74] Sully, *Studies of Childhood*, 70.

concepts as "becom[ing] real existences, like the hallucinations of the insane"[75]; in other words, children do not recognize mental phenomena for what they are, but take them for physical reality. Sully expresses a similar view. Once the intellect has begun developing, it is guided by fantasy: the child's "thought [...] grows out of the free play of imagination."[76] Standing, like Frazer, in the English tradition of associative psychology, he assumes that it is not the understanding so much as the imagination that answers the need for order by seeking out similarities in the welter of phenomena, which enable what is new to be assimilated into what is already known: "The child [...] is ever on the look-out for likeness."[77] He proceeds to say that the resulting "analogical" or "metaphorical" mode of "apperception" at work here leads to pictorial thinking, which is defined above all by concreteness, not abstraction.[78]

The only difficulty with this early form of thought, as Sully sees it, is that the imagination dominates empirical observation, which ultimately results in a faulty understanding of objects.[79] He elaborates that when contemplating an object, a child will pick out one attractive or interesting feature and disregard all others in order to connect it by association with another, already familiar object – in a manner that seems completely arbitrary by the standards of the adult observer. But instead of dismissing the child's fantasy-rich thought as a fundamental "falsehood" (as Frazer does when discussing 'primitive thinking'), he reframes this operation as part of play. In his estimation, the mismatch between fantastical thinking and reality leads to the child's fantasy life splitting in two different kinds of imagination: First, a playful activity that gives itself over to images of fantasy that are not subjected to any verification process, and second, a reflective attitude to reality that first tests and ultimately gives way to understanding.[80] In this manner, play becomes the site where taking-images-for-reality – (self-)deception, that is – can occur without further consequence or ill effect and where researchers can just as easily indulge their own fascination with this activity. This applies in any case to Sully, who seems quite charmed by the free play of fantasy. He celebrates the "selective activity in children's observation"[81] for its poetic quality and devotes the first chapter of his study to imagination and play in early life.

---

75 Preyer, *Mental Development in the Child*, 17.
76 Sully, *Studies of Childhood*, 70; cf. 29.
77 Sully, *Studies of Childhood*, 72.
78 Sully, *Studies of Childhood*, 72–73.
79 Sully, *Studies of Childhood*, 66–67; cf. 32.
80 Sully, *Studies of Childhood*, 115.
81 Sully, *Studies of Childhood*, 67.

Here, he offers another account of how sensory perception and imagination interact: the imagination assimilates or associatively links sensory data to what is already familiar. As a result, the object perceived seems to come alive or acquire a personality: "the child sees what we regard as lifeless and soulless as alive and conscious."[82] Sully underscores that this is a complete illusion that (prior to the split) is not tested against reality: "Children [...] quite seriously believe that most things [...] are alive and have their feelings."[83] The child's transformation in play into another person or thing is attended by their complete forgetting of the "real environment" and the "real me." Consequently, these "illusions," as Sully emphatically calls them, may last for days on end – far beyond the duration of a normal game.[84]

On the basis of his observations, Sully concludes that the creations of fantasy in general – that is, not just those arising from play – derive from cross-pollination between sensory perception and imagination: either the unknown world rouses curiosity and triggers the impulse to "understand" it by means of fantasy, or, alternatively, the intensity of images within makes the latter materialize in the outer world.[85] Being tricked by one's ideas results in the "enchantment" of the external world, which is especially pronounced in play. Later in life, play becomes its sole province. Throughout his study, Sully makes lavish use of the concept of "enchantment" – for instance, when he speaks of the "magic transmuting of things through [...] childish fancy."[86]

Compared to the fascination Sully exhibits for children's fantasy thinking in his logs, the standpoint adopted by later developmental psychologists is more sober, in keeping with the wish to emphasize the scientific nature of the new discipline. Yet these colleagues were also taken with the phenomena Sully described. Groos takes up delusion in *Das Seelenleben des Kindes* (The Mental Life of the Child, 1904) in a chapter on illusion, distinguishing between complete illusion, which tends to affect children much more than adults, and conscious self-deception. The latter he ties to aesthetic pleasure (a concept taken from Konrad Lange, as we will see in chapter 5). Groos considers conscious self-deception to be realized first of all in the hallucinations into which children sink when listening to stories or reading. Yet it is carried out above all in games of make-believe, when they "complete what is given by the senses in a twofold,

---

[82] Sully, *Studies of Childhood*, 30.
[83] Sully, *Studies of Childhood*, 32.
[84] Sully, *Studies of Childhood*, 38.
[85] Sully, *Studies of Childhood*, 53, 54.
[86] Sully, *Studies of Childhood*, 35.

illusory manner." Thus, a child will recognize a shape in an object (e. g., the form of a horse in the back of a sofa) and project a mental state onto it. Invoking Sully, Groos calls this process "personification,"[87] but unlike Sully, Groos distinguishes between the deception described here and actual delusion by affirming that, in addition to "incorrect apperception, the correct understanding is also present in consciousness."[88] The child actively seeks out deception and enjoys it. However, Groos cannot quite maintain the proximity between the child at play and the adult's reception of art because he is forced to acknowledge that children's illusions, especially in the act of personification, come very close to real error. Ultimately, he concludes that children occupy a middle position somewhere "between the mythological mindset of the 'primitive' and the aesthetic personification [enjoyed] by cultured adults,"[89] a location that would become momentous for the artistic appropriation of the child, e.g., by Walter Benjamin (see chapter 9).

As in Sully's work, fantasy is essential to Stern's concept of childhood. It is most active in play, which is ascribed a decisive function in childhood as a whole. Indeed, he calls childhood "the age of play."[90] In the central section of his study ("Fantasy and Play"), Stern distinguishes fantasies from other forms of concrete images, insofar as the spontaneity of the former sets them apart. At the same time, and like Sully, he stresses the connections between the two: the imagination, Stern observes, gains its material by means of contemplation and memory, and conversely, "the perception and reproduction of objective facts [...] are not without their subjective moment of imagination."[91] Among children, Stern argues, this intermingling is particularly pronounced, inasmuch as they cannot distinguish "between subjective and objective experiences." Here, in Stern's estimation, lies the "key to the most important characteristics of the child's psychic life."[92] In contrast to Sully, he does not leave this peculiarity unexamined, and he traces it back to the child being a "creature of the moment": children measure reality by the intensity of experience. "'Real' for this early stage of life is simply what is keenly felt [...]. The child is engrossed in an imaginary concept, and whilst it lasts its content is no less real for him than, at other times perhaps, his food."[93] Over time, he notes, children take distance from such

---

[87] Groos, *Das Seelenleben des Kindes*, 175.
[88] Groos, *Das Seelenleben des Kindes*, 165.
[89] Groos, *Das Seelenleben des Kindes*, 176.
[90] Stern, *Psychology of Early Childhood*, 265.
[91] Stern, *Psychology of Early Childhood*, 267.
[92] Stern, *Psychology of Early Childhood*, 273.
[93] Stern, *Psychology of Early Childhood*, 273–274.

wholesale illusion and come to infer the "idea of reality possessed by [...] adults."[94] In his account of the resulting condition, Stern also enlists Konrad Lange's notion of conscious self-deception;[95] in contrast to Groos, however, he affirms that the latter does not exist alongside actual deception but replaces it at a subsequent phase of development.

Another feature specific to child fantasy, for Stern, is its "untrammelled" nature, that is, children's ability to spin their fantasies out of little or no outside material.[96] He illustrates this quality by attending to the dynamic, fleeting, and quickly changing nature of fantasy images, as well as the child's budding sense of symbolism, which takes the raw stuff of experience and bends it at will. Discrete fantasy images are chained together in a purely associative fashion, Stern observes, which is why they demonstrate singular "caprice" and "perseveration."[97] Though a "determining impulse" is said to develop here over time – which counteracts the passive principle of association – it is much less pronounced and sets in later than other mental activities.[98] After Stern's extensive exposition on the "conscious condition" during play, he takes up play's "personal function" in the child's life, which, like Groos, he equates with self-training.[99] By his own account, Stern's attitude to child fantasy falls in the middle between criticism ("nonsense, lack of method and judgment") and celebration ("wonderful, almost creative power").[100]

Like the ethnologists discussed in chapter 2, Preyer and Sully are convinced that a complete deception is brought about by the imagination during play, whereas Groos and Stern, in their developmental psychology, sway between that conjecture and one of a "self-aware" deception. Bühler – despite his claims about the child's "anthropocentric" perspective and the occurrence of "hallucinations and illusions"[101] – eventually assumes that imaginary events possess a wholly illusory character (*Scheincharakter*) in play: "When [a child] [...] treats a piece of wood as a mother does her child, we can see in this treatment of the object [...] an act of interpretive pretending [*Scheindeutung*]."[102] Though he introdu-

---

94 Stern, *Psychology of Early Childhood*, 274.
95 Stern, *Psychology of Early Childhood*, 274–275.
96 Stern, *Psychology of Early Childhood*, 268.
97 Stern, *Psychology of Early Childhood*, 285.
98 Stern, *Psychology of Early Childhood*, 283.
99 Stern, *Psychology of Early Childhood*, 296.
100 Stern, *Psychology of Early Childhood*, 295.
101 Bühler, *The Mental Development of the Child*, 155.
102 Bühler, *Die geistige Entwicklung des Kindes*, 325; cf. 326, 327, 334, 337.

ces repeated examples that suggest the opposite,[103] he explains them as (pathological) deviations from normal (healthy) behavior. For instance, "when a child [...] asks whether a blade of straw can talk, or if both the grandmother and Little Red Riding Hood have enough room in the wolf's belly."[104] Bühler does not follow the middle course of conscious self-deception that Groos and Stern had taken, then. Likewise, Jean Piaget, in *La représentation du monde chez l'enfant* (1926; *The Child's Conception of the World*, 1929), excludes from consideration "all that belongs strictly to play" because, even though such activity is "continuously interwoven with participations,"[105] it lacks the dimension of conviction. Since he does not believe that children take ludic thought and activities very seriously, his research focuses on the non-playful sphere of early life.

Studies from the first decades of the twentieth century dedicated exclusively to the theory and history of play disagree with that decision, pointing out that *no* area of the child's life lies beyond the sphere of play. In *The Play of Man*, Groos himself contends that "the child's whole existence [...] is occupied by play"; indeed, it represents "the single, absorbing aim of his life."[106] Against the background of this totality of play, it is suggestive to think that children's peculiar way of thought also first develops by means of play.[107] But this does not mean for Groos, that these thoughts are only 'feigned.' Though in this book he distinguishes between illusions that "appear as a substitute for reality" and those that are "products of conscious self-deception," he nevertheless postulates various "transitional stages" between the two forms of illusion. He also locates children's play in such a stage of transition. Thus he writes that "illusion is often so strong for playing children [...] that it forms a perfect substitute for reality"[108]; "even in half-grown children the power of detachment is much greater than in adults."[109] In contrast to colleagues who assume that children's perceptions occupy positions oscillating between appearance and reality, Groos declares that the child enters a state similar to hypnosis, in which the awareness of unreality is only present in the sense of a "subtile [sic] consciousness of free, voluntary

---

103 Bühler, *Die geistige Entwicklung des Kindes*, 331, 337.
104 Bühler, *Die geistige Entwicklung des Kindes*, 337; cf. 334.
105 Piaget, *The Child's Conception of the World*, 133.
106 Groos, *The Play of Man*, 369.
107 The conception of the game at issue can be described as mimicry, in the sense it is used by Roger Caillois (*Man, Play, and Games*, trans. Mayer Barash [Champaign, IL: University of Illinois Press, 2001], where *paida* predominates and elements of *ludus* are excluded. Caillois points to split personality disorder as one of the dangers that mimicry poses and thus points the way to pathology and the process of pathologization, which I will explore in the next chapter.
108 Groos, *The Play of Man*, 131.
109 Groos, *The Play of Man*, 134.

acceptance of the illusion."¹¹⁰ Groos' considerations thus mean to say that children's play is on the one hand always already every bit as serious as claimed by developmental psychologists about children's 'primitive thinking.' And on the other hand this other thinking is always already invested with the seed of demystification, which the illusory character of the game brings with it. To put it differently, play brings about both belief and disbelief in the reality that the child's own 'primitive thinking' has created.¹¹¹

With this conception of children's play, the 'child-primitive' of developmental psychology and pedagogical discourse became associated with the figure of the artist and the reception of art.¹¹² This process involved the resurrection of old ideas, particularly those of Friedrich Schiller,¹¹³ to affirm the relationship between art and play. Groos places the two activities in analogy by claiming that "aesthetic behavior" only concerns a "partial phenomenon [*Teilerscheinung*] out of the realm of games of illusion."¹¹⁴ In *Der ästhetische Genuss* (Aesthetic Pleasure, 1902), he even writes that "aesthetic pleasure" should be understood "directly as play."¹¹⁵ Groos concentrates on the analogy between children who are partially deceived while at play and *recipients* of art. Nonetheless, he also mentions that the "joy of being the cause," which Stern and Freud would later posit in connection to the child's destructive act, is relevant to artistic *production*.¹¹⁶ Similarly, in "Der Dichter und das Phantasieren" (1908; "Creative Writers and Day-dreaming," 1959), Freud asks,

> Should we not look for the first traces of imaginative activity as early as in childhood? [...] Might we not say that every child at play behaves like a creative writer, in that he creates a

---

110 Groos, *The Play of Man*, 368.
111 Rainer Maria Rilke found a fitting expression for such experience in a fragmentary elegy where the child's subjectivity develops as it plays with an animated doll until it is finally recognized as a lifeless object ("Unvollendete Elegie 'Lass dir, daß Kindheit war,'" *Werke II* [Frankfurt am Main: Fischer, 1987], 459–460). Chapter 9 will also return to this notion of play in relation to Benjamin.
112 On Groos in the context of other theorists of play at the time and in relation to literature, cf. Thomas Anz, *Literatur und Lust. Glück und Unglück beim Lesen* (Munich: DTV, 2002), 33–76.
113 E.g., Groos, *Der ästhetische Genuss* (Giessen: J. Ricker, 1902), 19.
114 Groos, *Das Seelenleben des Kindes*, 172. In his earlier book *The Play of Man*, the author adopts a narrower perspective, claiming that the relationship exists primarily for "artistic pleasure" and less for "artistic production" (Groos, *The Play of Man*, 390).
115 Groos, *Der ästhetische Genuss*, 24.
116 Groos, *Der ästhetische Genuss*, 19, 21.

world of his own, or, rather, re-arranges the things of his world in a new way which pleases him?[117]

These suggestions in turn are informed by the conviction that children do not consider the world of play to be real. Instead, Freud, like Groos, believes they are partially aware of their creative activity in shaping this world.[118]

Against these backgrounds, contemporary works of pedagogy credited the child with a particular capacity for appreciating and creating art. An entire branch of Reform pedagogy, the so-called art-education movement (*Kunsterziehungsbewegung*), was based on this premise.[119] One representative of this line of thought was Gustav Friedrich Hartlaub's *Der Genius im Kinde* (The Genius within the Child, 1922), which celebrates the child's "unsuspecting superiority [...] to competent but mediocre art" by adults. Simultaneously, the reverse argument is also carried out and the artist is described as a grown-up child. As Hartlaub puts it, "only the poet and artist preserves the general, imaginative vigor of the child. [...] Only the artist is able to salvage, to varying degrees, the immense inner life of childhood."[120] In chapter 5, I will discuss at length how art reception and above all production were modeled after the play-based thinking and behavior of children.

Jean Piaget and the Magical Thinking of Children

Piaget's concept warrants discussion in some detail here because he has superceded almost all the authors I have been examining, both in terms of method

---

[117] Sigmund Freud, "Creative Writing and Day-Dreaming," *The Freud Reader*, ed. Peter Gay (New York: Norton, 1995), 437.
[118] Groos observes that enjoying art also involves "the pleasure of being the cause" insofar as "the state that emerges is itself, in a certain sense and in part, an effect we ourselves produce." In his estimation, this is implied by the very term "conscious self-deception" (Groos, *Der ästhetische Genuss*, 21).
[119] In its first phase, this movement sought above all to train the child's aptitude to appreciate art; in its second phase, it promoted artistic creativity. For an impressive array of documentation of the movement, cf. *Kunsterziehung. Ergebnisse und Anregungen der Kunsterziehungstage in Dresden, Weimar und Hamburg* (Leipzig: Voigtländer, 1906); as well as the following exhibition catalogs: Carl Götze, *Das Kind als Künstler* (Hamburg: Kunsthalle zu Hamburg, 1898); and *Die Kunst im Leben des Kindes. Katalog der Ausstellung im Hause der Berliner Secession, März 1901* (Leipzig and Berlin: E.A. Seemann, 1901). On teaching children to draw, cf. Wittmann, *Bedeutungsvolle Kritzeleien*, 141–186; and Götze, *Das Kind als Künstler*, 214–222.
[120] Hartlaub, *Der Genius im Kinde*, 69, 30.

and theory. In contrast to his forebearers, Piaget remains an influential figure in the field of developmental psychology. On a structural level, his approach resembles that of theorists focused on the maturation process, inasmuch as his own explanation of the development of thought during childhood dismisses the influence of external factors. In lieu of hereditary programming, however, Piaget gravitates toward constructivism. He premises that by means of discovering and structuring activities, the child constructively uses the stimuli from its surroundings, not with conscious intentionality (the precondition for a fully constituted subject), but in an ongoing process of modifying the boundaries between the self and the world. This modification is initiated through confrontations with the environment, which call the operative models gained from prior experience into question. In contrast to most of the developmental psychologists before him, Piaget held that the 'magical thinking' of children is furthermore not guided by an epistemological interest. Instead, this thinking consists first of all in a belief "in the automatic realisation of our desires."[121] This connects to the psychoanalytic notion of the primary function of the pleasure principle, which I will treat in greater detail in chapter 4.

Already in 1920, Piaget's article, "La psychanalyse dans ses rapports avec la psychologie de l'enfant" ("Psychoanalysis in Its Relations With Child Psychology"), enlists Freud's dream theory to propose the idea of another way of thinking that consists of an "inextricable network of symbol-associations whose only logic is that of the emotions,"[122] shared by neurotics, dreamers, artists, mystics, and indigenous peoples alike. At the same time, he refers to Lévy-Bruhl, whom he credits with having investigated thought of this kind under the label of "prelogical thinking." In doing so, Piaget makes clear that the difference between the magic practiced among indigenous peoples and the symbolism invented by children concerns content alone: one violates the laws of reality, the other those of logic. More important is his view of what they share structurally: "they all are governed by the laws of the dream itself." Following Eugen Bleuler's lead, Piaget describes such dreamlike mental activity as "autistic thought" insofar as it is (in contrast to scientific thinking) "strictly personal and incommunicable"[123] and distinguishes it as also still serving an essential role in adult life.

This early article can be regarded as the nucleus of Piaget's foundational studies of the following decade: *Le langage et la pensée chez l'enfant* (1923; *The Language and Thought of the Child*, 1932), *Le jugement et le raisonnement*

---

121 Piaget, *The Child's Conception of the World*, 152.
122 Piaget, "Psychoanalysis in Its Relations with Child Psychology," in *The Essential Piaget*, ed. Howard E. Gruber and J. Jacques Vonèche (London: Routledge, 1977), 56.
123 Piaget, "Psychoanalysis in Its Relations with Child Psychology," 56.

chez l'enfant (1924; *Judgment and Reasoning in the Child*, 1928), *The Child's Conception of the World*, and *La causalité physique chez l'enfant* (1927; *The Child's Conception of Physical Causality*, 1929). All of them orbit around the "egocentric thinking" of children and its consequence for their sense of logic and conception of causality and reality. These books abandon the diary-writing approach of early developmental psychology with its literary and philological overtones. Instead, they are oriented in the conventions of the natural sciences, based on the so-called "clinical method" whereby large samples of children were questioned in detail about their thoughts and mental images.[124]

Judgment and Reasoning in the Child maps out the essential principles of egocentric thinking. These include, for instance, the principle of "juxtaposition," whereby only one of two contradictory pieces of information is processed, and the principle of "syncretism," which is at work when two things appear to be connected in arbitrary fashion. Each of these operations, Piaget argues, takes the place of the child's inability to synthesize data and also demonstrates that the law of non-contradiction does not apply to children's thought.[125] Piaget recognizes the first and most spontaneous manifestation of this thinking in children's play, which he also refers to as the "quasi-hallucinatory form of imagination which allows us to regard desires as realized as soon as they are born."[126] It follows for Piaget that children consequently operate on two different levels of reality: First, on the level of play, where the child is not concerned with adapting to outer reality but only with satisfying its needs and interests. This satisfaction is achieved through the child's transformation of its external reality: "reality is infinitely plastic for the ego, since autism is ignorant of that reality shared by all, which destroys illusion and enforces verification."[127] Second, on the level of "true" reality, where the child does not play so much as observe. In keeping with the principle of juxtaposition, the two levels and modes of engagement exist side by side. No hierarchy exists between them because they are not present at the same time.[128]

For this reason Piaget considers Groos' thesis of "conscious self-deception" inadequate, since it presupposes that the child is simultaneously aware of both levels – i.e., that it has an adult's awareness of fiction. For his own part, Piaget posits that the reality of child's play is autonomous and that even "true" reality has only a very slight dependence on the principles of observation and experi-

---

[124] See the introductory remarks by Gruber and Vonèche, in *The Essential Piaget*, 63–64.
[125] Piaget, *Judgment and Reasoning in the Child*, 209–232.
[126] Piaget, *Judgment and Reasoning in the Child*, 202.
[127] Piaget, *Judgment and Reasoning in the Child*, 244–245.
[128] Piaget, *Judgment and Reasoning in the Child*, 242.

ence. This is because this reality too "is made up almost in its entirety by the mind and by the decisions of belief."[129] Referring to studies on children's drawings, Piaget calls this phenomenon the child's "intellectual realism." In other words, the child's reality is intellectually determined: populated by phenomena of mental origin that are considered to be real. Piaget defines such realism in terms of "precausality," the mentality "most in agreement with ego-centrism of thought," that is to say, the child's tendency to believe motifs stemming from its own psyche are the cause of phenomena. It is quite possible that Piaget here found inspiration in Lévy-Bruhl's notion of the "pre-logical mentality." Indeed,

> [i]t is [...] our belief that the day will come when child thought will be placed on the same level in relation to adult, normal, and civilized thought, as "primitive mentality," as defined by Levy-Bruhl [sic], as autistic and symbolical thought as described by Freud and his disciples.[130]

Whereas Piaget's first two studies are primarily dedicated to the formal characteristics of children's thinking, *The Child's Conception of the World* takes up matters of content: ideas about dreams, names, and life. This work simultaneously marks a shift away from explaining egocentric thinking in terms of social psychology, as the expression and result of lacking communication with others. Instead, it now represents a primary feature of the still undeveloped thinking of children, of which the communication deficit is only a secondary result.[131] Three characteristics of the child's worldview grow out of its egocentrism: "realism," "animism," and "artificialism."

With the first term, Piaget refers to the concept of "intellectual realism" previously discussed in *Judgment and Reasoning in the Child*. He shows, for example, that the young child is convinced of the realism of names, that is, that the name for an object is part of that object and belongs to it just as any of its other features do (e.g., color and shape). As the mindset that Piaget considers "most in agreement with ego-centrism of thought," realism occupies the central position in the child's worldview and provides the explanatory groundwork for the phenomena of artificialism and animism.[132] For both, the concept of "partic-

---

129 Piaget, *Judgment and Reasoning in the Child*, 248.
130 Piaget, *Judgment and Reasoning in the Child*, 255.
131 Cf. the introduction by Hans Aebli in Jean Piaget, *Das Weltbild des Kindes* (Stuttgart: Klett-Cotta, 1978), 9.
132 Piaget, *Judgment and Reasoning in the Child*, 255.

ipation" plays an essential role because it represents a sense of causality suited to the developing mind.

> Following the definition of M. Lévy-Bruhl, we shall give the name "participation" to that relation which primitive thought believes to exist between two beings or two phenomena which it regards either as partially identical or as having a direct influence on one another, although there is not spatial contact nor intelligible causal connection between them.

As Piaget conceives it, participation is closely related to "magic" – in other words, "the use the individual believes he can make of [...] participation to modify reality."[133] Whereas not every instance of participation implies magic, every act of magical thinking requires participation and in one of four possible variants: the participation of actions and things, of thoughts and things, of substances, or of intentions (which often amount to magical commands).[134] Hence, as Piaget writes, magical acts often evince a "tendency towards symbolism."[135] Accordingly, Piaget observes, this tendency follows the law that governs the child's linguistic development.

> Signs begin by being part of things or by being suggested by the presence of the things in the manner of simple conditioned reflexes. Later, they end by becoming detached from things and disengaged from them by the exercise of intelligence which uses them as adaptable and infinitely plastic tools. But between the point of origin and that of arrival there is a period during which the signs adhere to the things although they are already partially detached from them.

This is the "magical stage":

> What the magical stage itself shows [...] is precisely that symbols are still conceived as participating in things. Magic is thus the pre-symbolic stage of thought. From this point of view the child's magic is a phenomenon of exactly the same order as the realism of thought.[136]

The child's realistic ideology encompasses both causality determined by participation as well as the use of symbols that are magically intended to influence those participatory connections.

The child clings to magical thinking for a relatively long time, according to Piaget, because of its seeming success. To understand this supposition, he maps

---

[133] Piaget, *The Child's Conception of the World*, 132.
[134] Piaget, *The Child's Conception of the World*, 133–134.
[135] Piaget, *The Child's Conception of the World*, 134.
[136] Piaget, *The Child's Conception of the World*, 161.

out two structures of gratification: For one, the "social environment" or parents play a decisive role, who respond to the child's screams from birth onward: "Every cry of the baby leads to an action on the part of the parents, and even the desires it can least express are always foreseen."[137] Accordingly, the child becomes convinced that, by means of sounds, or even thoughts, it can influence the surrounding world. A "class of things" that obey its wishes ("the parents, like the parts of its own body, like all the objects that can be moved by the parents or by its own actions" – in other words, what most interests the child) becomes the model for organizing the rest of the universe, so that from the child's perspective everything is subject to the law of magic. Piaget also contends that magical gestures are "simply ritual." Thus, mistaking signs for causes, "the child makes sure the bed-clothes are tucked in"[138] and takes this fact as the source of its security. Yet this inference is not to be explained as mere madness because the satisfaction is considered real. Even among rational thinkers, the performance of pure rituals during a state of anxiety provides the longed-for reassurance because they are signs of normality.

Compared to developmental psychologists whose work was familiar to him, Piaget generally assumes – especially in this work and his next study – that an initial unified state precedes any separation of the self and the world: "During the first stage, the self and things are completely confused."[139] "During the early stages the world and the self are one: neither term is distinguished from the other."[140] In regard to this unified phase, there is no need for Piaget to address the question of deception that preoccupied earlier theorists. This is because the label of deception would not be applicable to the first (and at that point only) state of being. The same argument applies to the peculiarity of thought (e.g., participation) during the phase that follows the first but incomplete division between the self and the world. The question of deception is not relevant for Piaget at this stage either because no alternative to participation is available to the child's mind. It would only make sense to speak of the child being mistaken if they had another way of thinking available to them. In brief, magical thinking lies beyond the realms of truth and error. Along lines similar to how Lévy-Bruhl parted ways with English ethnologists, Piaget abandons the "adult standpoint," which can always only recognize the unity of self and world, or the power of participation, in retrospect. For the same reason, Piaget

---

[137] Piaget, *The Child's Conception of the World*, 153.
[138] Piaget, *The Child's Conception of the World*, 156.
[139] Piaget, *The Child's Conception of the World*, 250.
[140] Piaget, *The Child's Conception of Physical Causality* (New York: Harcourt Brace & Company, 1930), 244.

does not take for granted the analogy between children's magical thinking and adult artistic activity taken up by many of the developmental psychologists named above. Unlike the artist, the child has no alternative to magical thinking at its command. It occurs beyond the space of deception and play – and therefore also beyond the realm of art.

Between Natural Science, Philology, and Literature: The Methodological Dilemma of Developmental Psychology

Developmental psychology others the child by means of the paradigm of the 'child-primitive' embedded in its methodology. Stern, for example, claims alterity to be a precondition for his scientific research: only by viewing children from a distant perspective do scientists feel motivated to study and explain their behavior and thought patterns. At the same time, this observation prompts him (like Groos) to call the methods of child psychology into question. He identifies the same dilemma here as the one confronting ethnology: How is it possible for adult minds to grasp a way of thinking that is so alien?[141] How can any judgment be made about a "psyche" accessible only indirectly through observation of an inarticulate body and by means of imperfect analogies with the psyche of "cultured adults"?[142]

Early developmental psychologists had answered this question through a combination of philological and natural scientific methods and a more properly literary approach to writing: journal writing. This genre indeed inaugurated the field of child psychology with the publication of Preyer's *Mental Development in the Child* in Germany, which researchers referred to, time and again, as the founding document of the profession.[143] This work is based on the author's systematic observation of his son, which he wrote down and interpreted daily.

---

141 Stern, *Psychology of Early Childhood*, 35–38.
142 Groos, *Das Seelenleben des Kindes*, 12. Cf. Fritz Mauthner, "Kinderpsychologie," in *Wörterbuch der Philosophie* (Leipzig: Meiner, 1923).
143 E.g., Stern, *Psychology of Early Childhood*, 12. However, many others preceded him, for instance, Dietrich Tiedemann, "Beobachtungen über die Entwickelung der Seelenfähigkeit bei Kindern," *Hessische Beiträge zur Gelehrsamkeit und Kunst* 2 (1787): 313–333; 486–502; other figures who recorded their children's development include Johann Heinrich Pestalozzi, Charles Darwin, and Hyppolite Taine. Cf. Wittmann, *Bedeutungsvolle Kritzeleien*, 122–125; Cavanaugh, "Cognitive Developmental Psychology before Preyer"; Jaeger: "The Origin of the Diary Method in Developmental Psychology," in *Contributions to a History of Developmental Psychology*.

> I have [...] kept a complete diary from the birth of my son to the end of his third year. Occupying myself with the child at least three times a day [...] and guarding him, as far as possible, against such training as children usually receive, I found nearly every day some fact of mental genesis to record. The substance of that diary has passed into this book.[144]

As Stern notes, many others followed Preyer's lead: "America, above all, was flooded with descriptive records of little children; of these, by far the most comprehensive are the studies of Miss Shinn, but the records of Moore, Major, Chamberlain are deserving of mention."[145] Indeed, Stern's own *Psychology of Early Childhood*, as the title page indicates, is "supplemented by extracts from the unpublished diaries of Clara Stern," documenting her children's activities; to this day, the work sets the standard for the diary method in developmental psychology.[146]

The labeling of such records as diaries is as vexing as it is revealing. Counter to what one would expect after 1800, these diaries are not a medium of self-analysis and contain hardly any reflections on the writer's thoughts and feelings.[147] As a rule, this diarist only observes others (children), and as matter-of-factly as possible – the very opposite of the soul-searching that the word implies today.[148] A similar paradox is evident in the text's status as readerly. Whereas the modern diary primarily serves to bring about self-understanding in the author, most examples of this genre in child psychology were intended from their inception to be read by others.

For those reasons, the diaries at issue more closely resemble an earlier form of the modern diary, namely the private chronicle, a chronologically ordered and factual record of information and events of potential interest to a family and its descendants but with little reflection on the writer's inner life.[149] Still, one might

---

144 Preyer, *Mental Development in the Child*, x.
145 Stern, *Psychology of Early Childhood*, 27.
146 Heike Behrens and Werner Deutsch stress "differences from earlier diary entries" in the context of a "method [that] already had a more than 100-year history in Germany" ("Die Tagebücher von Clara und William Stern," in *Theorien und Methoden psychologiegeschichtlicher Forschung*, ed. Helmut E. Lück and Rudolf Miller [Göttingen: Hogrefe, 1991], 68). Stern himself, the authors note, distinguished between two traditions of journal keeping: that of professional educators interested in the child from age six onward, and that of psychologists focused on development prior to this age. Behrens and Deutsch credit the Sterns with moving beyond this convention and breaking with the rigid scheme of observation dictated by Preyer (whose method was considered authoritative) (68–69).
147 Cf. Martin Lindner, *ICH schreiben*, Chapter 1–2: Definition der Textsorte Tagebuch, n.p.
148 Cf. Lindner, *ICH schreiben*, Chapter 1–2: Definition der Textsorte Tagebuch, n.p.
149 Cf. Lindner, *ICH schreiben*, Chapter 1–2: Definition der Textsorte Tagebuch, n.p.

read this lexical choice – and the fact that it was made in the years before and after 1900 – as a hint that the writing subject is possibly more involved than they purport to be. At play here is an implicit conflict between an impersonal and scientific bearing and personal involvement, which is reflected in the preliminary remarks to many studies of developmental psychology. Sully, for example, discusses the need for sympathetic insight[150] into the child's mind, which is why he views the mother (or nanny) as a particularly suitable observer.[151] Stern also declares that observation should be conducted by familiar parties, especially the mother and close relations:[152] "Inner understanding," a general "atmosphere" of being "on intimate terms,"[153] is required for the child to feel at ease and for the observer to "interpret" actions correctly. At the same time, however, these texts point to a problem arising from the close relationship between the observer and the observed: although mothers and other persons close to the child are granted hermeneutic superiority over strangers, it is also feared that the observer's relation to the child may distort the interpretation – for instance, when typical behavior is mistaken for precocious talent, or when facial movements that are merely reflexive are taken to represent an early attempt at communication.[154] Accordingly, it is recommended that a scientist attend the observer. Sully describes the mother as an assistant to a scientifically trained father.[155] Stern recommends that she has training herself.[156] The purpose is clear enough: "The observation which is to further understanding, which is to be acceptable to science, must itself be scientific."[157] In this spirit, at the beginning of his book Stern formulates rules that are intended to equip lay observers with the right methodological tools: First, observers should distinguish between factual matters and interpretations. Second, interpretations ought to be at a level appropriate to the child's stage of development. And third, general statements without sufficient empirical evidence are to be avoided.[158] Stern also indicates techniques

---

[150] Sully, *Studies of Childhood*, 14, 16.
[151] Sully, *Studies of Childhood*, 236.
[152] Stern, *Psychology of Early Childhood*, 37.
[153] Stern, *Psychology of Early Childhood*, 38.
[154] Sully, *Studies of Childhood*, 11.
[155] Sully, *Studies of Childhood*, 17, 18.
[156] Stern, *Psychology of Early Childhood*, 37. On the division of roles in the bourgeois family and the father's "function as an observing, controlling third party" supervising the "mental development of children," cf. Wittmann: from the late eighteenth century on, an increasing number of fathers "kept literal records of the development of their children's behavior and mental development" (*Bedeutungsvolle Kritzeleien*, 123).
[157] Sully, *Studies of Childhood*, 11.
[158] Stern, *Psychology of Early Childhood*, 8.

that will grant the diary a proper, "scientific" status – for instance, recording raw data promptly and according to a strict chronology, all the while ensuring that the child remains unaware of the proceedings.[159]

Bound to the scientification of diaries was the hope that the individual case could be generalized into the exemplary one. The diaries tended to be treated as (collections of) case histories; that is, from the observations on an individual child, readers drew conclusions about child development in general. To this end, Groos calls for combining "individual-" and "mass observation" so that particular details might be verified in light of overall trends.[160] Sully and Stern acknowledge the virtues of statistical research, even though they preferred individual observation for its advantageous "close rapport" with the subject.[161] Small-scale experiments, with outcomes noted in the diary, contributed to the diary's drift into case history, that is, into the typical genre for reporting human experiments.[162] Such experiments had already been performed by Preyer, whose authority Sully invokes when recommending the same.[163] Groos also calls for observations to be conducted under both natural and artificial (experimental) conditions, while hoping that an ideal balance between the two might be struck so that the young child would behave normally without noticing.[164] Piaget, as I have noted, employed the "clinical interview" method, which is based on exact observation of a broad sample of children by means of questions, interviews, and tests. Stern alone evinces skepticism, especially about large-scale and longitudinal experiments likely, in his estimation, to falsify the child's behavior.[165] In sum, the ambivalence I have already noted is once again evident in respect to statistical surveys and experimentation, practices that play a central role in authenticating the scientific status of the diaries yet stand at odds with the ideal of maintaining a close relationship with the child and the 'natural' conditions for observation (which the term 'diary' implies).

That the 'case-historicization' of the diary resulted in more than its scientification offers a glimpse of the proximity of these case-studies to literary texts.

---

159 Sully, *Studies of Childhood*, 10.
160 Groos, *Das Seelenleben des Kindes*, 14.
161 Cf. the American Child Study Movement, whose "prophet" was G. Stanley Hall (Wittmann, *Bedeutungsvolle Kritzeleien*, 207) and whose propagandists mobilized "veritable legions of teachers and parents" to "collect as much, and as varied, data on childhood as was possible" (205).
162 A clear effort at scientificity in the form of tables and statistics can also be found in Siegfried Levinstein, *Das Kind als Künstler* (Leipzig: Voigtländer, 1905).
163 Sully, *Studies of Childhood*, 19.
164 Groos, *Das Seelenleben des Kindes*, 14.
165 Stern, *Psychology of Early Childhood*, 39–40.

They follow a narrative scheme to the extent that they present a narrator (the observer), a protagonist (the child), and a notable event, often presented in the rhythm of exposition–climax–resolution. Take the following example from Preyer (chosen more or less at random):

> The twenty-third month brought at length *the first spoken judgment*. The child was drinking milk, carrying the cup to his mouth with both hands. The milk was too warm for him, and he set the cup down quickly and said, loudly and decidedly, looking at me with eyes wide open and with earnestness *heiss* (hot). This single word was to signify "The drink is too hot!" In the same week [...] the child of his own accord went to the heated stove, took a position before it, looked attentively at it, and suddenly said with decision, *hot* (*heiss*)! Again, a whole proposition in a syllable.[166]

The three-tiered ambivalence that results – personal involvement, scientificity, and literariness – matches the reaction provoked in the readers. They entertain a distance from the portrayed research and its objects of study, in keeping with the scientific nature of the text. In spite of this, the child whose multistage development the readers are following comes closer and closer to them, in keeping with the details provided in the diary and their literary cast whereby the child is given a distinct character, interests, emotions, and a story. All of this means that the reader's perspective is constantly shifting between analytical distance and personal involvement. Reading Stern, one soon gets to know his children, Hilde and Günter, and responds to the events in their lives and their overall development in an emotional manner. In the case of Preyer's son, it is hard not to feel pity for the boy, inasmuch as his father uses him as an object of research without showing him affection.

The approach taken by developmental psychologists to the observations they record in their diaries also wavers between that of the natural sciences (experimentation, statistics) and philology. Time and again, the authors reflect with approval on their ways of interpreting the events they've narrated. Sully even calls "the observer [...] a sort of clairvoyant reader of [children's] secret thoughts."[167] In such statements, we can recognize the hermeneutic operations of philology. This proximity is also evident in frequent references to literary examples, which are often mixed in with purportedly authentic diary entries. Groos cites scenes from Gottfried Keller and Goethe, among others, for example, when discussing childish destructiveness.[168] Indeed, he also advocates including poet-

---

[166] Preyer, *Mental Development in the Child*, 144.
[167] Sully, *Studies of Childhood*, 14.
[168] Groos, *The Play of Man*, 98.

ry and artists' autobiographies because such works combine self-observation with the observations of others. Artists, he believes, are far more perceptive when it comes to children.

> Although the artist's imagination, even for purely biographical purposes, leads to many deviations from reality, he has the ability, more than other people, to recall the emotions of childhood as though they happened yesterday and to express their characteristics most fully.[169]

The argument positing a relationship between the child and the artist – which is typical of discourse about the 'primitive' around 1900 (cf. chapter 5) – upends the hierarchy that subordinates works of imagination to scientific observation. Literature and art are granted greater accuracy because they bridge the gap to authentic (self-)observation.[170]

Charlotte Bühler adopts a different, but still largely philological approach. She seeks to acquire information on the way children think by studying the books they prefer to read (or hear), which she assumes therefore have an affinity with the child's mind. Accordingly, her work explores children's imagination by analyzing fairy tales. To ensure scientific soundness, she incorporates a statistical survey on the ages at which children are most interested in these stories.[171] Generally this takes place through a recognition of the fairy tale's typical features (characters, setting, plot, and their representation) and simply correlating them with the way children's minds work.

> This naive concatenation of the everyday, even profane, with the extraordinary and miraculous is a peculiarity inherent only in folk tales, and one that expresses a unique simplicity. Such an approach must be very close to the childlike view of life. It accepts the profane and the sacred without distinction, unbiased and with innocence; reality and wonder are not yet separated by an unbridgeable gap. The fairy tale world may be natural to the child to the same extent as it is unreal to the adult.[172]

In sum, early developmental psychology presented itself through mixed writing methods drawn from the fields of natural science, philology, and literature. Unlike texts published in ethnology at the time, those written by developmental

---

169 Groos, *Das Seelenleben des Kindes*, 18.
170 Sully (*Studies of Childhood*, 11) and Stern (*Psychology of Early Childhood*, 41–42) also value the authenticity of poets' memory more highly than that of other people; however, they doubt its usefulness for science because of the very "poetry" such recollections contain.
171 Charlotte Bühler, *Das Märchen und die Phantasie des Kindes* (Leipzig: Barth, 1918), 5.
172 Bühler, *Das Märchen*, 11.

psychologists start to reflect on this structure. Indeed they hint critically at the *poème*-like character of these texts, but at the same time they view the proximity to creative works and the emotional involvement of the scientists as a seal of quality insofar as they imply greater authenticity in observation and interpretation. The 'primitive,' here configured as the child, stands not only on the border between an othered self and a nostrificated other, but also on the boundary between competing scientific methods – to say nothing of the much-debated 'two cultures' of literature and science. Much the same holds for ethnological writings, but works of developmental psychology bring out the tension much more, since emotional connections to the object of study and, thus, skepticism about the natural scientific approach inspired researchers to look for alternatives to standard scientific writing techniques.

Chapter 4
Psychopathology in the Paradigm of the 'Primitive'

Like the fields of developmental psychology and ethnology reviewed in the previous chapters, psychiatric studies of the early twentieth century relied on analogical thinking. However, instead of equating indigenous peoples or children with humanity in its original state, the analogy concerned the mentally ill, and in this way it constructed the third figuration of the 'primitive' addressed in this book. The premise that the 'primitive' was present in schizophrenics – who suffered from "*the* artistic malady of the 1920s"[1] – and that their experience and thought corresponded to that of prehistoric humans formed the cornerstone of mental health research during the 1910s and 1920s.

Paul Schilder's *Wahn und Erkenntnis* (Delusion and Knowledge, 1918), for example, aims to "delineat[e], in sharper fashion, the essential lines common to the delusion of the [mentally] ill and the thinking of 'primitive man.'"[2] In regard to the latter, the book's scope is limited to beliefs in magic and animism, which – in light of the works of Frazer, Wundt, Preuss, and Vierkandt – Schilder considers essential aspects of 'primitive thinking.' Accordingly, he sets out to identify such beliefs and the thought mechanisms underlying them in schizophrenics and paranoiacs (who suffer from a "closely related disease"[3]).

Schilder proceeds by presenting an array of case histories, each of which is followed by a summary and analytical remarks stressing the patient's proximity to 'primitive thinking.' The first case concerns a woman identified as Anna H., who suffers from the delusion that she is under the spell of a black hand exercising control through "little wishes" (*Wünschelchen*); this condition, Schilder contends, exemplifies the indigenous belief in "magic-as-substance" (*substantiell gedachte zauberische Substanz*) – that is, *mana* or, in this instance, the "*orenda* of the Iroquois."[4] In the next case, the patient Helene K. is convinced that her thoughts are all-powerful, which Schilder declares to conform to 'prim-

---

[1] Bettina Gockel, *Die Pathologisierung des Künstlers. Künstlerlegenden der Moderne* (Berlin: Akademie Verlag, 2010), 13.
[2] Paul Schilder, *Wahn und Erkenntnis* (Berlin: Springer, 1918), 87. The author refers to Felix Krueger's 1911–1912 lectures in ethnology at the University of Halle, as well as to the writings of Freud and Jung (see below).
[3] Schilder, *Wahn und Erkenntnis*, 60.
[4] Schilder, *Wahn und Erkenntnis*, 62.

itive thinking.' The third case, of Hans Felix K., illustrates the author's claim that the mentally ill share the indigenous belief in the magic power of words. In the fifth case, Schilder takes the wish expressed by the next patient, Rudolf B., to undergo a "testicular cross section" as a resurrection of archaic puberty rites.[5]

On the whole Schilder's analysis does not focus on mental structure so much as content. He devotes little analysis to the logic of his patients' delusions (for instance, the way heterogeneous entities are grouped together on the basis of chance similarities). Instead, he traces them back to a putative imaginary complex from long ago. Only in his concluding remarks does he acknowledge that an emotional logic is at work in 'primitive' and pathological thought.

> The worldview that places the magical in the foreground, whether among primitives or the mentally ill, is a worldview constructed in decisive measure by the affective element. [...] One could say that part of the primitive or insane person's drives have turned into his object.[6]

Schilder describes correlations between 'primitive' and paranoid (or schizophrenic) thinking in detail. However, he does not offer nuanced explanation of what kind of correspondence is at issue.

Addressing this shortcoming is the aim of Alfred Storch in *Das archaischprimitive Erleben und Denken der Schizophrenen* (1922; *The Primitive Archaic Forms of Inner Experiences and Thought in Schizophrenia*, 1924). Storch also adopts a purely "phenomenological" approach at the outset, revealing the peculiarities of schizophrenic experience and thought. In turn, he places these observations in a developmental psychological context. This added step involves taking a "genetic psychological viewpoint" meant to reveal how "corresponding to all the processes and structures in adult man, lower and less perfect forms are met with in men at lower cultural levels, in children, and in animals."[7] In other words, Storch sets up a series of developmental stages. At the bottom of the scale are animals, "peoples of nature" (*Naturvölker*), and children; the top is occupied by "civilized man" (*der Kulturmensch*). This scaling enables him to understand correspondences between the thinking of "peoples of nature" and schizophrenics as evidence that they belong to the same (low) rung of development. For Storch, schizophrenics think on the same level as indigenous peoples

---

5 Schilder, *Wahn und Erkenntnis*, 76–77.
6 Schilder, *Wahn und Erkenntnis*, 100.
7 Alfred Storch, *The Primitive Archaic Forms of Inner Experiences and Thought in Schizophrenia: A Genetic and Clinical Study of Schizophrenia*, trans. Clara Willard (New York: Nervous and Mental Disease Publishing Company, 1924), ix.

(or as children even, as he repeatedly declares): "We have thus, in [...] analysis [...], stumbled upon an abundance of peculiar tendencies and motivations [...] which all alike have parallels in the primitive levels of thought."[8]

The *Poèmes* of Psychology

Storch hardly reflects on how or why this state of affairs has come to be. In lieu of theoretical discussion, he uses four complexes of metaphors to attempt to figuratively capture the relationship between the mentally ill and 'primitive thinking.' Storch states at some points that the patient's thinking "sinks back"[9] to a lower stage and at others that the 'primitive' mental and emotional world is "breaking forth."[10] Both representations of the psyche are based on an imaginary topography relegating earlier forms of consciousness to the bottom and newer ones to the top, where a step downward can also represent a movement backward.[11] Significantly, however, Storch's figurative representations remain incongruous with one another inasmuch as the first case pictures activity emanating from the subject, and the second from the 'primitive' world of emotion. Accordingly, resistance appears comparatively low in the first case and quite high in the second: the image of a soft surface of water is set in opposition to a hard crust that must be broken through by force. In terms of psychoanalysis, the first metaphorical complex comes quite close to regression, and the second approaches that of repression (he also speaks of "a breaking forth of emotional currents which had been dammed back"[12]) – I will address both in due course.

The two other metaphorical complexes employed by Storch, which occur less frequently but likewise contradict each other, concern "substitution" and "undercurrents." At times, archaic experience is viewed in terms of "the undercurrent of the waking thoughts of the day"[13] affecting all human beings at all times. Alternatively, it is described as a building block that fills in the gaps that result when a developed consciousness falls apart: "Using this comparative genetic method we discover [...] that in schizophrenia certain mental conditions which are stable in highly developed minds [...] are replaced by more primitive

---

8 Storch, *Primitive Archaic Forms*, 4.
9 E.g., Storch, *Primitive Archaic Forms*, 25, 96.
10 Storch, *Primitive Archaic Forms*, 59, 83.
11 Storch, *Primitive Archaic Forms*, 99.
12 Storch, *Primitive Archaic Forms*, 59.
13 Storch, *Primitive Archaic Forms*, 105.

mental conditions."[14] Storch clearly reserves greater sympathy for the first notion. It lays the foundation not just for the metaphorical complexes of sinking and eruption, but also for the hymn-like rhapsody concluding his study:

> All the dams which reason has erected [...] give way and the psychic experiences unfold themselves unimpeded in the boundless sphere of the unconditioned. From the substrata archaic elements swell up, an intoxicating Dionysiac cosmic consciousness, a grandiose world phantasy; the person [...] becomes [...] God.[15]

Storch exposes this bearing as "Promethean temerity" and prophesies its imminent collapse, but all the same he grants the patient the status of an ancient hero. However, in contrast to the standard narrative of Enlightenment, rebellion against the gods does not represent an appeal to innate human reason so much as a liberation from it and an abandonment of oneself to experience and thought guided by emotion.[16]

The positive evaluation of schizophrenic thought resounding in this hymnic conclusion contradicts the study's scientific claim. At the same time, however, it agrees with Storch's use of metaphors, which do not act as mere rhetorical ornamentation but constitute an integral element of his reflections on operations of the 'primitive mind.' On this score, his reflections also express a certain affinity with 'primitive thinking' as he himself defines it, namely as being based on literal interpretations of figurative language.[17]

This poses the question of the *poème*-like character of the studies already mentioned, a quality that is also pronounced in the literary cast of Schilder's case studies. Inasmuch as typography sets the case studies apart from the rest of the text, a certain independence is already in evidence. Over the course of each chapter, the case studies grow longer and longer. The last two are some five pages each, which makes it easy for the reader to get lost in them – especially since the narrative distance between the doctor and the patient progressively

---

14 Storch, *Primitive Archaic Forms*, xii.
15 Storch, *Primitive Archaic Forms*, 106.
16 Doris Kaufmann speaks of a "pronounced concept of schizophrenia" in the cultural-scientific discourse of the 1920s ("Kunst, Psychiatrie und 'schizophrenes Weltgefühl' in der Weimarer Republik. Hans Prinzhorns Bildnerei der Geisteskranken," in *Kunst und Krankheit. Studien zur Pathographie*, ed. Matthias Bormuth, Klaus Podoll, and Carsten Spitzer [Göttingen: Wallstein, 2007], 57). Likewise, in her discussion of Binswanger and Jaspers, Gockel points out that during the same period mental illness went from being seen as a degenerative phenomenon to counting as a sign of election, especially in the context of art (Gockel, *Die Pathologisierung des Künstlers*, 83–103).
17 Storch, *Primitive Archaic Forms*, 98–99.

decreases. Thus, in the first case history, the grammatical mood switches from the subjunctive to the indicative: "She [thinks she] stands under the spell of the black hand" (*Sie stehe im Bann der schwarzen Hand*) becomes "The black hand has now, through the little wishes, stuck something in her throat" (*Die schwarze Hand hat ihr jetzt vermittels der Wünschelchen etwas in ihren Hals gesteckt*).[18] In subsequent case histories, more direct discourse from patients is included – their writings, for instance, or remarks made in conversation. A tripartite dramaturgy is also evident: the patient being admitted to care, progress (or lack thereof), and, finally, his or her discharge.

The tendency to give case histories a literary cast has a pendant in Storch's practice of not always distinguishing between real and fictional examples.[19] Thus, in the chapter entitled "The Schizophrenic Consciousness of Self: A Structure Belonging to a More Primitive Psychological Level" – he shares a dream from Gottfried Keller's *Der Grüne Heinrich* (1854, 1879; *Green Henry*, 1960) (an example previously used by Ludwig Klages) in order to shed light on the actual case of a patient from the St. Georg Hospital in Hamburg.[20] The literary nature of the study also comes out in the use of metaphor I have been discussing. Giving up analytical and conceptual terminology in favor of imagistic language to illustrate the relationship between schizophrenic and 'primitive thinking' corresponds to the author's advocacy (in line with the methodological considerations of developmental psychologists) for the researcher's "emotional participation and sympathetic understanding" of the schizophrenic mind, which, being engaged in an emotional and irrational process, is "only imperfectly accessible to rational analysis." Thus, Storch calls for "entering deeply [...] into the life of the schizophrenic, on the one hand, and into the ethnographical material, on the other."[21] This bearing does not represent a scientific approach so much as a literary aesthetics of empathy.

---

18 Schilder, *Wahn und Erkenntnis*, 60–61.
19 Breuer and Freud's *Studien über Hysterie*, (1895; *Studies on Hysteria*, 1936) is well known for the literary cast given to cases. For a recent discussion, see Achim Geisenhanslüke, *Das Schibboleth der Psychoanalyse* (Bielefeld: transcript, 2008).
20 Storch, *Primitive Archaic Forms*, 21–22.
21 Storch, *Primitive Archaic Forms*, x. Cf. Werner, *Einführung in die Entwicklungspsychologie*, 30, who writes that one should mentally assume the psychopath's position.

The Analogy of Regression

The theoretical paradigm for schizophrenia advanced by Schilder and Storch follows the principle of analogy insofar as schizophrenics and prehistoric humans are equated under the category of the 'primitive.' Both authors refer to Freud, whose *Totem und Tabu* (1913; *Totem and Taboo: Some Points of Agreement between the Mental Lives of Savages and Neurotics*, 1919) is one of the first works to have pointed out affinities between the mentally ill (neurotics, in this case) and prehistoric humanity. But unlike Schilder and Storch, Freud's theory of regression seeks to account for how such affinities come about, or more precisely, why more than a mere analogy is at stake.[22] Regression (not survival or recapitulation) is the temporal model that his argument follows.

Freud first developed his ideas about a different way of thinking – one that is possibly archaic and perhaps also found in children – by attempting to explain the phenomenon of dreaming. A few comments on this aspect of his studies are in order before I turn to the theory of regression, properly speaking.[23] In *Traumdeutung* (1899; *The Interpretation of Dreams*, 1913), Freud develops a model of "dream-work" that informs subsequent theorists' (e.g., Piaget and Jung) conceptions of another way of thinking. Freud stresses that dream-work differs from waking thoughts in that the former does not form new thoughts so much as transform existing dream-thoughts[24] in a manner that is "irrational"[25] by the standards of daytime life. The process occurs by means of "condensation" (when two images are fused into one), "displacement" (when one image is replaced by another that is similar or connected by association), and "conditions of representability" (i.e., visualizing things or making use of symbols). Moreover, logical connections can transform into temporal ones (e.g., rendering a logical connection as simultaneity or a causal relation as succession) or can simply

---

22 For an "analysis of the emergence and diffusion of the theory of schizophrenic regression," cf. Andreas Heinz, *Anthropologische und evolutionäre Modelle in der Schizophrenieforschung* (Berlin: VWB, 2002), 5. The author provides an overview from the end of the nineteenth century to the end of the twentieth, but often devotes too little space to individual theories. See also Peter Geissler, *Mythos Regression* (Giessen: Psychosozial-Verlag, 2001), who deals with theories both by psychoanalysts and researchers in other fields but also in too cursory of a manner.
23 Cf. Gardian (*Sprachvisionen*, 94–121) for a discussion of psychopathological primitivism in the writings of Freud, Jung, and Kretschmer.
24 Freud, *The Interpretation of Dreams*, 510.
25 Freud, *The Interpretation of Dreams*, 601.

be disregarded.[26] Through these processes (compounded by "secondary revision"), the dream-work defamiliarizes latent dream thoughts into the dream's manifest content, which, bypassing mental censorship, finally gains admission to consciousness.

Significantly, Freud avoids speaking of dream-work as a different, other, or alien way of thinking. For the most part, *The Interpretation of Dreams* discusses mental operations in the sense of waking life, and his subsequent studies do too. Thus, in "Formulierungen über die zwei Prinzipien des psychischen Geschehens" (1911; "Formulations on the Two Principles of Mental Functioning," 1925), Freud defines thought as a reaction to the formation of the so-called reality principle: conjectural activity that serves to postpone the immediate gratification ("motor discharge" of excited states) that the so-called pleasure principle demands.[27] The state of sleep, on the other hand, is described as the "likeness of mental life as it was before the recognition of reality"; accordingly, in *The Interpretation of Dreams*, the processes of dream-work are described as "primary" processes of the "psychical apparatus,"[28] serving the pleasure principle exclusively – hence, they cannot be considered thinking or thought at all.[29] Freud nevertheless still refers to dreams as "a particular *form* of thinking, made possible by the conditions of the state of sleep" in a footnote added to *The Interpretation of Dreams* in 1925.[30] However, since he is discussing dream-work as an autonomous process, it is clear that the dream *itself* does not think; rather, it *presents* thoughts in a foreign form.

---

26 The only exception to the overall distortion of logic, according to Freud, is the treatment of similarity, which "is capable of being represented in dreams in a variety of ways" (Freud, *The Interpretation of Dreams*, 335).

27 Sigmund Freud, "Formulations on the Two Principles of Mental Functioning," in *The Freud Reader*, ed. Peter Gay (New York: W.W Norton, 1989), 303. Fantasy stands apart as a form of thinking devoted to the pleasure principle. Before the reality principle sets in, Freud argues, the object of desire (or thought) is hallucinatory.

28 Freud, *The Interpretation of Dreams*, 601.

29 Cf. Carl Gustav Jung on Freud's definition of thinking: "Freud finds that the hallmark of waking thought is *progression:* the advance of the thought stimulus from the systems of inner or outer perception through the endopsychic work of association to its motor end, i.e., innervation. In dreams he finds the reverse" (Jung, *Collected Works of C.G. Jung*, vol. 5, *Symbols of Transformation*, trans. Gerhard Adler and R.F.C. Hull [Princeton, NJ: Princeton University Press, 1976], 21). Emphasis in original.

30 Freud, *The Interpretation of Dreams*, 510.

In *Totem and Taboo*, his examination of the 'primitive worldview,'[31] Freud again takes up the hallucinatory premises of desire along lines bound to the pleasure principle – the same structure that governs dreams (or, more precisely, hallucinatory dreams[32]) and the mental lives of children[33] and psychotics.[34] Here, he focuses on animism as a "system of thought"[35]: as the "doctrine of spiritual beings" teeming everywhere in the world, which are held to be responsible for natural processes and to infuse not only animals, plants, and inanimate things with life, but also human beings by means of those entities.[36] Unlike the English ethnologists he cites, Freud does not trace such belief back to prescientific curiosity so much as look for its psychological cause. As he argues apropos of magic (which he deems a technique of animism), animism follows from attaching excessive value to purely mental processes, which is expressed in magic by satisfying a wish through "motor hallucinations" (which represent the wish in question as having been fulfilled). Hereby, "things become less important than ideas of things: whatever is done to the latter will inevitably also occur to the former": "the principle governing magic, the technique of the animistic mode of thinking, is the principle of the 'omnipotence of thoughts.'"[37] Freud goes on to observe that this principle shapes the worldview of neurotics, whose compulsive actions are likewise "magical."[38] What's more, neurotics and "savages" share the condition of being stuck at an early stage of sexual development, namely childhood narcissism.

For Freud, childhood narcissism is the actual source of the belief in the omnipotence of thoughts. The lacking reference to an external love-object, which defines narcissism, corresponds mentally to a devaluation of external reality compared to the products of inner life: "intellectual narcissism and the omnipotence of thoughts."[39] Indigenous or prehistoric cultures and children, he claims, occupy this same developmental level, and neurotics return to it through

---

31 For an examination of Freud in the context of artistic primitivism, see David Pan, *Primitive Renaissance* (Lincoln, London: University of Nebraska Press, 2001), 83–97. On Freud's reading of Frazer, cf. Ronald E. Martin, *The Languages of Difference: American Writers and Anthropologists Reconfigure the Primitive, 1878–1940* (Newark: University of Delaware Press, 2005), 91–131.
32 Freud, *The Interpretation of Dreams*, 544.
33 Freud, *The Interpretation of Dreams*, 533.
34 Freud, *The Interpretation of Dreams*, 114, 533.
35 Sigmund Freud, *Totem and Taboo: Some Points of Agreement between the Mental Lives of Savages and Neurotics*, trans. James Strachey (London: Routledge, 2001), 90.
36 Freud, *Totem and Taboo*, 88.
37 Freud, *Totem and Taboo*, 99.
38 Freud, *Totem and Taboo*, 98.
39 Freud, *Totem and Taboo*, 105.

regression. Or, more precisely, part of the native or archaic bearing has remained within them through fixation (i.e., an arrested development in the course of childhood), and another part of their mindset is reactivated by the sexualization of thought processes. In the paradigm of the 'primitive,' neurotics suffer from a double – ontogenetic and phylogenetic – regression: they repeat a behavior that is both childlike and archaic.

In this context, Freud speaks of animism as a "system of thought," of the "animistic mode of thinking," as well as of "sexualized thinking." He comes close to conjecturing that a different mode of mental life exists, which, however, is not characterized by another quality of thinking so much as by the attachment of a higher value to the products of thinking over actual reality. If thinking is actually supposed to obey the reality principle, then here its higher evaluation puts it into the service of the pleasure principle. A certain proximity of this thinking to fantasy results, which Freud, in "Formulations on Two Principles of Mental Functioning," defines as mental activity subordinate to the pleasure principle:

> With the introduction of the reality principle one species of thought-activity was split off; it was kept free from reality-testing and remained subordinated to the pleasure principle alone. This activity is *phantasying*, which begins already in children's play, and later, continued as *day-dreaming*, abandons dependence on real objects.[40]

Consequently, the main difference between fantastic thinking and non-fantastic thinking is that the former does not test its thoughts against reality, that is, it does not distinguish between imagination and reality – and inasmuch as it considers products of imagination to be real or constitutive of reality, it is less bound to the laws of logic. Sully had already argued along similar lines to explain children's magical thinking. And Jung will take up this affinity in *Wandlungen und Symbole der Libido* (1912; *Psychology of the Unconscious: A Study of the Transformations and Symbolisms of the Libido*, 1916), where he posits the existence of fantastic thought (as I will show in detail below).

The affinities of dream-work to theories of 'primitive thinking' developed in the fields of contemporary ethnology and developmental psychology are plain. Freud remarks as much in *The Interpretation of Dreams* when he describes dreams as "regression to the dreamer's earliest condition, a revival of his childhood, of the instinctual impulses which dominated it and of the methods of expression which were then available to him"; he expresses the hope of eventually discerning "behind this childhood of the individual [...] a picture of a phylogenetic childhood – a picture of the development of the human race":

---

40 Freud, "Formulations on the Two Principles of Mental Functioning," 303.

> Dreams and neuroses seem to have preserved more mental antiquities than we could have imagined possible; so that psycho-analysis may claim a high place among the sciences which are concerned with the reconstruction of the earliest and most obscure periods of the human race.[41]

Freud describes this double recourse as "regression." But what, exactly, does he mean by this term? The understanding presented in *Interpretation of Dreams* refers, for one, to a reversed course of motion within the psychic apparatus: the normal path leading from sensory organs to motor operations turns around so that the latter (or, at any rate, certain thoughts) provoke sensory stimuli – hallucinations, in other words. However, Freud goes on to stress that "regressive thought transformation" (mental activity that turns into physical sensation) can also occur in waking life under pathological conditions. This reversal is not brought about by the dream state but eased by it. Instead, it is triggered by the connection between thoughts and repressed or unconscious (for the most part infantile) memories.[42] Freud writes that these memories pull the thoughts associated with them into regression, as it were, for infantile memories generally resemble hallucinated or sensory perception. But in addition to the force of attraction exercised by memory, a force of resistance works against the penetration of such thoughts into consciousness. Regression is the result of these two conflicting aspects of mental life.

In a 1914 addition to *The Interpretation of Dreams*, Freud distinguishes between three kinds of regression: The first is topical, the reversed course of the psychic apparatus described above. The second is temporal, when *older* mental formations such as infantile memories are reactivated. The third is formal regression, when 'primitive' modes of expression replace those otherwise in place. All three are interrelated: "what is older in time is more primitive in form and in psychical topography lies nearer to the perceptual end."[43] In a sense then, Freud reconfigures the topography of the psychic apparatus along allochronic lines and formalizes its differences: the system of perception no longer stands at the beginning of a direction of motion only in a spatial sense; it also simultaneously re-establishes itself in the temporal distance, in the ontogenetic past, and therefore is distinguished by a less developed language of forms.

"General Theory of the Neuroses" (1917), one of Freud's *Introductory Lectures on Psychoanalysis*, revisits the topic of regression in the context of developmen-

---

41 Freud, *The Interpretation of Dreams*, 550.
42 Freud, *The Interpretation of Dreams*, 539–549.
43 Freud, *The Interpretation of Dreams*, 549.

tal abnormalities. Here, he at first distinguishes between regression and inhibited libido:

> I will therefore declare without more ado that I regard it as possible in the case of every particular sexual trend that some portions of it have stayed behind at earlier stages of its development, even though other portions may have reached their final goal. [...] Let me further make it clear that we propose to describe the lagging behind of a part trend at an earlier stage as a *fixation* – a fixation, that is, of the instinct.

Regression, on the other hand, does not involve getting stuck at a particular stage. Instead, more advanced components of the psyche *revert* to an earlier stage of development. Freud clarifies, however, that this process depends on fixations that have previously occurred: "The stronger the fixations on its path of development, the more readily will the function [i.e., attaining the means of gratifying the sexual urge] evade external difficulties by regressing to the fixations."[44] The latter have weakened the function and made the possibility of reversion more appealing. Freud locates "infantile sexual experiences" as the sites of fixation to which regression leads.[45] Two factors are at work here: For one, this is where inborn drives manifest themselves, which Freud deems "after-effects of the experience of an earlier ancestry"; accordingly, he also speaks of "prehistoric experience." Second, accidental experiences during childhood – external influences – are at least equally responsible for the emergence of fixations: "fixation of the libido in the adult [...] falls, for our purposes, into two further parts: the inherited constitution and the disposition acquired in childhood."[46]

Phylogenetic Regression

The works concerning regression treated up to this point assign a much greater role to childhood than to the archaic past. However, the theoretical reflections by Freud I have just discussed open the prospect of regression reaching back much further – a possibility Freud explicitly expresses in the already mentioned addition to the 1919 edition of *The Interpretation of Dreams*. For him, dreaming as a whole is understood as a regression not just to one's own childhood (that is, to instinctual stirrings and modes of expression stored in the unconscious), but to the phylogenetic past repeated during every childhood. It follows that dreams

---

[44] Sigmund Freud, *Introductory Lectures on Psychoanalysis*, trans. James Strachey (New York: Norton, 1989), 423.
[45] Freud, *Introductory Lectures on Psychoanalysis*, 451.
[46] Freud, *Introductory Lectures on Psychoanalysis*, 450.

provide the analyst with knowledge of the archaic inheritance of humankind. Freud emphasizes this point in his *Introductory Lectures on Psychoanalysis* when discussing the "archaic traits and infantilism" that constitute both the formal and the material properties of dreams. Not only is a "primitive" form of expression realized in dreams, but, as in childhood, the "dominance of the id" and early (from an adult perspective, "perverse") sexual impulses are also reestablished. At the same time, symbolic relations, which comprise the "intellectual endowment" of early humans, are revived.[47]

In *Totem and Taboo*, Freud had already hinted that "agreements" between the mental lives of neurotics and "savages" amount to more than a mere analogy. While wary of drawing the conclusion of "any internal relationship"[48] between them, he simultaneously stresses that neurotics "may be said to have inherited an archaic constitution as an atavistic vestige,"[49] associating it thus with development that was suspended at an earlier phylogenetic stage. Elsewhere in the study, Freud describes the behavior of neurotics in terms of regression to a narcissistic stage shared with indigenous peoples and children.[50] Finally, at the end of the work, he ventures the "bold" claim that a mass-psyche is handed down from one generation to the next, which he justifies by arguing that otherwise "there would be no progress in this field and next to no development."[51] Nonetheless, he leaves open the question of how transfer of this kind might take place. Freud's thesis here is more idealistic than materialistic, i.e., he assumes that later generations take up the feelings of earlier ones by inheriting customs, ceremonies, and statutes that carry deposits of their emotional life.[52]

A manuscript from 1915, *Übersicht der Übertragungsneurosen* (*Overview of the Transference Neuroses*, 1987) returns to the themes of *Totem and Taboo*. This document, rediscovered by Ilse Grubrich-Simitis, makes it perfectly clear that regression extends back to phylogeny for Freud. In contrast to the former publication, this text provides a precise biological explanation for this extension. Describing regression as "the most interesting factor and instinctual vicissitude," he once again discourses on "problems of fixation and disposition" and traces regression back to "a fixation point in either ego or libido development,"

---

47 Freud, *Introductory Lectures on Psychoanalysis*, 262.
48 Freud, *Totem and Taboo*, 158.
49 Freud, *Totem and Taboo*, 77.
50 Freud, *Totem and Taboo*, 120.
51 Freud, *Totem and Taboo*, 183.
52 Freud, *Totem and Taboo*, 31.

which for him represents the tendency toward neurosis.[53] As before, Freud stresses that fixation can be ascribed either to experiences from early childhood or genetic constitution, which he understands as the "acquisitions of our ancestors."[54] In so doing, he arrives at the hypothesis that the "phylogenetic disposition" can be used to elucidate neurosis. From here, he sketches a "phylogenetic playlet in two acts and six scenes"[55]: In the first act, "prehistoric man" must contend with the catastrophe of the Ice Age, which had abruptly ended their carefree existence under paradisiacal conditions. In reaction to the catastrophe, human beings grow anxious and introduce a prohibition on reproduction; consequently, their unused libido is sublimated and applied to intellectual tasks, specifically to an "animistic world view and its magical trappings."[56] Freud then associates these three types of behavior to the three transference neuroses that he identifies as plaguing his contemporaries (anxiety hysteria, conversion hysteria, and obsessional neurosis). By explaining the neuroses as regressions to corresponding phases of human history, he assigns a phylogenetic disposition to them. The same scheme holds for the narcissistic neuroses (dementia praecox, paranoia, melancholia-mania) that he assigns to the next three stages of development in the culture of the "second generation."[57] Here, Freud contends, the decisive events are the jealous patriarch's castration of his sons, the sons forming a homosexual fraternity, the communal grief experienced when they murder the primal father, and then the joy when the latter is resurrected. That these relations between primordial culture and modern neurotics constitute more than a mere analogy for Freud becomes especially prominent in his extensive discussion of how inheritance could be conceived in a homosexual society: "It is evident that the castrated and intimidated sons do not procreate, therefore cannot pass on their disposition."[58] Freud solves this problem with the youngest son, who (in the scenario envisioned) has not been castrated by the father or banished, but who witnesses the fate of his older brothers and their alternative community. Through this son, Freud surmises, acquisitions originally belonging only to men are passed on: "next to those men who fall by the wayside as infertile, there may remain a chain of others, who in their person go through the vicissitudes of

---

[53] Sigmund Freud, *A Phylogenetic Fantasy: Overview of the Transference Neuroses*, ed. Ilse Grubrich-Simitis (Cambridge, MA: Harvard University Press, 1987), 9.
[54] Freud, *A Phylogenetic Fantasy: Overview of the Transference Neuroses*, 10.
[55] Ilse Grubrich-Simitis, "Metapsychology and Metabiology," in Freud, *A Phylogenetic Fantasy*, ed. Ilse Grubrich-Simitis (Cambridge, MA: Harvard University Press, 1987), 89.
[56] Freud, *A Phylogenetic Fantasy: Overview of the Transference Neuroses*, 15.
[57] Freud, *A Phylogenetic Fantasy: Overview of the Transference Neuroses*, 17.
[58] Freud, *A Phylogenetic Fantasy: Overview of the Transference Neuroses*, 19.

the male sex and can propagate them as dispositions." Through this line, acquisitions initially made only by men pass down to women, too: "We are spared the grossest difficulty by observing that we should not forget human bisexuality. Thus women can assume the dispositions acquired by men and bring them to light in themselves."[59]

Passages like this – especially given that Freud stood in active correspondence with Sándor Ferenczi during the First World War – indicate that he, like Ferenczi, was convinced that the psychic maladies of his contemporaries could be explained in phylogenetic terms and that he assumed traumatic experiences and their cultural consequences could be inherited. It follows that regression reaches back to phylogenetic points of fixation. Indeed, according to Freud it even extends into the inorganic realm, that is, to the earliest state of phylogenesis. In *Jenseits des Lustprinzips* (1920; *Beyond the Pleasure Principle*, 1922), he defines instincts as regressive:

> It seems, then, that an instinct is an urge inherent in organic life to restore an earlier state of things. [...] Let us suppose, then, that all the organic instincts are conservative, are acquired historically and tend towards the restoration of an earlier state of things. It follows that the phenomena of organic development must be attributed to external disturbing and diverting influences. The elementary living entity would from its very beginning have no wish to change. [...] Every modification which is thus imposed upon the course of the organism's life is accepted by the conservative organic instincts and stored up for further repetition. [...] It would be in contradiction to the conservative nature of the instincts if the goal of life were a state of things which had never yet been attained. [...] 'the aim of all life is death.'[60]

Here, regression appears as the necessary outcome of the death instinct.[61]

One of Freud's letters to Ferenczi credits the latter as the originator (*Urheberrecht*) for developing the phylogenetic theory elaborated in *Overview of the Transference Neuroses*.[62] As Grubrich-Simitis has documented in detail, the two psychoanalysts maintained close contact when Freud was writing his meta-

---

59 Freud, *A Phylogenetic Fantasy: Overview of the Transference Neuroses*, 20.
60 Sigmund Freud, *Beyond the Pleasure Principle*, trans. James Strachey (New York: Norton, 1961), 31–32. Emphasis in original.
61 In contrast to the texts discussed earlier, Freud's *Beyond the Pleasure Principle* does not explain neuroses as regressions to an earlier phylogenetic stage. Instead, he develops a new theory of instinct, in which regression is a necessary expression of the death instinct. In this respect, it is not to be classified as pathological per se – it only becomes so if it lacks its counterpart, the urge to develop, which is triggered by the life instincts.
62 Freud to Ferenczi, 12 July 1915, in *A Phylogenetic Fantasy: Overview of the Transference Neuroses*, 95.

psychological studies. In particular, Freud appreciated Ferenczi's *Entwicklungsstufen des Wirklichkeitssinnes* (1913; *Stages in the Development in the Sense of Reality*, 1916), which contains initial speculations on the formative role of geological catastrophes for anthropogenesis and the genetic transmission of collective memories to individuals.[63] Still greater influence, in my estimation, was exercised by Ferenczi's efforts to devise a "bio-analysis" on the basis of a new theory of coitus. The results would appear in *Versuch einer Genitaltheorie* (1924; *Thalassa: A Theory of Genitality*, 1968), but Ferenczi and Freud had already been corresponding about the matter at length for years. The works of Lamarck occupy a central position in these exchanges, and Freud followed his colleague's lead and immersed himself in them. The two men even planned a joint work on Lamarck with which, as Freud wrote to Ferenczi, psychoanalysis would "[leave] its calling card with biology." The direction the project would have taken is indicated by Freud's remark that research by "psycho-Lamarckists such as [August] Pauly" risked leaving them with "little to say that is completely new."[64] (According to Pauly, physiological demands and the organism's efforts to meet them are what actually fuel the course of evolution, and organic adaptations are inherited.[65]) Although the collaborative work never materialized, Ferenczi's *Theory of Genitality* represents the outcome of the exchange. As Grubrich-Simitis has shown, it contains numerous reflections developed together, as revealed, for instance, in Ferenczi's request to include "assumptions about Lamarckism" in his own book that had been constructed jointly with Freud.[66]

The second part of *Theory of Genitality* traces the "individual experience of the catastrophe of birth and its repetition in the act of coitus" back to the emergence of humankind's distant forebears from the water: "What if […] birth itself [were] nothing but a recapitulation on the part of the individual of the great catastrophe which at the time of the recession of the ocean forced so many animals, and certainly our own animal ancestors, to adapt themselves to a land existence?" In this way, Ferenczi invokes Haeckel's fundamental biogenetic law to explain the repetition of phylogeny not just in the embryo but also in the "devel-

---

[63] Cf. Grubrich-Simitis, "Metapsychology and Metabiology," 79–81. The following draws on Grubrich-Simitis's reflections in this essay.
[64] Freud to Ferenczi, 28 January 1917, in *A Phylogenetic Fantasy: Overview of the Transference Neuroses*, 94.
[65] Cf. August Pauly, *Darwinismus und Lamarckismus. Entwurf einer psychophysischen Teleologie* (Munich: E. Reinhardt, 1905).
[66] Ferenczi to Freud, 25 July 1917, in *A Phylogenetic Fantasy: Overview of the Transference Neuroses*, 95.

opment of the means of protection of the embryo."⁶⁷ Nascent life in the amniotic fluid of the uterus repeats the conditions of existence of human beings' most remote ancestors: fish in the primeval sea. Ferenczi also enlists Lamarck's thesis that characteristics, acquired during the organism's adaptive reaction to internal needs and external forces, can be inherited – and does so in order to formulate a theory of inherited trauma. He begins with the premise that "memory traces of all the catastrophes of phylogenetic development accumulated in the germ-plasm." In other words, experiences of disaster in the early history of humankind might be stored in biological material and passed on to future generations:

> What we call heredity is perhaps [...] only the displacing upon posterity of the bulk of the traumatically unpleasurable experiences in question, while the germplasm, as the physical basis of heredity, represents the sum of the traumatic impressions transmitted from the past and handed on by the individual.

In subsequent generations, this genetic material causes the "perpetual repetition of the painful situation" on a physical level, albeit in mutated and weakened form so that, over the course of time, "unpleasurable tension" diminishes.⁶⁸ Applied to Ferenczi's hypothesis, this means not just that each individual birth repeats the traumatic expulsion of humankind's primordial ancestors from the water but also that this expulsion represents the biological cause for the emergence of uterine and natal conditions. The germ plasma, with its mnemic charge, is compelled to repetition, forming an organ and physical process that renews traumatic experience, which can be dismantled over the generations.

For Ferenczi, the womb, intrauterine existence, and the process of reproduction provide biological proof of a "thalassal regressive trend," i.e., a "striving towards the aquatic mode of existence abandoned in primeval times."⁶⁹ This view reproduces Freud's radical thesis that life seeks to return to an inorganic state. Ferenczi considers rest in the womb and in orgasm not only to be a return to life in the sea, but at the same time also to the "repose of the era before life originated, in other words, the deathlike repose of the inorganic world."⁷⁰ More importantly, this claim represents the extreme of a biological view of regression – as implied by Freud's discussion of heredity in *Overview of the Transference Neuroses* and made explicit in his effort in *Beyond the Pleasure Principle* to prove

---

67 Sándor Ferenczi, *Thalassa: A Theory of Genitality*, trans. Henry Alden Bunker (New York: W. W. Norton, 1968), 45.
68 Ferenczi, *Thalassa*, 66.
69 Ferenczi, *Thalassa*, 52.
70 Ferenczi, *Thalassa*, 63.

that the death instinct already prevails at the level of germ cells. Ferenczi takes on directly what Freud leaves relatively open-ended, namely how fixations are passed from one generation to the next. His answer is radically materialistic: this inheritance is stored in the germ plasm and takes place through the formation of certain organs and the physical processes tied to them. Thus, for Ferenczi, it is ultimately the body that permits regression to earlier phylogenetic conditions and simultaneously the repetition and overcoming of ancient *traumata*. Ferenczi conceives of regression biologically then. In terms of the paradigm of the 'primitive,' this means that schizoprenics are viewed as 'primitives' because they regress to an earlier phylogenetic stage where a fixation once took place, which has been passed on biologically from generation to generation ever since.

Ferenczi's speculations go far beyond what Freud had imagined. At the same time, they display a tendency that, as we have seen, is also evident in the latter's work. In *Freud, Biologist of the Mind,* Frank J. Sulloway points out, "from the discovery of spontaneous infantile sexuality (1896–1897) to the very end of his life, Freud's endorsement of biogenetic and Lamarckian viewpoints inspired many of his most controversial psychoanalytic conceptions."[71] Among other things, Sulloway shows how Freud ties neuroses and regression to one another through phylogenetic scenarios:

> Freud resolved the problem of the choice of neurosis in the following manner. Ontogenetically, a particular illness was linked to a particular stage of libidinal fixation, to which the libido has later regressed. Freud assumed that both the initial fixation point and the later process of regression were favored by organic predispositions – neuroses once experienced by the race. Such inborn predispositions served, he concluded, as the basic "schema" for ontogenetic development, remodeling many childhood experiences in phantasy according to the universal guidelines of phylogeny.[72]

In *Ontogeny and Phylogeny*, Stephen Jay Gould describes Freud's belief in biogenetic constitution and stresses the difference between the latter's notion of a *mental* recapitulation and Haeckel's understanding of a *physical* one: "Physical recapitulations are transient stages [...]. But the stages of mind can coexist. [...] The earlier stages are characteristically repressed in the healthy adult, but they need not disappear."[73] This difference forms the precondition for Freud's theory of neurosis, which is based on the possibility of regression that is built into mental recapitulation.

---

71 Frank J. Sulloway, *Freud, Biologist of the Mind* (London: Burnett, 1979), 498.
72 Sulloway, *Freud, Biologist of the Mind*, 391.
73 Gould, *Ontogeny and Phylogeny*, 157.

The manuscript, "Overview of Transference Neuroses," substantiates Sulloway and Gould's claims about the influence of biogenetic theory on Freud. In her essay accompanying the published manuscript, Grubrich-Simitis also documents in detail Freud's close collaboration with Ferenczi as well as the heavy impact of Haeckel and above all Lamarck on his work. She explains how Lamarckian schemes – to which Freud adhered all his life – form a "bracket between two stages in the development of Freudian theory."

> The traumatic real experience in Freud's early conception of the etiology of hysteria appears in fully developed psychoanalytic theory set back into the distant past of the prehistory of the species, that is, transposed from the ontogenetic to the phylogenetic dimension.

According to Grubrich-Simitis, the Lamarckian-Haeckelian

> postulate helped Freud bridge "the gulf between individual and group psychology." At the same time he hoped to bridge the gulf that "earlier periods of human arrogance had torn too wide apart between mankind and the animals," because he saw in the archaic inheritance of *Homo sapiens* the analogue to the instinctual equipment of animals. And he probably harbored the hope of overcoming yet another gulf, the one between the natural sciences and the humanities.[74]

After all, Grubrich-Simitis justly observes, Freud was

> radical in two directions: in the impetus of his analysis of civilization, critical of society and religion, and in his relentless insistence on the final anchoring of all human behavior in the pleasure-creating, mortal biological-organic substrate.[75]

Finally, Laura Otis argues that Freud's writings represent one of the most important lines of transmission for the biological theory of organic memory: "By the 1940s, relatively few reputable biologists relied on Lamarck's and Haeckel's thinking [...]. In psychoanalysis, however, the claim that one 'remembered' not only one's infancy, but the experiences of one's ancestors, appeared much more reasonable." Accordingly, she points to a conflict. On the one hand, Freud was convinced of the theory of organic memory:

> The individual acquired new characters: those stimuli that had sufficient impact or that were repeated frequently enough created an impression upon the nervous system and eventually upon the germ plasm and thus could be passed on to subsequent generations. Each

---

[74] Grubrich-Simitis, "Metapsychology and Metabiology," 99. The quotes within this passage are from Freud, *Moses and Monotheism*.
[75] Grubrich-Simitis, "Metapsychology and Metabiology," 105.

individual bore in his or her unconscious the heritage of ancestral impressions and added to it with personal experiences.[76]

On the other hand, Otis notes, Freud was more aware than contemporary biologists that the theory of organic memory was based on a mere analogy, i.e., on the psychologist's interpretation.

The biogenetic and Lamarckian bias that scholars have noted in Freud's works holds implications for my purposes here not only because it made the model of regression appeal to authors like Robert Müller and Gottfried Benn (whose works will be examined in detail later on), but also because it connects so clearly with the two models discussed in the previous chapters. The analogies that ethnology, developmental psychology, and psychoanalysis drew between 'prehistoric man' and indigenous peoples, children, and the mentally ill are based on three temporal models: survival, recapitulation, and regression. Survival involves the persistence of the archaic into the present, and it is modeled after atavism. Just as organs that no longer serve a purpose may be retained in rudimentary form, this theory holds that aspects of ancient cultures may be preserved even if cultural evolution is believed to have long since reached a far 'higher' level of development.

In the process of recapitulation, by contrast, a primal development is repeated. This model, with its return to the biogenetic foundation, is even more obviously understood in biological terms. Already for Haeckel, physical development proceeds hand in hand with psychic development. Thus he assigns single-cell organisms with a "cell-soul" whose function is to store the memory of earlier sensations. This memory is materially realized in a specific change to the germ plasma and then all the way to the development of certain organs. The hypothesis suggests that phylogenesis is repeated through ontogenesis both physically and mentally, and from this derives the work of many developmental psychologists and psychopathologists. But if, in this course of development, inhibitions arise – that is, a developmental standstill at a given phylogenetic/ontogenetic level – this may trigger a survival phenomenon whereby an adult behaves like a child or a contemporary 'civilized' individual exhibits archaic behavior. Accordingly, Freud's example for the inhibition of development is atavism:

> As you know, [...] in the highest mammals the male sex-glands, which are originally situated deep in the abdominal cavity, we find in a number of male individuals that one of these paired organs has remained behind in the pelvic cavity, or that it has become permanently

---

[76] Laura Otis, *Organic Memory: History and the Body in the Late Nineteenth and Early Twentieth Centuries* (Lincoln, NE: University of Nebraska Press, 1994), 183.

lodged in what is known as the inguinal canal, through which both organs must pass in the course of their migration.[77]

Finally, regression does not refer to the persistance of an archaic substance, nor to the repetition of phylogenesis; instead, it involves a turn back from a later stage of development to an earlier one. Whereas in the case of survival the enduring element is intrinsically old and never has reached a higher level, Freud applies the concept of regression to his contemporaries, whose pathologies result from their fall back to earlier stages. Freud sees inhibition as one reason for such backsliding, i.e., a fixation at an earlier level of development, which in the case of phylogenetic fixation can be passed on through organic memory to later generations.

The models of survival, recapitulation, and regression are thus interrelated. Against the background of the theory of recapitulation, developmental inhibitions can arise in the course of ontogenesis, corresponding to phylogenetic fixations or reactivating the latter. These inhibitions give rise to survivals, which are perceived in successfully evolved contexts as atavisms, rudiments of the ancient past. Conversely, regression can pass from a later stage of development to an earlier one, where an inhibition had occurred during the course of phylogenesis (or in the process of ontogenesis, which recapitulates the latter) and was then transmitted in the organic memory. Three figures – indigenous people, children, and the mentally ill – are understood to be the expressions of these three models.

From the differences between these models arise the various fantasies attached to the three figures: Even though survival presents something archaic, it at the same time is delegated to the realm of indigenous peoples outside of Europe, who are thought to embody the origins of European culture but are seen by the latter as being at a significant temporal and developmental remove. The model of recapitulation in turn draws attention to the conviction that those who belong to 'civilized' societies repeat phylogenesis ontogenetically; however, they are assured that the developmental heights they have achieved separate them from these beginnings. In brief, when looking at indigenous peoples and children, the European spectator says, "we" can see who "we" were and also confirm who "we" no longer are.[78] In contrast to both of these models, the concept of regression means that for even the healthiest adult there is apparently still a possibility of slipping back to the primal state. Consequently, analogies be-

---

77 Freud, *Introductory Lectures on Psychoanalysis*, 473.
78 On the history of this figure of thought, see Gess, "Sie sind, was wir waren," *Jahrbuch der deutschen Schillergesellschaft* 56 (2012). Schiller, "On Naïve and Sentimental Poetry," 180–181.

tween schizophrenic thinking and dream-thoughts and other, scattered states of consciousness were especially unsettling for those invested in distancing themselves from the 'primitive.'[79] For these analogies make clear that everything the schizophrenic does applies in milder form also to the self-declared rational Modern and that therefore categorical borders cannot be drawn between the two.[80] The same holds for the opposite evaluation of 'primitive thinking.' Those critics of rationality who lamented the loss of such thinking or who searched for origins and deeper modes of thought found encouragement in the model of regression because it offered hope that origins and 'primitive thinking' could be recuperated by not only schizophrenics but anyone. In this way, the distinction between the ill and the healthy was questioned, and the schizophrenic was depathologized and framed instead as the discoverer of true being.

Ontologization

Freud understands regression as a psychic process that initially provides subjective relief to the person affected, but ultimately proves highly detrimental to his or her mental health. The aim of therapy must therefore be to eliminate the current causes and historical preconditions of the underlying regression.[81] A healthy psyche does not regress. The Swiss psychiatrist Carl Gustav Jung parted ways with his teacher and arrived at a different – and, in the end, an appreciative – view of regression. For Freud, the phylogenetic and ontogenetic development proceeds in determinate stages, and, if undisturbed, this procedure ultimately guarantees mental equilibrium. None of the phases is judged as having greater value than any other, but each one must occur at the right time and give way to the next. Jung thinks otherwise and assigns a different value to developmental stages. He interprets their course of development ontologically; in other words, he seeks the essence of a person all the way back in the origins of his or her de-

---

79 Cf. Kretschmer, *Textbook of Medical Psychology*, 125.
80 Accordingly, Eugen Bleuler points out the disastrous consequences that the direct translation of "autistic thinking" into action can have on healthy people and its ruinous effects in world history (for instance, "hounding peoples and classes against each other into a gruesome struggle for annihilation") ("Das autistische Denken," *Jahrbuch für psychoanalytische und psychopathologische Forschungen* 4.1 [1912]: 34).
81 On the "harmful aspects [of regression], as a dangerous form of resistance, as a symptom of the compulsion to repeat, and finally as the most important clinical example of the death instinct," cf. Michael Balint, "Freud und das Regressionsthema," Chapter 19, in *Therapeutische Aspekte der Regression: Die Theorie der Grundstörung* (Stuttgart: Klett-Cotta, 1970).

velopment in order to find indications of that person's true *type*. In this way, regression comes to mean a path to truth and therefore acquires a positive value for Jung.[82]

It is conceivable that Freud's "Formulations on the Two Principles of Mental Functioning" prompted Jung to call a specific mode of thinking described in *Symbols of Transformation* "fantasy-thinking."[83] At any rate, the work refers frequently to Freud's writings – especially *The Interpretation of Dreams* – in order to propose a theory of "two kinds of thinking."[84] The first is "directed or logical": "thinking that is adapted to reality, by means of which we imitate the successiveness of objectively real things, so that the images inside our mind follow one another in the same strictly causal sequence as the events taking place outside it." Such mental operations require language; their substance is material, linguistic, and oriented toward communication; evaluating and reworking propositions is also a matter of language. Jung also calls it *"thinking in words."*[85] The other, opposite mode he calls "dreaming or fantasy-thinking." Such thought does not follow a directed course – or, if it does, its objective remains unconscious – so much as it proceeds by association, quickly leading from reality to fantasy. Thinking along these lines largely defies language. Instead, it relies on a rapid succession of images and feelings to gratify desires:

> We have, therefore, two kinds of thinking [...]. The [first] operates with speech elements for the purpose of communication, and is difficult and exhausting; the [second] is effortless, working as it were spontaneously, with the contents ready to hand and guided by unconscious motives. The one produces innovations and adaptation, copies reality, and tries to act upon it; the other turns away from reality, sets free subjective tendencies, and, as regards adaptation, is unproductive.[86]

Jung traces back to the origins of this thinking by way of the products of the unconscious mind,[87] childhood, and even the medieval and ancient worlds, all the way to the prehistoric past: "infantile thinking and dream-thinking are simply

---

82 Cf. Martin, *The Languages of Difference*, 91–131.
83 Jung, *Symbols of Transformation*, 18, 28, 29.
84 Jung, *Symbols of Transformation*. The chapter, "Two Kinds of Thinking," had already appeared independently (1911) in *Jahrbuch psychoanalytischer und psychopathischer Forschungen*, the same journal that published Freud's "Formulierungen über die zwei Prinzipien des psychischen Geschehens" (1911) and Bleuler's "Das autistische Denken" (1912).
85 Jung, *Symbols of Transformation*, 11. Emphasis in original.
86 Jung, *Symbols of Transformation*, 18.
87 Jung, *Symbols of Transformation*, 24, 30: "All this shows how much the products of the unconscious have in common with mythology."

a recapitulation of earlier evolutionary stages."[88] Thus he applies what holds for infantile thinking to the "fantasy-thinking" available to adults "all through [their] lives": it "corresponds to the antique state of mind."[89]

However, primordial thinking and its products are not directly accessible. Jung takes a metaphor from geology and speaks of the stratification of the psyche. The oldest mental strata, which correspond to the unconscious, come to light in the course of regression (such as occurs in schizophrenia, which Jung understands to be a process of "introversion").

> We should therefore have to conclude that any introversion occurring in later life regresses back to infantile reminiscences which, though derived from the individual's past, generally have a slight archaic tinge. With stronger introversion and regression the archaic features become more pronounced.[90]

Tying fantasy back to archaic thought, Jung advances the thesis that the symbols current in fantasy-thinking possess merit collectively and independently of history, since they come from a vanished age when they had held "legitimate truth."[91] At the same time Jung assumes that such "products" have persisted because they "express the universal and ever-renewed thoughts of mankind."[92] Thus, when a person in the present day is confronted with a desire that she or he cannot give conscious form, fantasy-thinking will step in and reach for an appropriate archaic symbol to express the wish and facilitate the person's indirect reflection on it. Like the human beings of mythical prehistory who thought in fantastic terms, Jung takes symbols much more seriously than Freud – that is, in a sense, more literally. Ultimately, he attributes them not to individuals but to human groups with distinct ethnic psychologies and histories ("every Greek of the classical period carries in himself a little bit of Oedipus, and every German a little bit of Faust").[93] Possessing authority underwritten by antiquity and collective experience, the symbol expresses "the universal [...] thoughts of mankind." In this framework, Freud's notion that interpreting dreams means dispelling illusion and accounting for mechanisms of distortion retreats to the background. Thus, Jung concludes his analysis of the Abbé Oegger's Judas fantasy (found in Anatole France's *Le Jardin d'Épicure* [1895; *The Garden of Epicurus, 1908]*) by de-

---

[88] Jung, *Symbols of Transformation*, 23.
[89] Jung, *Symbols of Transformation*, 28.
[90] Jung, *Symbols of Transformation*, 31.
[91] Jung, *Symbols of Transformation*, 27.
[92] Jung, *Symbols of Transformation*, 31.
[93] Jung, *Symbols of Transformation*, 32.

claring, "*he* was the Judas who betrayed his Lord;"[94] in this case, identifying the fantasizing person with the fantasized symbol replaces interpretation.

Yet how are symbols like this handed down over generations? In an article that appeared in *Europäische Revue* in 1928, "Die Struktur der Seele" (and then revised and published as the volume *Seelenprobleme der Gegenwart* in 1931; *Structure and Dynamics of the Psyche*, 1960) Jung makes unmistakably clear that his contemporaries have "not acquired" the "immemorial patterns of the human mind" during their own lifetimes (by way of language, for instance). Instead, these features of the mind have been "inherited from the dim ages of the past,"[95] and people share them not just with other human beings, but also with animals.[96] Jung compares the psyche's geological structure – i.e., its levels of experience from different times, some of which are archaic – with that of the body:

> This whole psychic organism corresponds exactly to the body, which, though individually varied, is in all essential features the specifically human body which all men have. In its development and structure, it still preserves elements that connect it with the invertebrates and ultimately with the protozoa. Theoretically it should be possible to "peel" the collective unconscious, layer by layer, until we come to the psychology of the worm, and even of the amoeba.[97]

Jung is convinced that rudiments of both archaic physicality and archaic mentality remain present, and by examining those remaining elements, one can trace both the mind and the body back to the time of origins. As a result, his ideas concerning heredity and the connection between organic and psychic strata reflect his belief that psychic phenomena always rest on a physical substrate. He concludes his discussion by referring to the collective unconscious as "the whole spiritual heritage of mankind's evolution, born anew in the brain structure of every individual."[98] In his later work, *Die Bedeutung von Konstitution und Vererbung für die Psychologie* (1929; *The Significance of Constitution and Heredity in Psychology*, 1960), he proceeds typologically, on the basis of parallels supposedly

---

[94] Jung, *Symbols of Transformation*, 31.
[95] Jung, *Structure and Dynamics of the Psyche*, trans. R.F.C. Hull (Princeton: Princeton University Press, 1975), 201.
[96] Jung, *Structure and Dynamics of the Psyche*, 204. However, the contents of mental representations do not stand at issue so much as their "possibilities." In a later text (*Constitution and Heredity*), Jung speaks of "forms without content" when referring to the reactive schemata of the imagination.
[97] Jung, *Structure and Dynamics of the Psyche*, 212.
[98] Jung, *Structure and Dynamics of the Psyche*, 212.

brought to light by physiology and psychology: just as bodily constitution is a matter of heredity, the collective unconscious and its symbols are passed on genetically.

For Jung, regression leads the patient back to an archaic inheritance: fantasy-thinking and its symbols. In contrast to Freud, he considers that such thinking and its elements hold the key to the truth.[99] Instead of leading to points of fixation that might be mitigated by an analytic cure, regression terminates at the core of being, an essence that has remained hidden away in the unconscious until now – or has even been combated. The malady from which the patient suffers, according to Jung, is the result of not accepting (or being unaware of) the core of one's own being; true health can only be achieved inasmuch as it is brought to the level of consciousness and accepted, or lived. In this sense, regression represents to Jung a vital step not just in the process of recovery but in the overall path toward every healthy existence. Only people who know to which archaic type they belong can live in such a way that they will avoid illness. Here the archaic substance is ontologized and thereby also de-temporalized. It serves as the core of being *now* as much as it did *then*.[100] And with that, Jung discards the model on which Freud had based regression. Well-being is not a matter of passing through stages of development correctly so much as a project of self-realization along lines drawn long ago.

In Jung's interpretation of regression, primitivizing the mentally ill means depathologizing them. Freud had already set out on this course by showing that the thought processes of the mentally ill and of those deemed to be in good health do not differ as much as the latter might wish to believe. Indeed, such thinking haunts the dreams of healthy individuals and plays a vital role in both ontogenetic and phylogenetic development. In Jung's writings this depathologization works up to ontologization in that mental illness is understood as the first step, as it were, on the road to recovering an archaic essence. At the same time, he hints that the mentally ill could serve as a model for the collective,

---

99 As Otis stresses, Jung "stands out among Freud's followers as the one who paid the most attention to the 'anthropological' or phylogenetic dimension of psychoanalysis" (*Organic Memory*, 205).

100 In this sense, Gould writes that "Jung's appeal is not to recapitulation (an ontogenetically ordered series of ancestral stages), but to a general notion of racial memory (the static possession by adults of a complete racial history). As McCormick puts it, 'For Freud, the later problems of life arise during the early period of recapitulation when stages of advance are blocked. But for Jung the important stage is long after this period [...] Recapitulation ceases to be a question of research for Jung because the archetypes exist independently of any individual's development'" (*Ontogeny and Phylogeny*, 162–163).

including the one to which he belonged: Germans, according to this diagnosis, have the Faust myth as a collective inheritance. Jung speaks of "typical myths which serve to work out our racial and national complexes,"[101] thereby adhering to nationalist and racist stereotypes and equating individual and collective destiny.[102] In mental illness, the individual finds the symbol that permits him or her to steer the destined course in life, and so does the nation. Jung appreciatively quotes the words of Jacob Burckhardt: *"Faust* is a genuine myth, i.e., a great primordial image, in which every man has to discover *his* own being and destiny in his own way."[103]

These tendencies show up clearly, for example, in Hermann Hesse's references to Jung's conception of regression, which Hesse utilized to justify the First World War as part of a collective destiny.[104] In the novel *Demian. Die Geschichte einer Jugend* (1919; *Demian. The Story of Emil Sinclair's Youth,* 1923), which was written toward the end of Hesse's analytical sessions with Jung's pupil J. B. Lang, the protagonist Emil Sinclair causes problems at school and grows increasingly isolated from others ("the change [...] did not bring me any closer to [...] anyone – it only made me lonelier"); in so doing, the young man turns more and more toward a primal, maternal principle of being ("Eve! The name fits her perfectly, she really is like the mother of us all"),[105] led by a series of mythical symbols including the biblical figure of Cain and the Gnostic deity Abraxas. Ultimately he embraces his "destiny" by enlisting to fight in the First World War, which is interpreted as the beginning of a general "world-transformation": "The [...] remarkable thing was that my 'destiny,' this private and solitary thing, would now be shared with so many other people, with the whole world."[106] War is hailed as the proving ground on which the "primal feelings" of humankind can run riot so that the soul, hitherto "divided," will perish before undergoing a miraculous "rebirth": "The bird fights its way out of the egg. The egg is the

---

101 Jung, *Symbols of Transformation*, 32.
102 On Jung's affinities with National Socialism and his anti-Semitism, see Heinz Gess, *Vom Faschismus zum Neuen Denken: C.G. Jungs Theorie im Wandel der Zeit* (Lüneburg: Klampen, 1994).
103 Jung, *Symbols of Transformation*, 32.
104 For a critical reading of Hesse, see Nicola Gess, "Kunst und Krieg. Zu Thomas Manns, Hermann Hesses und Ernst Blochs künstlerischer Verarbeitung des Ersten Weltkriegs," in *Imaginäre Welten im Widerstreit. Krieg und Geschichte in der Literatur seit 1900*, ed. Lars Koch and Marianne Vogel (Würzburg: Königshausen und Neumann, 2007), and "Musikalische Mörder. Krieg, Musik und Mord bei Hermann Hesse," in *Literatur und Musik in der klassischen Moderne*, ed. Joachim Grage (Würzburg: Ergon, 2006).
105 Hermann Hesse, *Demian: The Story of Emil Sinclair's Youth*, trans. Damion Searls (New York: Penguin, 2013), 71–72, 116.
106 Hesse, *Demian*, 131.

world. Whoever wants to be born must destroy a world."[107] In this light, the young, eccentric Sinclair finds his way back to his own destiny, which is also representative of the collective's. This narrative recurs in many variations in Hesse's works, often in connection with the theme of the artist. *Klingsors letzter Sommer* (1919; *Klingsor's Last Summer*, 1970), for instance, describes how the title character – a "mentally ill"[108] and alcoholic painter – encounters in his states of intoxication and madness visionaries and spiritual leaders such as the "Armenian astrologer," who teaches him that downfall and rebirth are not only one and the same, but necessary for individuals and the collective alike (during war).[109] In becoming the painter of his own self, Klingsor embodies the "dying European man who wants to die," "at once Faust and Karamazov," for whom rebirth is indissolubly fused with downfall, just as "progress" is with "retrogression."[110]

The Schizophrenic Artist

The importance of the paradigm of the schizophrenic as a figuration of the 'primitive' not only for depth psychology but also for psychiatry is evident in one of the most influential books of the field (especially for literary authors): Ernst Kretschmer's *Medizinische Psychologie* (1922; *A Text-book of Medical Psychology*, 1952). As Christoph Gardian observes, Kretschmer "vividly summarizes the accounts of his predecessors (among others, Freud, Wundt, Jung, Bleuler, Storch, Schilder, and Preuss) and combines them in a consistent narrative,"[111] in which the relationship between "primitive races"[112] and schizophrenics and the affinity of 'primitive thinking' to art and the creative process play a central role.

Kretschmer traces an "Evolution of the Psyche," which he divides into "imagery," "affectivity," and "means of expression."[113] To do so, he draws on research in the fields of ethnology and developmental psychology. He gains insights into the development of pictorial processes by analyzing 'primitive language'. Such language, he emphasizes, lacks both abstract notions and com-

---

107 Hesse, *Demian*, 76.
108 Hermann Hesse, *Klingsor's Last Summer*, trans. Richard and Clara Winston (New York: Farrar, Straus and Giroux, 1970), 147.
109 Hesse, *Klingsor's Last Summer*, 139.
110 Hesse, *Klingsor's Last Summer*, 213.
111 Gardian, *Sprachvisionen*, 113.
112 Kretschmer, *A Text-book of Medical Psychology*, 82.
113 Kretschmer, *A Text-book of Medical Psychology*, 81–110.

plex grammatical structures, since only lexical sequence and deictic interjections establish relationships between the "picture-words." In this process, many discrete images are required to express even a simple thought. "New concepts," he writes, must be created by the "agglutination of already existent picture-words."[114] This is the same process at work in mythology and indigenous art, and it follows certain laws. The first concerns "condensation," or "*complex thought* (Preuss)," when, for example, one recognizes animal shapes in geometrical patterns.[115] Like others before him, Kretschmer stresses that, from the modern European perspective, these condensed images may be described as symbols, but "primitive minds" are not aware of their symbolic dimension; instead they are convinced of the *identity* of the image and its meaning. The second law is that of "displacement": the possibility for a part to stand in for the whole.[116] The third law, "stylization," is evident when forms are simplified or repeated in order to underscore what is essential.[117] Finally, the law of "imaginal projection" prevails when distinctions in terms of categories of mental representation and perception grow vague or go missing altogether.[118]

For "primitive people," Kretschmer contends, relationships between objects emerge where strong affect invests mental images:

> Sex, war, and conflict, the longing for rain or the spoils of hunting, above all, illness, fear of death and death itself – these are *foci* for the production of those psychic phenomena termed 'magical thinking.' From these *foci* magical thinking extends to objects, and, later, gradually covers the whole phenomenal world.[119]

In other words, this worldview operates in a "catathymic" manner, that is, by a "transformation of the psychic content by affective influences."[120] Like the German ethnologists, Kretschmer stresses that "the projection of affect" is how the living beings and objects in the "primitive man's" presence are imbued with a soul.[121]

Kretschmer finds many analogies between the early stages of psychic development and the adult mind. In "dreams, hypnosis, hysterical twilight-states, and the disordered thinking met with in schizophrenia," he identifies mental "types

---

114 Kretschmer, *A Text-book of Medical Psychology*, 84.
115 Kretschmer, *A Text-book of Medical Psychology*, 86–88. Emphasis in original.
116 Kretschmer, *A Text-book of Medical Psychology*, 88.
117 Kretschmer, *A Text-book of Medical Psychology*, 89–92.
118 Kretschmer, *A Text-book of Medical Psychology*, 92–94.
119 Kretschmer, *A Text-book of Medical Psychology*, 95–96.
120 Kretschmer, *A Text-book of Medical Psychology*, 96.
121 Kretschmer, *A Text-book of Medical Psychology*, 96–99.

of functioning [...] represent[ing] phylogenetic remnants," which he categorizes as "hypnotic mechanisms."[122] Like Freud, Kretschmer understands "dreams"[123] to involve imaging processes that regress to a lower level of development, that is, from the abstract to the concrete, from grammatical propositions to asyntactic series of pictures, from concepts to agglutinated word-images, and from logical to associative connections, all of which are guided by affect.[124] Hereby, the barriers of space, time, and causality are suspended. Also, according to Kretschmer, the boundary between the ego and the outside world dissolves, which is accompanied by the disintegration or splitting of personality (often in the form of identification with outside beings or objects) as well as the inability to distinguish between the inner and outer worlds.[125] On this basis, Kretschmer concludes that our "dream thought" is closer to the "waking thought" of early humans than it is to our own "waking thought." In order to explain what takes place in the unconscious mind, he invokes dreaming:

> Dream events allow us to divine much which occurs in our waking thought in the '*sphaira*' on the frontiers of consciousness, i.e. in those obscure shifting zones which are the wellsprings of all thought, especially intuitive, creative, and artistic thought.

The "sphaira's" productions also take shape through other states of altered consciousness – when one is unfocused or distracted, for instance, or, conversely, in the event of "hypnoidal over-concentration on a single focus."[126] In order to illustrate the proximity of these states to "primitive phylogenetic tendencies," Kretschmer points to poetry, which is created in such states and displays the traits of "stylization," "rhythm," concretion, "imaginal agglutination," an absence of logic, and "strong affective currents."[127] Therefore, according to Kretschmer, such poetry does not move readers' intellects so much as their "sphaira."

Likewise, hypnosis and the "*hysterical twilight state*" bear comparison to the "imaginal mechanisms" of dreams for Kretschmer.[128] The only point of difference is that they are more affect-laden and unfold in a more intense and dramatic manner. Comparable states, in milder form, occur in the process of free associ-

---

122 Kretschmer, *A Text-book of Medical Psychology*, 114.
123 Kretschmer, *A Text-book of Medical Psychology*, 122.
124 Kretschmer, *A Text-book of Medical Psychology*, 124.
125 Kretschmer, *A Text-book of Medical Psychology*, 122–124.
126 Kretschmer, *A Text-book of Medical Psychology*, 125.
127 Kretschmer, *A Text-book of Medical Psychology*, 125–126.
128 Kretschmer, *A Text-book of Medical Psychology*, 131.

ation or when one exerts oneself mentally when tired and half-asleep. Under such conditions, the mind tends toward "orderly 'picture-strip thinking'" (as in film) and "fantastically disoriented [thoughts]." Kretschmer describes people able to easily experience such phenomena as "*day-dreamers*" who have a particularly creative potential: his example is the writer E.T.A. Hoffmann.[129]

"Schizophrenic thinking" finally represents an extreme case of regression, to which Kretschmer devotes special attention. Here,

> the imaginal processes are often broken up in such a regressive way that [...] large cohesive features of the primitive world-pictures are made to live again before our eyes [...]. There are no important imaginal or affective mechanisms of the kind found amongst primitive peoples which cannot be found extensively in schizophrenics. As a matter of fact many of the terms used [...] are not derived from folk-psychology but from the psychopathology of schizophrenia and neurosis.

According to Kretschmer, schizophrenics do not exhibit "thought based on causality" so much as "thought based on magic" because, as in fairy tales, whatever is desired immediately takes place.[130] Even though patients often realize that this world of wishes is different from the real world – one patient calls it the "surreal" world – they still grant it a higher degree of truth.[131]

From this, Kretschmer draws further parallels to the production and reception of art, as well as to religion. In his eyes, art and pathological states of mind both arise from the "sphere of the unconscious"; therefore, they necessarily share common features: "Consequently, excessive psychic clarity and logical awareness are often fatal for mental creativity which flourishes best in the sphairal twilight. These matters are of special importance for the understanding of the neuroses and psychoses."[132] The difference between the activity of a healthy person's "sphaira" and the "magical thinking" of the schizophrenic can prove to be rather slight. Kretschmer only insists that, in schizophrenia, "magical thinking" moves in "the central point of the psychic field of vision" instead of remaining hidden at the outer edges of consciousness.[133] The products of the "sphere" emerge from the margins as art. Accordingly, Kretschmer identifies links between the "magical thinking" of schizophrenics and Expressionism: "If we think of our patient's inner 'picture show' as a painting with the title, 'The Infinity of Space,' underneath, we can exactly understand the principles underlying expressionistic

---

129 Kretschmer, *A Text-book of Medical Psychology*, 147. Emphasis in original.
130 Kretschmer, *A Text-book of Medical Psychology*, 134.
131 Kretschmer, *A Text-book of Medical Psychology*, 135.
132 Kretschmer, *A Text-book of Medical Psychology*, 128.
133 Kretschmer, *A Text-book of Medical Psychology*, 102.

pictures in which the artist seeks to set down inner feelings and ideas."[134] He writes of one patient's figurative thinking:

> We can immediately observe how the abstract line of thought disintegrates the emerging "Infinity of Space" [...] into the imaginal make up of its sphaira, i.e., into asyntactic, obliquely thrown together conglomerations of images, that [...] symbolize in a dreamlike manner the infinity of space [...].[135] A single example of this kind suffices to provide a clear explanation of the modern tendency in art known as 'expressionism.'[136]

In his discussions of schizophrenia, Kretschmer frequently makes connections to artists' creativity. Even more than in the writings of developmental psychologists, then, it is evident in Kretschmer's study that for him 'primitive thinking' – which schizophrenia is supposed to manifest most fully – is the key to understanding the creative process. The implicit thesis is that artistic activity takes up this thinking's typical procedures, such as image-agglutination through condensation, displacement, and stylization, as well as the projection of affect and catathymia. Artistic genius, according to Kretschmer, enlists "primitive phylogenetic tendencies" that have persisted in dreams and "psychic twilight":

> Men and women of creative genius, especially artists and poets, have so frequently drawn analogies between dreams and the way in which their creative works came into being, that we may regard that relationship as definitely established. Such creative products tend to emerge from a state of psychic twilight, [...] providing an entirely passive experience, frequently of a visual character, divorced from the categories of space and time, and reason and will. [...] The dreamlike phases of artistic creation evoke primitive phylogenetic tendencies toward rhythm and stylization with elemental violence.[137]

---

134 Kretschmer, *A Text-book of Medical Psychology*, 103. Kretschmer's statement that affinity exists between the art of the mentally ill and that of Expressionists was shared by the latter, for instance the artists associated with *Der Blaue Reiter* and *Die Brücke*; cf. John MacGregor, Chapters 14 and 16 (on Expressionism and Surrealism, respectively) of *The Discovery of the Art of the Insane* (Princeton: Princeton University Press, 1989).
135 Kretschmer, *Medizinische Psychologie*, 12th ed. (Stuttgart: Georg Thieme, 1963), 142. This passage is not included in the English translation, which was based on the 10th edition (1950). Therefore, this and any other passage where the German edition is cited have been translated for this volume.
136 Kretschmer, *A Text-book of Medical Psychology*, 137.
137 Kretschmer, *A Text-book of Medical Psychology*, 95.

Part Two: **Art, Language, and 'Primitive Thinking'**

Chapter 5
The Origins of Art

Around the turn of the century, books devoted to the birth of art began to multiply. Titles included Ernst Grosse's *Die Anfänge der Kunst* (1894; *The Beginnings of Art*, 1897), Yrjö Hirn's *Origins of Art* (1900), Carl Stumpf's *Die Anfänge der Musik* (1909; *Beginnings of Music*, 1911), Ludwig Jacobowski's *Die Anfänge der Poesie* (The Beginnings of Poesie, 1891), Francis B. Gummere's *Beginnings of Poetry* (1901), Erich Schmidt's "Die Anfänge der Literatur und die Literatur der primitiven Völker" (The Beginnings of Literature and the Literature of Primitive Peoples, 1906), and Heinz Werner's *Die Ursprünge der Lyrik* (The Origins of Lyric, 1900). On the one hand, we can understand this trend in the context of empirical aesthetics, which was formed by Gustav Theodor Fechner in the last third of the nineteenth century.[1] On the other hand, it can be viewed alongside the search in the human sciences for the origins of Europe's own culture. For while Fechner pursued empirical psychology, the studies listed above combined an empirical,

---

[1] This at any rate is the argument made by Sebastian Kaufmann in *Ästhetik des 'Wilden.' Zur Verschränkung von Ethno-Anthropologie und ästhetischer Theorie 1750–1850. Mit einem Ausblick auf die Debatte über 'primitive' Kunst um 1900* (Basel: Schwabe, 2020). His study draws extensively on my own research (Nicola Gess, ed., *Literarischer Primitivismus* [Berlin: De Gruyter, 2013]), as well as Priyanka Basu, "Die 'Anfänge' der Kunst und die Kunst der Naturvölker: Kunstwissenschaft um 1900," in *Image Match. Visueller Transfer: "Imagescapes" und Intervisualität in globalen Bildkulturen*, ed. Martina Baleva, Ingeborg Reichle, and Oliver Lerone Schultz (Paderborn: Fink, 2012), and Ingeborg Reichle, "Vom Ursprung der Bilder und den Anfängen der Kunst. Zur Logik des interkulturellen Bildvergleichs um 1900," in the same volume. Also of note are the following articles by Doris Kaufmann, which were also very important to my work on the intersection of primitivism and the theory of art as I was writing the German version of this book: "Kunst, Psychiatrie und 'schizophrenes Weltgefühl' in der Weimarer Republik. Hans Prinzhorns Bildnerei der Geisteskranken," in *Kunst und Krankheit. Studien zur Pathographie*, ed. Matthias Bormuth, Klaus Podoll, and Carsten Spitzer (Göttingen: Wallstein, 2007); "Zur Genese der modernen Kulturwissenschaft. 'Primitivismus' im transdisziplinären Diskurs des frühen 20. Jahrhunderts," in *Wissenschaften im 20. Jahrhundert. Universitäten in der modernen Wissenschaftsgesellschaft*, ed. Jürgen Reulecke and Volker Roelcke (Stuttgart: Steiner, 2008); "'Pushing the Limits of Understanding': The Discourse on Primitivism in German *Kulturwissenschaften*, 1880–1930," *Studies in History and Philosophy of Science* 39 (2008); "Die Entdeckung der 'primitiven Kunst.' Zur Kulturdiskussion in der amerikanischen Anthropologie um Franz Boas, 1890–1940," in *Kulturrelativismus und Antirassismus. Der Anthropologe Franz Boas (1858–1942)*, ed. Hans-Walter Schmuhl (Bielefeld: transcript, 2009); and "'Primitivismus': Zur Geschichte eines semantischen Feldes 1900–1930," in *Literarischer Primitivismus*, ed. Nicola Gess. For an early (and largely uncritical) discussion, see also Thomas Munro, *Evolution in the Arts* (Cleveland: Cleveland Museum of Art, 1967), especially chapter 10.

inductive approach with a search for human prehistory – even though barely any hard data existed for it.

The paradigm of the 'primitive' promised a way out of this impasse. As we have seen in earlier chapters, it understood certain categories of people in the present day – children, the mentally ill, and indigenous communities – as survivals of prehistoric humanity. In such a framework, empirical-inductive projects examined the linguistic, visual, and musical productions of native peoples, the mentally ill, and children in order to understand the nature and function of art in its 'primal state.' Studies of this kind included Richard Wallaschek's *Primitive Music* (1893), Herbert Kühn's *Die Kunst der Primitiven* (The Art of Primitives, 1923), Alfred Vierkandt's *Das Zeichnen der Naturvölker* (The Drawings of Primitive People, 1912), Hans Prinzhorn's *Bildnerei der Geisteskranken* (1922; *Artistry of the Mentally Ill*, 1972), and Gustav Friedrich Hartlaub's *Der Genius im Kinde* (The Genius within the Child, 1922). Another good example is Karl Lamprecht's "Einführung in die Ausstellung von parallelen Entwicklungen in der bildenden Kunst" (Introduction to the Exhibition of Parallel Developments in Visual Art, 1913), a speech delivered at the *Kongress für Ästhetik und Allgemeine Kunstwissenschaft* (First International Congress of Aesthetics [ICA]), where "numerous lectures complemented the research on non-European and prehistorical 'primitive' art by considering [works] by children."[2] Lamprecht contends that "the artistic development of lower cultures in the present day and prehistory alike has proceeded according to the principles of development found in children's art."[3]

As Max Dessoir emphasizes in *Ästhetik und Allgemeine Kunstwissenschaft* (1906; *Aesthetics and Theory of Art*, 1970), the task of a systematic and empirically based study of the arts was to research their "genesis and divisions." To this end, it was considered necessary to study "the art of peoples in a state of nature, children, and prehistory," and more and more frequently also that of the mentally ill. Together, these works were to be "viewed as interconnected elements in the research field of 'primitive art.'"[4] It was only logical, then, for the field of art

---

[2] Sebastian Kaufmann, *Ästhetik des 'Wilden,'"* 680.
[3] Karl Lamprecht, "Einführung in die Ausstellung von parallelen Entwicklungen in der bildenden Kunst," in *Kongress für Ästhetik und allgemeine Kunstwissenschaft, Berlin 7.-9. Oktober 1913* (Stuttgart: Enke, 1914), 78.
[4] Basu (summarizing Max Dessoir), "Die 'Anfänge' der Kunst," 117. On the other hand, Barbara Wittmann points out that recapitulation theory was increasingly losing significance for cultural history from about 1910 on. This can be seen, for example, in Max Verworn ("Kinderkunst und Urgeschichte," *Korrespondenz-Blatt der Deutschen Gesellschaft für Anthropologie, Ethnologie und Urgeschichte* 27 [1907]; *Die Anfänge der Kunst: Ein Vortrag* [Jena: Fischer, 1909]; *Ideoplastische Kunst: Ein Vortrag* [Jena: Fischer, 1914]); and Wilhelm Wundt ("Die Zeichnungen des Kindes und die zeichnende Kunst der Naturvölker," in *Festschrift Johannes Volkelt zum 70. Geburtstag*

studies (*Allgemeine Kunstwissenschaft*) to draw on the findings of ethnology, developmental psychology, and psychopathology, which offered considerations of their own on the linguistic, visual, and musical works of their objects of research, often in close conjunction with theories of 'primitive thinking' (see Chapters 2–4).

Without intending to, these disciplines thus provided historians and theorists of art a possible answer to one of their most important questions: the enigma of creativity. As Ernst Meumann observes in *Einführung in die Ästhetik der Gegenwart* (Introduction to Contemporary Aesthetics, 1908), "aesthetics' most difficult problem" was "genius" – in other words, the reasons underlying creative activity. To date, answers had proven "quite unsatisfactory"; as he puts it, "we are far from having said the final word [...] on the essence of artistic creation."[5] The paradigm of the 'primitive' promised to shine light into this black box, as it regarded artistic creativity as recourse to a 'primitive thought' that was deeply buried but not completely inaccessible to men and women of the time.[6]

Nonetheless, research into the origins of art did not merely provide a justification for the study of art and new stimuli for aesthetics. It also legitimized modern art itself through a spectrum of arguments. For evolutionary thinking, the topos of origins made it possible to demonstrate the extent to which modern European art supposedly stood at the summit of the 'evolution of the arts.' The topos also granted a sounder footing to *critical* views of progress inasmuch as it could be used to establish *general* laws of artistic activity that would be valid for *all* places and times and which now might be studied *ab ovo*. Finally, the topos of origins also had an important function, where, with a gesture critical of 'civilization,' contemporary art was denounced for having grown estranged from an anthropological 'essence.' Or, conversely, contemporary art was identified as the last refuge of and sole access to an origin from which modern society had alienated itself to its detriment.

[Munich: C. H. Beck, 1918]). However, the "ghostly power" it held for the avant-garde was unaffected (Wittmann, *Bedeutungsvolle Kritzeleien*, 241).
5 Ernst Meumann, *Einführung in die Ästhetik der Gegenwart* (Leipzig: Quelle & Meyer, 1908), 85.
6 As Susanne Leeb points out, at the end of the nineteenth century artistic production was increasingly "viewed as a generic trait of human nature acquired through evolution or activity prompted by drives," while "intelligence, talent, creativity, and the creative drive" were deemed "biological and genetic capacities" above all (*Die Kunst der Anderen*, 19).

Justifying the Study of Art

The proliferation of books around 1900 dealing with the 'beginnings' of art stood in the larger context of the search for origins that shaped the "Age of History" (Foucault). Indeed, the paradoxical relation between the foreign and one's own culture attending the emergence of ethnology and the paradigm of the 'primitive' (see Chapter 2) also informed the discipline of art studies. 'Primitive art' was thought of as the historical source of modern European art. In this way, it was not regarded merely as foreign, but rather its otherness proved to be the basis for one's own artefacts.[7]

However, art studies dealt with this paradox quite differently than the human sciences. Instead of unintentionally destabilizing standard notions of cultural identity, they projected the basic features of modern European art back onto a foreign past.[8] Often, such undertaking did not concern the beginnings of art so much as seek out the fundamental principle thought to shape art's further evolution.[9] In this framework, whatever is supposed to stand at the beginning does not become obsolete over time, but carries on in ulterior stages of development. In its most extreme form, such reasoning gives rise to an ontologizing view in which the first beginnings of art *are* its essence. Accordingly, in "The Origin of the Work of Art" (1935), Martin Heidegger declared,

> Origin means here that from where and through which a thing is what it is and how it is. That which something is, as it is, we call its nature [*Wesen*]. The origin of something is the source of its nature. The question of the origin of the artwork asks about the source of its nature.[10]

In what follows, I will expand on these theses by examining some representative studies in the fields of art history, musicology, and literary studies. In doing so, I will show how these disciplines shared the goal of justifying both their own existence and that of contemporary art. At the same time, we will see that their lines

---

[7] See also Leeb: "What is decisive is that modern art formed its self-understanding in the first place through both the included and the excluded Other, e.g., through the art of 'primitives'" (*Die Kunst der Anderen*, 16).
[8] On the mechanism of projection and its significance for theories of culture at the turn of the century, see Müller-Tamm, *Abstraktion als Einfühlung*.
[9] The argument draws on Alexander Rehding, "The Quest for the Origins of Music in Germany circa 1900," *Journal of the American Musicological Society* 53.2 (2000), especially 346–347.
[10] Martin Heidegger, "The Origin of the Work of Art," in *Off the Beaten Track*, trans. Julian Young and Kenneth Haynes (Cambridge: Cambridge University Press, 2002), 1. Also quoted in Rehding, "The Quest for the Origins of Music in Germany circa 1900," 347.

of argument proceeded quite differently and also shifted significantly from the late nineteenth century to the late 1920s as a positivistic orientation gave way to speculativism and the evolutionary paradigm was replaced by cultural critique.

Art Studies (Allgemeine Kunstwissenschaft)

With its intensive thematization of origins, the discipline of *Allgemeine Kunstwissenschaft* sought to justify itself. Whether focused on music, literature, or visual art, scholars sought to secure the scientificity of their approach by examining the underlying essence and "laws of development" governing aesthetic production and by enlisting empirical data, especially from ethnology. Grosse's *The Beginnings of Art* represents a case in point. Here the author criticizes art studies for having thus far neglected to "begin at the beginning" and therefore for failing to identify the laws of development for art, as befits a serious science:[11]

> If we are ever to attain a scientific knowledge of the art of civilized peoples, it will be after we have first investigated the nature and condition of the art of savages. [...] The first and most pressing task of the social science of art lies, therefore, in the study of the primitive art of primitive peoples. In order to compass this object, the study of the science of art should not turn to history or pre-history, but to ethnology.[12]

Grosse seeks to remedy the lack of data from prehistoric cultures by performing an allochronic turn, relocating indigenous peoples of the present to a point earlier in time. Inasmuch as "savages" are thought to have no history, their works still display the qualities of those produced by the first human beings to inhabit the earth.

At the same time, however, Grosse acknowledges that the "highest [...] mastery"[13] is evident in the "artistic achievements [of] primeval men."[14] The reason for such sophistication lies with the "exercise of two faculties"[15] that archaic communities had to cultivate in their struggle for existence (that is, not for purely aesthetic purposes): skilled observation and manual dexterity. In other words, Grosse relativizes the evolutionistic standards used until that point by evaluating

---

11 For a thorough discussion of Grosse's work, see Basu, "Die 'Anfänge' der Kunst"; Reichle, "Vom Ursprung der Bilder"; and Kaufmann, *Ästhetik des 'Wilden,'* 664–674, who observes that Grosse was hardly the first to make this claim (665).
12 Grosse, *The Beginnings of Art*, 21.
13 Grosse, *The Beginnings of Art*, 197.
14 Grosse, *The Beginnings of Art*, 164.
15 Grosse, *The Beginnings of Art*, 198.

'primitive art' as masterful.¹⁶ Moreover, he contextualizes creative activity in cultural and historical terms, revealing a tendency for cultural relativism (as the American anthropologist Franz Boas was doing in the same period). Grosse grants to each culture artistic forms of its own and stresses that it is impossible to judge them as having higher or lower value on a universal scale of development. Art can only be more or less suitable to its own community.

The novelty of Grosse's view of 'primitive art' is plain in light of more traditional perspectives from the time. Heinrich Schurtz's *Urgeschichte der Kultur* (Prehistory of Culture, 1900), for instance, baldly declares that European culture occupies the summit of developments to date; it follows that the cultures of non-European peoples would lag far behind. Although he also calls for research on them, this serves only to gain information about the beginnings of European culture. Schurtz invokes Haeckel's biogenetic law and advises researchers "to come closer to achieving insight into the past through self-observation."[17] Inasmuch as he regards people as passing through the stages of human development over the course of their childhood and youth, he believes traces of ontogenetic and phylogenetic antiquity still exist in adults and can be investigated. Consequently – and in contrast to Grosse – Schurtz does not deem these childlike 'primitives' to be great artists so much as a strange combination of wild children and philistines:

> The "bad behavior" of our children, which comes out in seemingly inexplicable fits of defiance, stubbornness, and destruction and has its counterpart in eruptions of tempestuous tenderness, is found among members of peoples living in the state of nature, just in more dangerous form. [...] A Philistine learns what is necessary for his station, and this is enough for him to live his life without needing to learn anything new. Primitive peoples occupy the same position: their period of apprenticeship lies far in the past, and they have in a sense retired and need nothing more.[18]

Therefore, Schurtz is convinced that because they are unwilling to evolve, they have remained at one incipient phase for thousands of years.

However, Schurtz agrees with Grosse that general rules of art can be inferred from examinations of 'primitive art.'[19] Even though he cautions against trying to

---

[16] See Kaufmann, "Zur Genese der modernen Kulturwissenschaft," 43; for her, the "fundamental shift in the conception of primitive art" starts with Alois Riegl (43).
[17] Heinrich Schurtz, *Urgeschichte der Kultur* (Leipzig, Vienna: Bibliographisches Institut, 1900), 24.
[18] Schurtz, *Urgeschichte der Kultur*, 66, 75.
[19] The same holds for other representatives of *Allgemeine Kunstwissenschaft*. August Schmarsow, for instance, invoked Grosse when he called for researchers to take a cue from ethnology

extract "the essence of a phenomenon from its sprouts,"[20] he maintains that looking at the "most primitive peoples" will facilitate "deeper understanding" of art's "root," which still fuels modern art's creative force. This root reaches "so far back that it was there before any awareness" of it existed; "fundamentally, and still today, art derives its true creative power from the process [*Treiben*] unconsciously at work."[21]

In similar fashion – and contradicting the cultural relativism he endorses elsewhere – Grosse, at the end of his study, declares that "primitive forms of art" are suitable for formulating the laws of art in general, since they show that what now exists was already there at the beginning. In acknowledging this state of affairs, he calls for scientific aesthetics to acquire an empirical footing:

> Strange and inartistic as the primitive forms of art sometimes appear at the first sight, as soon as we examine them more closely, we find that they are formed according to the same laws as govern the highest creations of art. And not only are the great fundamental principles of eurhythm, symmetry, contrast, climax, and harmony practised [...]. Our investigation has proved what aesthetics has hitherto only asserted: that there are, for the human race at least, generally effective conditions for aesthetic pleasure, and consequently generally valid laws of artistic creation.[22]

Grosse looks for a starting point where basic principles determining future development had been cultivated. In so doing, he projects features of modern European art back onto 'primitive artforms' in order to then recognize them as its supposed source. To take just one example, the eurhythm he identifies as a universal principle evident in 'primitive art' is a term that would only come into fashion in the early twentieth century in the context of anthroposophy.

Musicology

The same pattern is evident in specific disciplines of the study of the arts, for example in musicology. As Alexander Rehding has shown, scholars sought to

---

in order to explain the "nature of art" and its "genesis" – that is, to retrace its evolution ("Kunstwissenschaft und Völkerpsychologie," *Zeitschrift für Ästhetik und Allgemeine Kunstwissenschaft* 2, no. 3 [1907]: 309). Ultimately, Schmarsow considered art to be based in affect and expressive motion (327, 337–339), in keeping with the rules governing human physiology.

20 Schurtz, *Urgeschichte der Kultur*, 493.
21 Schurtz, *Urgeschichte der Kultur*, 494.
22 Grosse, *The Beginnings of Art*, 307.

prove that their field was a true science by adopting a genealogical perspective; in so doing, they would disclose the beginnings of music and demonstrate its "essential constitution" (*Wesensbeschaffenheit*).²³ Carl Stumpf's *Die Anfänge der Musik* (1911; *Origins of Music*, 2012) exemplifies that undertaking. The author begins by taking issue with the assumptions on music guiding the thought of Charles Darwin, Herbert Spencer, and Karl Bücher, who, in his estimation, had failed to explain the origins of an art "whose material consists essentially of fixed and transposable tonal steps."²⁴ For his own part, Stumpf considers that music has its origin in "acoustic signals" consisting of consonants shouted together. "Primordial humans [...] may have noticed this uniformity and may have particularly liked using simultaneous pitches [...] while having the impression of singing the self-same note, i.e., a strengthened note."²⁵ Such calls would have served the purpose of "signalling to people" and "the invocation of gods" (or, more precisely, "the demonic magical powers of air and water").²⁶ From here, intervals, polyphony, and tonality were gradually discovered.

Stumpf's observations are marked by a rhetorical move that projects basic principles of Western music back to the origin as supposed universals. Rehding notes that "the categories Stumpf privileged as universals are in fact particularly fitting for Western music, with its emphasis on the harmonic and polyphonic structure."²⁷ Accordingly, the author views the compositions of his own day as the fulfillment of the essence of music in general:

> Our present European music [...] is now, by contrast, entirely built on the chordal system which is derived by consistently and exclusively carrying through the principle of consonance. Since this is the primordial phenomenon out of which music arose, which forms its flesh and bones, and since it has brought this elementary fact most purely and perfectly

---

23 Guido Adler, "Antrittsvorlesung an der Universität Wien, Musik und Musikwissenschaft," *Jahrbuch der Musikbibliothek Peters* 5 (1898): 29; also quoted in Rehding, "The Quest for the Origins of Music in Germany circa 1900," 345–385. On the debate (especially Wallaschek's position), cf. Alexandra Hui, "Origin Stories of Listening, Melody and Survival at the End of the Nineteenth Century," in *Music and the Nerves, 1700–1900*, ed. James Kennaway (Basingstoke: Palgrave Macmillan, 2014); Gernot Gruber, "Das 'Archaische' in der Musikkultur der Wiener Moderne. Eine Skizze," in *Kunst, Kontext, Kultur. Manfred Wagner 38 Jahre Kultur- und Geistesgeschichte an der Angewandten*, ed. Gloria Withalm, Anna Spohn, and Gerald Bast (Berlin: De Gruyter, 2012); Basu, "Die 'Anfänge' der Kunst."
24 Carl Stumpf, *The Origins of Music*, trans. David Trippett (Oxford: Oxford University Press, 2012), 43.
25 Stumpf, *The Origins of Music*, 46.
26 Stumpf, *The Origins of Music*, 47.
27 Stumpf, *The Origins of Music*, 56. Rehding, "The Quest for the Origins of Music in Germany circa 1900," 382.

into being and thereby established the stylistic principle for the whole imposing design [Bau], we may regard it as the highest form of music so far, without being narrow-minded from the perspective of either ethnopsychology or developmental history.[28]

Stumpf was not alone in holding this belief, which led early ethnomusicologists to find the foundations of – and justification for – modern European music among indigenous peoples. Richard Wallaschek's *Primitive Music* (1893), draws on travelogues to advance such a claim:

> When at the beginning of the last century [Friedrich Wilhelm] Kolbe travelled among the Hottentots he found them playing different gom-goms in harmony. They also sang the notes of the common chord down to the lower octave [...], thus producing a harmonious effect. [William John] Burchell, who repeatedly assures us that he probably was the first European who ever touched the African soil in that part where he travelled, describes the harmonious singing of the Bachapin boys [...] guided only by their own ear, [...] in correct harmony. The Bechuana [...] have a sufficient appreciation of harmony to sing in two parts.[29]

Accounts like these demonstrate for Wallaschek the "naturalness" of harmony and, with that, the naturalness of European music in the present day.[30] Indigenous music that did not display such characteristics was attributed to a different genetic disposition (that is, "racial" difference).[31] Alternatively, it was ignored or dismissed. Hugo Riemann exemplifies the latter attitude:

> The striking congruencies of the division of the octave into twelve semitones, which completes the seven-step scale by interspersing a semitone between alternatively two and three tones [i.e., the diatonic scale] – found likewise by the Chinese, Greeks, and the nations of the European West in the space of many centuries – is a historical fact, which cannot simply be overthrown by a couple of pipes with faulty bores from Polynesia or by the questionable vocal achievements of colored women.[32]

---

28 Stumpf, *The Origins of Music*, 64–65; also quoted in Rehding, "The Quest for the Origins of Music in Germany circa 1900," 353–354.
29 Richard Wallaschek, *Primitive Music: An Inquiry into the Origin and Development of Music, Songs, Instruments, Dances, and Pantomimes of Savage Races* (London: Longmans, Green and Co., 1893), 139.
30 This judgment does not concern expanded tonality, much less atonality, but the status quo of classical and Romantic music. In fact, the line of argument at issue lent itself to a dismissal of avant-garde compositions as "pathological," if not "degenerate."
31 Wallaschek, *Primitive Music*, 144. Cf. Rehding, "The Quest for the Origins of Music in Germany circa 1900," 359.
32 Hugo Riemann, *Handbuch der Musikgeschichte* (Leipzig: Breitkopf & Hartel, 1904). 1: vi; quoted in Rehding, "The Quest for the Origins of Music in Germany circa 1900," 355.

Compared to the *Allgemeine Kunstwissenschaft* at the turn of the century, musicology exhibited even more strongly the paradox of simultaneously taking distance from and identifying with supposed origins – that is, retrojecting Western musical principles while denying or disqualifying all others. First beginnings counted not as the expression of a primordial essence so much as an element that unfolds over time and culminates in modern European music as its fulfillment and highest form.

Literary Studies

Literary studies at the turn of the century were similarly motivated to cite ethnological findings in order to replace the merely hypothetical structure of earlier claims with "unbroken chain[s] of evidence obtained by empirical means"[33] and thus to demonstrate the scientific nature of their undertaking. In *Anfänge der Poesie* (1891), for example, Ludwig Jacobowski calls for a "poetics [...] based strictly on empiricism, [that is,] the natural sciences" since this alone would be able "to validate the scientific nature of literary studies in the future."[34] On the basis of Haeckel's biogenetic law, he seeks to prove the historical priority of lyric over epic. To that end he shows how for children, "subjective (i.e., lyrical) moments of feeling precede the objective (i.e., epic) moments of perception" and how children from their earliest days of life can express such sentiments in the sounds they utter.[35] The equation of "subjective" and "lyrical" as well as "objective" and "epic" bears the mark of contemporary poetic theory – the entire process is once again informed by a projective mechanism.

For Jacobowski, this insight into the child's psyche applies to "primitive man" as well:

---

[33] Karl Bücher, "Arbeit und Rhythmus," *Abhandlungen der philologisch-historischen Classe der königlich sächsischen Gesellschaft der Wissenschaften*, 17 (1897): 80. Grosse also plays an important role in this context, insofar as in 1887 he presented the most comprehensive (and earliest) plan for turning literary history into literary science (*Literaturwissenschaft*): "The task of the modern science of literature is determining laws," which include the "law of poetic evolution in general" (Grosse, *Die Literatur-Wissenschaft*, quoted in Klaus Weimar, "Die Begründung der Literaturwissenschaft," in *Literaturwissenschaft und Wissenschaftsforschung*, ed. Jörg Schönert [Stuttgart: Metzler, 2000], 139). Since no data is available for prehistory, Grosse advises scholars to take a "detour" via "similar but less complicated [...] phenomena" – for instance, children's games (quoted in Weimar, "Die Begründung der Literaturwissenschaft," 140).

[34] Ludwig Jacobowski, *Anfänge der Poesie. Grundlegung zu einer realistischen Entwickelungsgeschichte der Poesie* (Dresden: E. Pierson's Verlag, 1891), v.

[35] Jacobowski, *Anfänge der Poesie*, 7.

> Inasmuch as we must deem primitive man [...] approximately equal to the child intellectually and psychically, because, according to Haeckel's biogenetic law, we have a miniature image of primitive man in the child's development, we carry over results obtained on an ontogenetic basis to the phylogeny of primitive man and find, for him, what holds for the newborn child and highly developed animals: that his "worldview" is a matter of epistemological sensualism. With that, it is proven that perceptions represent the first, decisive moment in the psychic life of primitive man. And since I conceive of primordial lyricism [*Urlyrik*] as the transposition of perceptions into vocalizations, its priority stands beyond doubt.[36]

The historical priority of lyric – that is, the (supposed) fact that it developed before epic and drama – implies a positive value judgement: Jacobowski considers contemporary poetry that stands closest to that of human origins to be "the actually 'true'" lyric – in particular, "intimate confessional or occasional lyric in the highest Goethean sense."[37] Many literary scholars would follow him in affirming the priority of lyric. Thus, Erich Schmidt's "Die Anfänge der Literatur und die Literatur der primitiven Völker" still considers the point of origin to lie in spontaneous vocalizations of sentiment and choral expression.[38]

Movement and rhythm competed with affective vocalization in theories of literary origins.[39] *Arbeit und Rhythmus* (Work and Rhythm, 1897) by Karl Bücher – a work regularly invoked as authoritative by his contemporaries – identified "energetic, rhythmical physical movement, especially that motion we call work," as

---

36 Jacobowski, *Anfänge der Poesie*, 10.
37 Jacobowski, *Anfänge der Poesie*, 11.
38 Erich Schmidt, "Die Anfänge der Literatur und die Literatur der primitiven Völker," in Erich Schmidt, Adolf Erman, Carl Bezold, et al., *Die orientalischen Literaturen* (Berlin and Leipzig: Teubner, 1906), 7–8. Yrjö Hirn (*The Origins of Art. A Psychological and Sociological Inquiry* [London: Macmillan, 1900]) also derives "artistic drive" from conventional psychological notions of expressing emotion.
39 Heinz Werner did not take sides in the debate between rhythm/motion and vocalization/affect (although his sympathies lay with the latter view) so much as he identified two equally valid "primitive types": "The first primitive type is distinguished by its senselessness, following from the predominance of the motoric component of vocalization. The second primitive type is brief, extemporal interjection, which derives immediately from overall mood dictated by feeling; this is the primal form of logical poetry, from which higher types evolve" (*Die Ursprünge der Lyrik* [Munich: Reinhardt, 1924], 8). Werner's study attains a greater level of sophistication than the others discussed here because the author reflects on and defines the operative conception of primitiveness (e. g., "[its] essence [lies] in a significant lack of differentiation, diffuseness, and [...] much lower degree of centralization and subordination" [5]); the main part of the book delineates the developmental course of major "poetic elements" [42], for instance, allegory, repetition, ellipsis, rhythm, and rhyme instead of undertaking a sweeping survey of literary genres.

having "led to the development of poetry."[40] Another example is Francis B. Gummere's *The Beginnings of Poetry* (1901), which starts out by warning against equating indigenous peoples, children, and prehistoric ancestors and draws attention to the speculative nature of many sources. However, he then goes on to enlist these sources himself and to point to the behavior of children to substantiate his thesis about literary origins.[41] Gummere focuses on the ballad or "communal song," arguing that it emerged from "choral rhythm" as a means of creating collective identity: "In rhythm, in sounds of the human voice, timed to movements of the human body, mankind first discovered that social consent which brought the great joys and the great pains of life into a common utterance." With this insight into its beginnings, he makes the demand that contemporary poetry not neglect rhythm, lest it lose its community-building power.[42]

Finally, a third position affirming the historical and normative priority of lyric deemed figurative language (not affective expression or rhythmic movement) to be the primordial form of human expression (see Chapter 6). In contrast to theorists who favored affect and rhythm, scholars such as Alfred Biese adopted an ontologizing perspective: since metaphorical language is non-arbitrary and originary, it counts as true in a fundamental sense and offers a privileged means for disclosing reality. Poetic language opens the way for gaining insight into the world-in-itself (*Welt an sich*).[43]

Art History

In the 1910s and 1920s, the tendency to ontologize origins was most pronounced among art historians. In *Die Kunst der Primitiven*, Herbert Kühn – an art historian and authority on prehistory – does his best to liberate 'primitive art' from evolutionistic prejudices. Like Grosse had done two decades earlier, he stresses the interconnection of art and worldview, pointing out that the 'primitive art' in question is not underdeveloped so much as it has emerged from a different way of looking at things. In spite of this relativizing perspective, however, 'primitive art' continues to play the part of a timeless ideal for him.

Kühn identifies two styles – the sensory/naturalist and the imaginative/abstract – that have alternated time and again throughout the history of art in keeping with dominant lifestyles and social structures. In so doing, he takes

---

[40] Bücher, "Arbeit und Rhythmus," 80.
[41] Francis Gummere, *The Beginnings of Poetry* (New York: Macmillan, 1901), 11–29; 102.
[42] Gummere, *The Beginnings of Poetry*, 114, see also 473.
[43] Cf. Biese, *Die Philosophie des Metaphorischen*, 78–103.

up a broadly accepted art historical theorem of the time that was also influenced by contemporary art and its abstract or expressive modes, but he recasts it as an ahistoric universal.[44] He recognizes the ideal form of both styles in the productions of the earliest humans and thus posits that all subsequent art must revert to its 'primitive' antecedents:

> The sensory experience of paleolithic human beings, bushmen, and eskimos is thoroughly sensory, sensory without reserve: the imaginative life of mankind in the Neolithic and Bronze Ages is the very type of this sensibility [*Stilform*]. It is as if all later art looked back to these primal forms in unconscious recollection of these great works of art.[45]

Kühn, then, does not see the highest expression of a primordial principle in the art of his day, but he praises the latter for approaching an essence that found its fullest realization in the mythical past: "The modern artist and the artist of the Neolithic creates Law, Cosmic Force [*das Kosmische*], the Whole. Both possess the same will, the same thought, the same feeling of connectedness to the Universe and God."[46] Meanwhile it is clear that the summit of artistic development can never be reached again because it lies at its beginning. Kühn exhuberantly extolls the superiority of 'primitive art.' For instance:

> Here is a will for the extreme, the radical in art; later times, which always carry within them the inheritance of what is passed, cannot bring it forth again. In this early time, all is more defined, clearer, and unconditional. [...] This is what makes the primitive age so inwardly mighty for those who have eyes to see.[47]

In place of a distancing and deprecating treatment of 'primitive art,' Kühn's work takes an affirmative, ontologizing approach. No conflict emerges with the relativistic perspective that assigns different forms of art to different worldviews because Kühn distinguishes between two eternally recurring outlooks and as a mat-

---

[44] This is also the approach taken, e.g., by Max Verworn in his 1907 lecture, published as *Zur Psychologie der primitiven Kunst* (Jena: Fischer, 1908). Calling for a renewal of the psychology of art and ethnology (*Völkerkunde*), he starts with the opposition between paleolithic (authentic, true to nature and life) and later (stylized, ornamental, and distorted) art – that is, "physioplastic" and "ideoplastic" forms, which are respectively modeled on natural phenomena and what the human mind thinks or knows about them. On this basis, he contests the analogy between the art of primeval human beings and that of children.
[45] Herbert Kühn, *Die Kunst der Primitiven* (Munich: Delphin Verlag, 1923), 13.
[46] Kühn, *Die Kunst der Primitiven*, 78.
[47] Kühn, *Die Kunst der Primitiven*, 82; see also 24, 29.

ter of principle finds the fullest realization of the various forms of art at their origin.

In offering these reflections, Kühn is able to draw on a book published a decade earlier, Wilhelm Worringer's enormously influential *Abstraktion und Einfühlung* (1908; *Abstraction and Empathy*, 1957).[48] Worringer's search for art's origins does not concern its material or techniques. Rather, it involves speculations about an archaic psyche and the intentional "primal artistic impulse" (*Urkunsttrieb*) at work within it. The author recognizes this impulse in the push for abstraction as the sole possibility for "man [to] rest in the face of the vast confusion of the world-picture."[49] Worringer also finds its fullest realization – and with that "the highest, purest regular art-form" – in "primitive culture": "The less mankind has succeeded, by virtue of its spiritual cognition, in entering into a relation of friendly confidence with the appearance of the outer world, the more forceful is the dynamic that leads to the striving after [the] highest abstract beauty."[50] Worringer credits "primitive man" with an "instinct for the 'thing-in-itself,'" intuitive understanding of the "necessity" and "regularity" of phenomena beyond the coloration imparted by environment and subjective perception. Modern-day humans and their ancestors alike share this experience. But now what was once a matter of collective instinct has transformed into individual knowledge, which is why it can bear no fruit: "The individual on his own was too weak for such abstraction."[51] Worringer concludes that modern art's ideal lies out of its reach: 'primitive art' alone was able to achieve it. Yet here too it is clear that the ideal for 'primitive art' is derived from modern European works.[52] The tendency toward abstraction in the latter is once again projected into the past and declared to be the timeless and enduring essence of art itself.

Insofar as they see the highest level of art to have been achieved at its first beginnings, Worringer and Kühn's idealizations of 'primitive art' are critical of European civilization. As we will see below, this critical impulse is even more pronounced among works in art history concerning the paradigm of schizo-

---

[48] On Worringer in the context of literary studies, see Claudia Oehlschläger, *Abstraktionsdrang. Wilhelm Worringer und der Geist der Moderne* (Munich: Fink, 2005); Müller-Tamm, *Abstraktion als Einfühlung*, 249–286; in relation to primitivism, cf. especially Helmut Lethen, "Masken der Authentizität. Der Diskurs des 'Primitivismus' in Manifesten der Avantgarde," in *Manifeste: Intentionalität*, ed. Hubert van den Berg and Ralf Grüttemeier (Amsterdam: Brill, 1998).
[49] Wilhelm Worringer, *Abstraction and Empathy: A Contribution to the Psychology of Style*, trans. Michael Bullock (Chicago: Ivan R. Dee, 1997), 19.
[50] Worringer, *Abstraction and Empathy*, 17.
[51] Worringer, *Abstraction and Empathy*, 18.
[52] This is also the case for Carl Einstein's widely-read *Negerplastik* (1915; *Negro Sculpture*, 2016), which is primarily an engagement with Cubism based on an appreciation of primitive art.

phrenic 'primitives,' whose works were thought to protest against the alienated conditions of modern life.

The Enigma of Creativity

The nature of creativity often posed a mystery for earlier aesthetic theory. The question of what enables artists to produce original works was bypassed with references to their inborn genius. Kant's *Critique of the Power of Judgment* is exemplary in this regard:

> Genius is the talent (natural gift) that gives the rule to art. Since the talent, as an inborn productive faculty of the artist, itself belongs to nature, this could also be expressed thus: Genius is the inborn predisposition of the mind (*ingenium*) through which nature gives the rule to art. [...]
> From this one sees: [...] That it cannot itself describe or indicate scientifically how it brings its product into being, but rather that it gives the rule as nature, and hence the author of a product that he owes to his genius does not know himself how the ideas for it come to him.[53]

One can describe what genius does:

> [it] find[s] ideas for a given concept on the one hand and on the other hit[s] upon the expression for these, through which the subjective disposition of the mind that is thereby produced, as an accompaniment of a concept, can be communicated to others.[54]

But nothing, save for vague references to "talent,"[55] is said about what makes such activity possible in the first place. The "bourgeois myth of the artist"[56] could never have emerged without this air of mystery because it contributes to the aura of uniqueness so essential to it. The figure of the artist exhibits "an auratic structure in the Benjaminian sense: no matter how close he comes to his public, he remains at a remove from it." Both the nature and the capacities of

---

[53] Immanuel Kant, *Critique of the Power of Judgment*, trans. Paul Guyer (Cambridge: Cambridge University Press, 2000), 186–187.
[54] Kant, *Critique of the Power of Judgment*, 194–195.
[55] Kant, *Critique of the Power of Judgment*, 195.
[56] Andreas Reckwitz, "Vom Künstlermythos zur Normalisierung kreativer Prozesse," in *Kreation und Depression*, ed. Christoph Menke and Juliane Rebentisch (Berlin: Kulturverlag Kadmos, 2010).

the artist must remain unique and inaccessible for genius to exist as a special form of subjectivity (or "Spezialsubjekt").[57]

The rise of empirical aesthetics during the second half of the nineteenth century only partially changed this state of affairs. For example, Wilhelm Dilthey's *Die Einbildungskraft des Dichters* (1887; *The Imagination of the Poet*, 1985) presents an ambivalent picture. On the one hand, the author stresses that the point of departure for his theory necessarily lies "in the analysis of the creative capacity": "the poet's imagination and his attitude toward the world of experience provide the point of departure for every theory seriously directed to explaining the manifold world of poetry and literature in the succession of its manifestations."[58] In this spirit, Dilthey sets about examining psychological processes hitherto obscured by the designation of "poetic imagination."[59] He arrives at the insights that "the same processes" at work in the writer's mind "occur in every psyche"[60] and that poetic imagination is related to the psychic activities occurring in dreams, madness, and children's play.

At the same time, however, Dilthey is anxious to preserve the elect status of poets. Ultimately, he sets their imaginative activity apart from ordinary madness by granting them "the freedom of a creative capacity,"[61] that is, the ability to distinguish between fantasy images and reality. Likewise, he sets writers apart from children at play insofar as the latter have no alternative to the "freedom from purpose" that prevails in their fantasy worlds. The same principle of difference applies all the more to the general population of adults. Even if they possess the same psychological dispositions as poets, they remain miles away from them: "the creative imagination of the poet confronts us as a phenomenon totally transcending the everyday life of mankind." Indeed, the "great poet" "differs from every other class of human beings to a much greater extent than is usually assumed."[62]

---

57 Reckwitz, "Vom Künstlermythos zur Normalisierung kreativer Prozesse," 105.
58 Wilhelm Dilthey, *The Imagination of the Poet: Elements for a Poetics*, in *Selected Works*, vol. 5, *Poetry and Experience*, ed. Rudolf A. Makkreel and Frithjof Rodi (Princeton, NJ: Princeton University Press, 1985), 35.
59 Dilthey, *The Imagination of the Poet*, 5: 66.
60 Dilthey, *The Imagination of the Poet*, 5: 60.
61 Dilthey, *The Imagination of the Poet*, 5: 101.
62 Dilthey, *The Imagination of the Poet*, 5: 60. Cf. Sandra Richter: "The psychology of the extraordinary personality of the poet becomes a major part of Dilthey's poetics. According to Dilthey, the poet is different from ordinary men in the following respects, which result from his extraordinary 'imagination' (*Einbildungskraft*)" (*A History of Poetics: German Scholarly Aesthetics and Poetics in International Context, 1770–1960* [Berlin and New York: De Gruyter, 2010], 156).

Dilthey's observations are characterized by an antithetical movement: the artist's exceptional stature is diminished by psychological comparisons to children and the mentally ill, but at the same time his or her exclusivity and the black box protecting it are upheld. Resistance to shining light into the black box of creative genius was widespread and vigorous. In *Die dichterische Phantasie und der Mechanismus des Bewusstseins* (Poetic Fantasy and the Mechanism of Consciousness, 1869), Hermann Cohen therefore criticizes contemporary aesthetic discourse for its intense objection to uncovering the secret of creativity, stating that his colleagues deemed it "barbarous and unproductive" to voice "doubt in the grace of the moment, the divine cradle [*Götterschooß*] of genius." Accordingly he argues – inasmuch as "uncritical" belief in "the creations of genius" was still the norm – scholars had not come very far in "exploring the essence and origins of literature [*Dichtung*]."[63] Instead, time and again, they had gotten lost in tautologies (e. g., Friedrich Theodor Vischer, who merely "explains fantasy with fantasy"[64]).

Cohen, in keeping with principles of the journal *Zeitschrift für Völkerpsychologie und Sprachwissenschaft* (Journal for Folk Psychology and Linguistics), which was responsible for the original article's publication, sought to remedy this state of affairs by adopting a psychological approach to "poetic fantasy,"[65] which he derives from myth, its linguistic form, and the ways that children think and speak corresponding to mythic consciousness:

> Poetic fantasy [is based] on a mechanism that myth reveals to us. [...] Hereby, the first question concerning the *a priori* conditions of imaginative literature [*Dichtung*] has been solved. As inadequate and fabricated as [any particular instance of] poetic fantasy may appear, it nevertheless is drawn from myth. [...] Myth itself does not derive from a "creative fantasy," but rather is constituted by a group of apperceptions. The unity of consciousness in the first poet of all, the myth-making people [*das mythendichtende Volk*], is evident.[66]
>
> Mythical apperceptions are first practiced by the child, and therefore also by the poet in his childhood, and having penetrated the as-yet empty, receptive consciousness, they remain firmly lodged there.[67]

---

63 Hermann Cohen, *Die dichterische Phantasie und der Mechanismus des Bewusstseins* (Berlin: Ferd. Dümmler's Verlagsbuchhandlung, 1869), 2. First published as an article in the journal *Zeitschrift für Völkerpsychologie und Sprachwissenschaft*, 1869: 173–263.
64 Cohen, *Die dichterische Phantasie*, 7.
65 The same holds for others. For example, Wilhelm Scherer "[relied] on Darwin [...], Herbert Spencer and Edward Burnet Tyler [sic] to explore the origin of poetry" (Sandra Richter, *A History of Poetics*, 168). Hereby, "creative forces of the soul" drawing on "various 'empirical' contributions [to] the *Zeitschrift für Völkerpsychologie*" represented the "main area of interest" (171).
66 Cohen, *Die dichterische Phantasie*, 43.
67 Cohen, *Die dichterische Phantasie*, 68.

Cohen's reflections are already guided by what would become the obvious hypothesis in the paradigm of the 'primitive': the key to artistic creativity lies in 'primitive thinking,' which exists contemporaneously in the artist as much as it does in his or her individual past, which reenacts the development of the species. Thus, as Cohen writes, "The force of myth is not extinguished in modern man."[68] At the same time, this proposition entails the demystification of genius: there is no longer a black box. More still, even a person who is not a genius can do the same, provided that she or he is able to harness "the mythical force."[69]

Pathology or Heroization: Genius and Madness

However, aesthetic theories that saw the key to creativity in 'primitive thinking' differed on how best to understand and evaluate these origins. For example, in the late nineteenth century, the negative evaluation of the return of 'primitive thinking' and the related pathologization of artists (recalling the configuration of the mentally ill 'primitive') played a prominent role in studies of art. From a sociological perspective, this negative assessment can be explained as an effort to "delegitimize delegitimizers": The myth of the artist as an "individualistic ideal ego" (who "conveyed deviant ideas and images" and promoted bohemian lifestyles) questioned the "central patterns of bourgeois culture (morality, industriousness, marriage, rationality, etc.)."[70] But those questions were themselves now called into question as they were pathologized. This, as Bettina Gockel has shown, provided the starting point for a new "science of the artist" in the second half of the nineteenth century, which sought to demystify the genius of old in light of biologistic theories of degeneration.[71]

---

68 Cohen, *Die dichterische Phantasie*, 65.
69 From this changed conception, Leeb draws parallels to Joseph Beuys' dictum that "everyone is an artist. When 'man' takes the stage as a new epistemic figure, art becomes a human capacity" and a "generic trait" (*Die Kunst der Anderen*, 12).
70 Reckwitz, "Vom Künstlermythos zur Normalisierung kreativer Prozesse," 108, 106, 108.
71 Much has been written in recent years on the application of psychopathological theory to aesthetics; see John MacGregor, *The Discovery of the Art of the Insane*; Kaufmann, "Kunst, Psychiatrie und 'schizophrenes Weltgefühl'"; Gockel, *Die Pathologisierung des Künstlers*; Yvonne Wübben, *Verrückte Sprache. Psychiater und Dichter in der Anstalt des 19. Jahrhunderts* (Konstanz: UVK, 2012); Thomas Anz, "Schizophrenie als epochale Symptomatik. Eine Erinnerung – auch an die literarischen Anfänge von Gerhard Köpf," in *Feder, Katheder und Stethoskop – von der Literatur zur Psychiatrie*, ed. Corinna Schlicht and Heinz Schumacher (Frankfurt am Main: Peter Lang, 2008), and *Literatur der Existenz. Literarische Psychopathographie und ihre soziale Bedeu-*

Of particular influence was Cesare Lombroso's *Genio e follia* (1872; *The Man of Genius*, 1896), which explores the "resemblance between genius and insanity"[72] as well as the "art of the deranged."[73] While calling on astrological, racial, and familial complexes, he premises that "in the visible manifestation of their thoughts, the insane frequently revert (as also do criminals) to the prehistoric stage of civilization."[74] Around the same time, in *Die Ästhetik der Gegenwart*, Ernst Meumann discusses prominent "aesthetes" of the day who subscribed to Lombroso's position. These included Siegmund von Hausegger, "who compares the artist's work with dreamlike and hypnotic states," Max Dessoir, who emphasizes "how the increased nervous activity of the genius borders on the pathological," and Paul Julius Möbius, who "has tried to show, apropos of Schopenhauer, Nietzsche, Goethe, and others, how talent often occurs alongside a neurasthenic disposition, hereditary burdens, and various pathological traits."[75] How compelling the link between artistry and mental illness apparently was around the turn of the century is exhibited by the fact that Meumann numbers Dilthey among the advocates of Lombroso's line of argument, even though the philosopher was indeed ultimately interested in stressing the *difference* between genius and madness.

If the anti-bourgeois artist received a negative evaluation in this strain of discourse, some twenty years later – in the context of an increasingly pointed critique of civilization – the opposite tendency prevailed.[76] The 'special subject' of the artist continued to be associated with madness, but now the connection served to distinguish the artist as a heroic figure of protest whose thoughts and deeds resist bourgeois norms.[77] As I showed in Chapter 4, Alfred Storch,

---

tung im Frühexpressionismus (Stuttgart: Metzler, 1977); Thomas R. Müller, "Genie und Wahnsinn," *Soziale Psychiatrie* 141, no. 3 (2013): 11–13.
72 Cesare Lombroso, *The Man of Genius* (London: Walter Scott, Ltd., 1896), vi.
73 Lombroso, *The Man of Genius*, 185.
74 Lombroso, *The Man of Genius*, 191.
75 Meumann, *Einführung in die Ästhetik der Gegenwart*, 92.
76 Gockel writes, "in the discursive field that comes into view there are signs of a decisive change to [...] how the ideal figure of the artist is conceived during and after the First World War." The "image of the epileptic, degenerate genius" loses "more and more ground after 1900, yielding to schizophrenia as the paradigmatic affliction of artists at the end of the 1910s and into the 1920s." Thereby, the artist tends to be represented "as an exceptional human being triumphing over his illness," who "remains a mad genius" but, "as a schizophrenic, achieves a spiritual existence thanks to disciplined and self-disciplining work" (*Die Pathologisierung des Künstlers*, 22).
77 See also Anz, *Literatur der Existenz*, who points to the anti-bourgeois thrust of existential figures in literature around 1910 (39–45); as well as Anz, "Schizophrenie als epochale Symptomatik," 121–122.

for instance, took such a view. And as Gockel notes, it can also be observed in the existential psychology of the 1920s – for instance, in the works of Ludwig Binswanger, who, in discussing "the relationship of phenomenology [...] to psychology and psychopathology,"[78] declares

> there are people who know that, apart from sensory perception, there is another kind of more immediate and more direct way to know or experience things, that, besides conceptual analysis in discrete elements, another, more authentic and more complete mode of intellectual apprehension exists. Such people include, among others, the true artists.[79]

This line of argument may be observed not just among psychologists, but also among art historians. If Worringer and Kühn's idealizations of 'primitive art' serve as an implicit critique of Western civilization, then the same impulse is even more pronounced in the works of authors employing the paradigm of schizophrenic 'primitives.'[80] A prime example of that impulse is *Bildnerei der Geisteskranken* (1922; *Artistry of the Mentally Ill*, 1972) by the art historian and psychiatrist Hans Prinzhorn. A veritable sensation, this book made a lasting change to the reception of the art of the insane.[81] Prinzhorn concludes that "the differentiation of [patients'] pictures from those of the fine arts is possible today only because of an obsolete dogmatism" – he thus rejects artistic tradition and training as "external cultural embellishments of the primary configurative process."[82] The latter, he argues, is intrinsic "to all men," even if it has been "buried by the development of civilization."[83] Mental illness, combined with the isolation produced by institutionalization, leads to the reactivation of the primal creative drive.[84] Counter to received wisdom in reference works on psychopathology, Prinzhorn argues that regression is not at work in this process; this "natural-scientific" way of explaining things is too "causally directed" to be useful. Instead,

---

[78] Ludwig Binswanger, "Über Phänomenologie," *Zeitschrift für die gesamte Neurologie und Psychiatrie* 82 (1923): 11.
[79] Binswanger, "Über Phänomenologie," 12; quoted in Gockel, *Die Pathologisierung des Künstlers*, 94.
[80] Doris Kaufmann ("Kunst, Psychiatrie und 'schizophrenes Weltgefühl'") discerns a shift, after 1910, in popularity from the "native primitive" to the "schizophrenic primitive."
[81] On Prinzhorn, see also Doris Kaufmann, "Kunst, Psychiatrie und 'schizophrenes Weltgefühl.'"
[82] Hans Prinzhorn, *Artistry of the Mentally Ill*, trans. Eric von Brockdorff (New York: Springer, 1972), 274.
[83] Prinzhorn, *Artistry of the Mentally Ill*, 270.
[84] Prinzhorn, *Artistry of the Mentally Ill*, 270–271.

he calls for a "method of observation in which the creative factors of psychic life will be given their just place."[85]

In other words, Prinzhorn takes issue with his colleagues' allochronization of art by the mentally ill – their penchant for viewing it as an early point in the course of phylo- or ontogenetic development and deeming it 'less developed.' Against developmental logic of all stripes, he attributes ontological significance to the images he examines. They point to an essence at the core of human existence: the primary urge to create (*Gestaltungsdrang*), which stands fundamentally beyond history and can no longer thrive in modern civilization. Though it is impaired by the latter, this primary configurative urge can still be observed in works produced by those defined as outsiders – children, members of indigenous communities, and especially the mentally ill.[86] In Prinzhorn's eyes, such images represent the most suitable resource for studying the "primary configurative process" and all its "subconscious components" in "almost pure form."[87]

Moreover, Prinzhorn's study shows quite clearly that, in engaging with the works of the mentally ill, the justification of contemporary art stood at issue. The author asserts that images produced by the insane would be more closely related to those by modern artists than those by children and indigenous peoples would be. Prinzhorn bases his argument on a line of reasoning we have already heard from Kronfeld: that the behaviors to which the patient is driven – "renunciation of the outside world," "devaluation of [...] surface luster," and "a turn inward upon the self"[88] – and which nourish his creative drive are also sought by the artist, as "intuition and inspiration."[89] Therefore, the images of both groups resemble each other, even though in one case they have been created compulsively, and in the other deliberately.[90] This is also why so many painters hold the works of the mentally ill in high regard: "shaken to their foundation" by what they see, they believe "they [have] found the original process of all configuration, pure inspiration, for which [...] every artist thirsts."[91] Prinzhorn thus diagnoses among 'healthy' contemporaries a "longing for inspired creation" that is "denied to us." But he leaves it open as to whether he judges such a longing critically or pathologizes it. On the one hand, he speaks of "schizophrenic" feeling, declares that human beings "intoxicate themselves" with "primary configura-

---

85 Prinzhorn, *Artistry of the Mentally Ill*, 273.
86 Prinzhorn, *Artistry of the Mentally Ill*, 273.
87 Prinzhorn, *Artistry of the Mentally Ill*, 274.
88 Prinzhorn, *Artistry of the Mentally Ill*, 271.
89 Prinzhorn, *Artistry of the Mentally Ill*, 273.
90 Prinzhorn, *Artistry of the Mentally Ill*, 271, 273.
91 Prinzhorn, *Artistry of the Mentally Ill*, 271.

tions," and attributes to them a "craving [*Sucht*] for direct intuitive experience."⁹² On the other hand, he seems to share the views of Storch and Kronfeld, for he ultimately reaches the conclusion, at the end of the book, that an "original process of all configuration" is evident, in exemplary fashion, in the pictorial works of the mentally ill.⁹³

In reaction to Prinzhorn's book, the psychiatrist (and early contributor to the Expressionist movement) Arthur Kronfeld also examines "the process of artistic configuration in light of psychiatry" and identifies three factors in comparing artists and the mentally ill: For one, statistics indicate that schizophrenia is common among artists. Second, a similar creative situation prevails for both groups. Of the artist, Kronfeld writes, "The configurative process presupposes a psychic situation organized in such a way that it simultaneously represents its symbol and its solution, its outlet and its compensation."⁹⁴ The same holds for the schizophrenic, but here he adopts a more pathos-laden tone:

> Archaic strata of the psyche, magical and inspirational, projective modes of enormous vitality, summoned forth from primal urges [*Urtrieben*], give the self unconditional victory over what has prevailed until now, yielding in new but originary form, in hallucinatory, immediate experiences of a revelatory or inspired nature, in new intellectual processes of synthesis, original in kind, an immense, self-created reality, as it were, the "worldview of psychosis."

Third, Kronfeld claims that both artists and schizophrenics consider their products to be "intuitively evident." In affirming the kinship between "the creative element of works by psychotics and those of artists,"⁹⁵ he verges on heroizing the mentally ill – a tendency already evident in Storch's work.⁹⁶ Kronfeld describes both groups as freedom fighters, seeking to achieve liberation "from the effect of the world on the self."⁹⁷ In his estimation, the schizophrenic proceeds in a much more radical fashion and earns the distinction of being "the spiritually richer human being" possessed of "authentic life without compromis-

---

92 Prinzhorn, *Artistry of the Mentally Ill*, 272.
93 Prinzhorn, *Artistry of the Mentally Ill*, 271.
94 Arthur Kronfeld, "Der künstlerische Gestaltungsvorgang in psychiatrischer Beleuchtung," *Klinische Wochenschrift* 4.1 (1925): 29.
95 Kronfeld, "Der künstlerische Gestaltungsvorgang," 29.
96 Doris Kaufmann speaks of Kronfeld's "emphatic conception of schizophrenia," which "was widespread in scholarly discourse on culture in the 1920s" ("Kunst, Psychiatrie und 'schizophrenes Weltgefühl,'" 57).
97 Kronfeld, "Der künstlerische Gestaltungsvorgang," 29.

es."⁹⁸ Therefore, the analogy drawn between the schizophrenic and the artist does not entail the pathological depreciation of the latter; on the contrary, the artist is stylized on the model of the schizophrenic as a fearless loner living by his own concepts and laws.

Mystisches Denken, Geisteskrankheit und moderne Kunst (1923), by Walter Lurje, represents another attempt to understand the essence of modern art: "Why is it that, despite their best efforts, so many people fail to grasp modern art?"⁹⁹ The answer, Lurje maintains, is that the "mystical thinking" shared by "peoples living in a state of nature," children, and "psychically abnormal individuals" (including not only the mentally ill, but also "certain religious minds" and "true artists") defies the logic of modern European adults.¹⁰⁰ Lurje calls for the right measure to be implemented to evaluate such thinking. At the same time, he undermines his relativism by declaring that mystical thought affords insight into "primordial cause[s]," the "source of every essence," and "fount of being."¹⁰¹ The corollary of this position is his normative stance that only an artist "capable of mystical experience" is able to create "real works of art."¹⁰² Lurje invokes literary figures (Friedrich Huch, Alfred Kubin) whose writings represent to him a mystical perception of the world, and he offers examples from the fine arts (Marc Chagall, Alexander Archipenko, Oskar Kokoschka), where he sees mystical thinking expressed not by "content, but form."¹⁰³ In the latter, he traces how symbolic representation yields to (dream) images that evoke the fantasies of childhood and call its imaginative activity back to life. Such works, he argues, do not invite logical thought so much as its opposite, "innermost" feeling or "instinct." Once again, contemporary art is justified by its representation as the product of a primordial human endowment. The process of artistic production, which Lurje mystifies as "mystical thinking," is simultaneously projected back in time and detemporized – that is, it is declared to be still accessible (for some) as the "source of all Being."¹⁰⁴

---

98 Kronfeld, "Der künstlerische Gestaltungsvorgang," 30.
99 Walter Lurje, *Mystisches Denken, Geisteskrankheit und moderne Kunst* (Stuttgart: J. Püttmann, 1923), 3.
100 Lurje, *Mystisches Denken*, 10.
101 Lurje, *Mystisches Denken*, 14.
102 Lurje, *Mystisches Denken*, 15.
103 Lurje, *Mystisches Denken*, 21.
104 Lurje, *Mystisches Denken*, 14.

Normalizing the Artist

Regardless of whether heroicized or pathologized, the artist, understood as a 'primitive madman,' still qualifies as a 'special subject.' Even though artists' psychological capacities do not fundamentally differ from others in this line of thought, they possess (like the mentally ill) more courage to activate the other way of thinking lying dormant within them and with which they are perhaps more substantially equipped. As Lurje writes, "A true artist is no average person and cannot be understood or evaluated in terms of ordinary human beings."[105] Only "among true artists is the capacity for mystical-prelogical thought and experience present to a degree that is no longer the case for other adults."[106]

This rather exclusive conception of artistic identity contrasts, as Meumann had already observed in 1908 of contemporary aesthetics, with a "fundamentally different" perspective, which considered "artistic talent" in terms of "the science of talent in normal human beings,"[107] thus disregarding the creative individual's singularity. This gesture served to normalize artists, but it could also strike a utopian tone inasmuch as the creative potential for thinking and perceiving the world differently was now supposed to extend to the general population. In this line of thought, the artist was connected not so much to the 'schizophrenic primitive' but rather to figurations of the 'primitive' as represented by indigenous peoples or children.

Theories along these lines rested on one of two very different conceptions of artistic creativity. The *first group* focused on myth as a way to figure out the origin of creativity and art. Depending on the ethnological school in question, scholars either chose an individual-psychological explanation (the next two chapters will explore how influential this orientation was both for primitivist theories of language and metaphor and for literary figurations of the artist as 'primitive'), or they adopted a social-psychological mode of explaining the origin of myth and creativity. The latter approach was especially urgent for thinkers eager to connect politics and aesthetics in theory and practice. Examples include the Collège de Sociologie, active in Paris during the late 1930s, which sought to counter the fascists with their own weapons. The *collège* – whose members had strong ties to the milieus of literature and ethnology (and briefly included Walter Benjamin) – saw its activities as a continuation of the French sociological tradition,

---

105 Lurje, *Mystisches Denken*, 3.
106 Lurje, *Mystisches Denken*, 14.
107 Meumann, *Einführung in die Ästhetik der Gegenwart*, 92.

especially the Durkheim school, but with emphasis on the sacred.[108] As Stephan Moebius writes, its sociology of the sacred sought to "analyze, uncover, and renew vital elements [...] that were vanishing in the modern world, for example, collective experiences initiated by rituals, celebrations, or games"[109]. Activities involving "aspects of social bonds that are charged with energy, a-teleological, and experienced imaginatively and affectively" should be freed from their "secondary or supplementary status," and their revitalization should not be left to the fascists alone.[110]

Art as (Child's) Play

In the *second group*, which I will focus on in the remainder of this chapter, theoretical reflections on the origins of art and artistic creativity focused on play (not myth) and invoked the 'primitive' through the figuration of the child.[111] In 1895, James Sully remarks on a widespread belief "that children are artists in embryo, that in their play and their whole activity they manifest the germs of the

---

108 Moebius, *Die Zauberlehrlinge*, 134.

109 Moebius, *Die Zauberlehrlinge*, 135.

110 No project undertaken by members of the Collège makes this ambition as clear as Georges Bataille's secret society, Acéphale: "A communal myth (Acéphale/Dionysos) with assorted rules of conduct and rituals, the celebration of self-loss and self-sacrifice, and [...] a sense of transgressive, mystical-ecstatic 'joy before death' were supposed to create religious-magical cohesion and sound the depths of the sacred *in actu*" (Moebius, *Die Zauberlehrlinge*, 254). For Bataille was convinced, as he documents in his studies of Nietzsche, that "the formation of a new structure, of an 'order' developing and raging across the entire earth, is the only truly liberating act, and the only one possible, since revolutionary destruction is regularly followed by the reconstitution of the social structure and its head" (Georges Bataille, *Visions of Excess: Selected Writings, 1927– 1939*, ed. Allan Stoekl [Minneapolis: University of Minnesota Press, 1985], 198–199). Bataille's secret society and the project of the Collège de Sociologie as a whole have garnered criticism for attempting to combat fascism with its own weapons, that is, for instrumentalizing the power of myth to found community. Needless to say, the question is whether a myth-creating collective can act against fascism at all. Moebius quotes Philippe Lacoue-Labarthe and Jean-Luc Nancy, who ask if, "on the contrary, the mythical function with its national, *völkisch*, ethical, and aesthetic effects [...] is what a future politics must be reinvented *against*" (quoted in Moebius, *Die Zauberlehrlinge*, 154).

111 Parts of this section of the chapter were published previously in a longer version: Nicola Gess, "Vom Täuschen und Zerstören. Spiel und Kunst aus der Perspektive der Entwicklungspsychologie um 1900," in "*Sich selbst aufs Spiel setzen.*" *Spiel als Technik und Medium von Subjektivierung*, ed. Christian Moser and Regina Strätling (Paderborn: Fink, 2016).

art-impulse."[112] Endorsing this position, he studied the relationship between playing and art.[113] For Sully, children's play is determined by a fantasy-driven (re)shaping of the world,[114] and artistic activity is its continuation. Accordingly, he speaks of how "the impulse of the artist has its roots in the happy semi-conscious activity of the child at play" and states that "the play-impulse becomes the art-impulse."[115] For him, then, art is not just what had at one time been play. Instead, the productions of adult artists continue to draw their force from play. In his chapter on children's drawings, Sully indicates that "genuinely artistic work"[116] derives from an ability the small child possesses in full and the adult artist must endeavor to preserve: the "innocence" of seeing first introduced to critical discourse by John Ruskin.[117]

Sully's premise was still the consensus some thirty years later. In *Der Genius im Kinde*, the reformist educator, Gustav Friedrich Hartlaub, distinguishes between two "forms of experience" in the child, dreaming and playing, in terms of the media each involves: speaking on one side and doing and forming on the other. In either case, the "refusal to acknowledge the world as simply mechanical" – that is, an animistic bearing expressing a "'fairytale' relationship to nature,"[118] and "a world of invisible and magical relations"[119] – is closely allied with the "power to imagine/give form [*Ein-Bildungskraft*] in the proper sense." In this relationship to the world, the child repeats the "oldest, half-dreaming, and visionary state of man, in which [...] the fairytale is rooted,"[120] then gradually passes through all other levels of culture (from fairytale notions to heroic sagas, for instance).[121] Although Hartlaub invokes Haeckel's biogenetic law, he also justifies the parallel by affirming that the child possesses an "inextinguishable presentiment" that its fairytale world "somehow belongs to the sum-total of the cosmos, in a word, that it is also real [...] , and that one must [...] save it."[122] On this score, Hartlaub is making the child the mouthpiece of

---

112 Sully, *Studies of Childhood*, 318. On views in the field of child psychology concerning children's art and its relation to the adult artist, see also Boas, *Cult of Childhood*, 79–102. Cf. Wittmann, "Johnny-Head-in-the-air in America," for a discussion of children's drawings.
113 Sully, *Studies of Childhood*, 321.
114 Sully, *Studies of Childhood*, 326.
115 Sully, *Studies of Childhood*, 327.
116 Sully, *Studies of Childhood*, 398.
117 Cf. Wittmann, *Bedeutungsvolle Kritzeleien*, 84.
118 Hartlaub, *Der Genius im Kinde*, 24.
119 Hartlaub, *Der Genius im Kinde*, 25.
120 Hartlaub, *Der Genius im Kinde*, 24.
121 Hartlaub, *Der Genius im Kinde*, 27.
122 Hartlaub, *Der Genius im Kinde*, 25, see also 27.

his own views, which are critical of civilization: "human maturation is not just a step forward, but also a motion backwards, a loss."[123] Many passages speak of childhood as a "paradise,"[124] where the "golden age" at the beginning of humankind is repeated.[125] Ultimately, artists alone refuse to let this paradise (i.e., dreams, play, and their products) disappear. Hartlaub writes, "Only the poet and the artist preserve [...] this general imaginative potential [*allgemeine einbildungskräftige Möglichkeit*] of the child [...]. The 'artist' alone knows how to salvage, more or less, [what remains of] the immense inner life of childhood."[126] The only difference between dreaming/playing and art is that the former is an end in itself, whereas the latter is made for others and requires technical skill.[127]

Sully, Hartlaub, and many other child psychologists thus stress the role of play when drawing analogies between children and artists.[128] They understand play as the (re)shaping of the world through imagination that – in contrast to art, which is intended for others – occurs for its own sake and is believed to be real by the child engaged in it. As I already demonstrated in Chapter 3, child psychologists lent particular attention to the self-deception and destruction at work when children play – especially in activities violating the norms of moral and/or healthy conduct that prevail among adults. Deception borders on both lying and madness. Destruction borders on crime, which is regarded as either evil or pathological behavior, depending on one's interpretation of criminal responsibility. But both characteristics also interested child psychologists because they enabled them to connect play to art. Opposed to moralizing and pathologizing perspectives on deception and destruction, they offered affinities with the production and reception of art situated beyond moral and medical questions. A child deceived by play was no longer interpreted in pathological terms or as losing touch with reality, then. Instead child psychologists viewed such play as behavior that would later give rise to the adult's readiness to fully engage in the world of art. Likewise, destruction of a plaything did not

---

123 Hartlaub, *Der Genius im Kinde*, 29.
124 Hartlaub, *Der Genius im Kinde*, 25, 29.
125 Hartlaub, *Der Genius im Kinde*, 29.
126 Hartlaub, *Der Genius im Kinde*, 30.
127 Hartlaub, *Der Genius im Kinde*, 23, 30.
128 As noted in chapter 3, Freud's "Creative Writers and Day-dreaming" also acknowledges affinities between children's play and art: "A piece of creative writing, like a day-dream, is a continuation of, and a substitute for, what was once the play of childhood." In other words, literature emerges from play and provides the same pleasure that childhood games do. The wish that "finds its fulfillment in the creative work" arises when "a strong experience in the present awakens in the creative writer a memory of an earlier experience (usually belonging to his childhood)" (442).

count as a primordial destructive urge potentially leading to criminality, but as an indication of the mature artist's sovereign command of his materials. In what follows, I will explain in greater detail how these psychological theories of aesthetics ground art in either deception or destruction.

Art and Deception

In broad terms, early child psychologists viewed children's self-deception in one of three ways. According to the first model (e. g., William Preyer), the child's initial concepts are taken for something real in their own right.[129] Karl Bühler represents the second position: in *Die geistige Entwicklung des Kindes* (1918; *The Mental Development of the Child*, 1930), he argues that children at play are aware that what they are doing is illusory.[130] The third school of thought, which posits that the child wavers between belief and disbelief, opens the possibility for understanding play as a mode of aesthetic reception *avant la lettre*.

A good example of the latter is Karl Groos' *Das Seelenleben des Kindes* (The Spiritual Life of the Child, 1904), which distinguishes between full illusion, to which children are more susceptible than adults, and conscious self-deception, illustrated by games of make-believe, when the child "complete[s] what is given by the senses in a twofold illusory manner."[131] In this process, the child projects shapes onto objects (for instance, a horse onto the back of a sofa) and moreover attributes psychic states to them, which Sully, like Groos, terms "personification." Groos differentiates such experience from thoroughgoing deception because "alongside incorrect apperception, the correct conception is present to consciousness."[132] Moreover, this special type of deception is procured at will and enjoyed.

For Groos, conscious self-deception provides a key for understanding art: "aesthetic behavior is a partial phenomenon out of the realm of games of illusion."[133] His earlier study, *Die Spiele der Menschen* (1899; *The Play of Man*, 1901), identifies this bearing as common to play and art and centers less on the production than on the reception of art, or "aesthetic pleasure" (*Der ästhetische Genuss*), which is also the title of a work the author published in 1902.

---

129 Preyer, *Mental Development in the Child*, 17.
130 Bühler, *The Mental Development of the Child*, 91.
131 Groos, *Das Seelenleben des Kindes*, 175.
132 Groos, *Das Seelenleben des Kindes*, 165.
133 Groos, *Das Seelenleben des Kindes*, 172.

Groos's reflections tie in with other theories of art – and not just Schiller's *Über die ästhetische Erziehung des Menschen* (*Letters on the Aesthetic Education of Man*, 1795), a watered-down version of which had long since grown commonplace among the educated middle class. Connections include the writings of Theodor Lipps, the co-founder of *Einfühlungsästhetik*, or "the aesthetics of empathy," and the works of Konrad Lange.[134] In his inaugural lecture at Tübingen, "Die bewußte Selbsttäuschung als Kern des künstlerischen Genusses" (Conscious Self-deception as the Core of Artistic Pleasure, 1894), the latter declared "awareness of aesthetic self-deception" to be a "*conditio sine qua non* for the pleasure taken in art."[135] Both here and in the book he would publish half a decade later, *Das Wesen der Kunst. Grundzüge einer illusionistischen Kunstlehre* (The Essence of Art: Fundamentals of an Illusionist Doctrine of Art, 1901), Lange describes the dynamic as a constant "alternation between deception and non-deception, illusion, and recognition of reality."[136] To illustrate his point, he refers to Goethe's remark that, when reading a good novel, one wavers between emotion (indulgence in illusion) and admiration for the author's skill (which destroys illusion).[137] Such "continuous oscillation between reality and appearance, gravity and play"[138] constitutes the "core appeal of artistic enjoyment"[139] according to Lange.

Groos modifies Lange's thesis along the lines of Lipps's *Einfühlungsästhetik*. In his eyes, the essence of aesthetic enjoyment is "co-experience" (*Miterleben*), a specific form of aesthetic illusion (or conscious deception) that amounts to an "inner imitation."[140] Unlike Lange, Groos does not simply *describe* the process by which art is received. Rather, he seeks to *explain* the psychological and physiological factors responsible for conscious self-deception. Indeed, he goes so far as to offer reasons for what motivates such "inner imitation." Ultimately, he fills in this gap with a putative drive; that is, he contends that human beings are endowed with an inborn mimetic impulse that stands at the source of all learning

---

134 Konrad von Lange, *Die bewußte Selbsttäuschung als Kern des künstlerischen Genusses* (Leipzig: Veit, 1895), 18.
135 Lange, *Die bewußte Selbsttäuschung*, 22.
136 Konrad von Lange, *Das Wesen der Kunst: Grundzüge einer illusionistischen Kunstlehre* (Berlin: Grote'sche Verlagsbuchhandlung, 1907), 357.
137 Lange, *Das Wesen der Kunst*, 358.
138 Lange, *Die bewußte Selbsttäuschung*, 22.
139 Lange, *Die bewußte Selbsttäuschung*, 23.
140 Groos, *Der ästhetische Genuss*, 214, 198.

and that is particularly evident in children's play. Thus he provides the key to what Yrjö Hirn, whose example he follows, deems the "origins of art."[141]

Art and Destruction

Destructive play interested aesthetic theorists as much as it did child psychologists. Theorists of degeneration, in the wake of Lombroso, saw such activities in children as a portent of future criminality. Here, no path led to art – unless it was conceived in pathological terms. However, other psychologists viewed games of destruction along different lines, not as the symptom of biogenetic fatalism, but as the striving for free, self-determined action upon the world. For Freud, not only a basic drive, but also a quest for sovereignty are responsible for destructive games. Thus, the game of *fort-da* that he describes in *Jenseits des Lustprinzips* (1920; *Beyond the Pleasure Principle*, 1922)[142] manifests not just the so-called death drive (*Todestrieb*), but also the young child's effort to overcome an unpleasant situation (the mother's absence) by shifting from a passive to an active role. Likewise, William Stern recognizes children's destructive play as an effort to achieve mastery over unwelcome circumstances in *Psychologie der frühen Kindheit* (1914; *Psychology of Early Childhood*, 1924). In his estimation, "being the cause" procures pleasure, and it "can never be exhibited in a more elemental form than in destruction."[143] Stern thus clarifies that destruction is not sought for its own sake; instead, it represents the experience of being no longer the object but the subject of a situation. The child's impotence is replaced by a pleasurable destructive power. In contrast to Lombroso's claims of biogenetic determination, Stern regards children's destructive play as the first steps toward acquiring autonomy.

This view of destructive play held implications for a theory of artistic production that did not see the artist as the captive of his creation (i.e., in analogy to intoxication or madness), but as a self-aware creator forging a new order by sovereignly commanding his materials. Such a perspective had been topical since, at the very latest, Nietzsche's reading of Heraclitus, as follows:

> A Becoming and Passing, a building and destroying, without any moral bias, in perpetual innocence is in this world only the play of the artist and of the child. And similarly, just as

---

141 Groos, *Der ästhetische Genuss*, 192, 201.
142 Freud, *Beyond the Pleasure Principle*, 13–17.
143 Stern, *Psychology of Early Childhood*, 311.

> the child and the artist play, the eternally living fire plays, builds up and destroys, in innocence [...]. Not wantonness, but the ever newly awakening impulse to play [*Spieltrieb*], calls into life other worlds. The child throws away his toys; but soon he starts again in an innocent frame of mind. As soon however as the child builds he connects, joins and forms lawfully and according to an innate sense of order.[144]

Child psychologists took up this topos and, in a familiar manner, claimed to put it on positivistic footing in order to account for artistic activity and the sovereign gesture it represents. Along these lines, Freud asks in "Creative Writers and Daydreaming,"

> [s]hould we not look for the first traces of imaginative activity as early as in childhood? [...] every child at play behaves like a creative writer, in that he [...] re-arranges the things of his world in a new way which pleases him.[145]

In much the same way, Groos sees an analogy between artistic *production* and play primarily involving a shared "pleasure at being the cause."[146]

Above all, the politically committed avant-gardes of the early twentieth century embraced the attempt to derive art from the dynamic of destruction and creation; it is no coincidence that the "wild child" played a vital role in this context.[147] Examples include Dada, the Surrealists, the Futurists, and even cultural theorists such as Walter Benjamin, for whom, as I will discuss extensively in Chapter 9, the "grotesque, cruel, grim side of children's life" represents a model for artistic and revolutionary action.[148] But other art theorists were just as invested in the topos, for example Johan Huizinga's study, *Homo ludens* (1938; Eng. 1949), the roots of which extend as far back as 1903.[149] While it

---

144 Friedrich Nietzsche, "Philosophy During the Tragic Age of the Greeks" (1873), in *The Complete Works of Friedrich Nietzsche*, ed. Oscar Levy, vol. 2, *Early Greek Philosophy* (New York: Macmillan, 1911), 108.
145 Freud, "Creative Writers and Day-dreaming," 437.
146 Groos, *Der ästhetische Genuss*, 19, 21.
147 As Wittmann notes, the demise of recapitulation theory in science and scholarship did not lessen its "ghostly imaginative power" for figures as varied as Paul Klee, Walter Benjamin, Carl Einstein, and Georges Bataille (*Bedeutungsvolle Kritzeleien*, 241).
148 See also Nicola Gess, "Gaining Sovereignty: The Figure of the Child in Benjamin's Writing," *Modern Language Notes* 125.3 (2010).
149 Johan Huizinga does not restrict play and games to the child. Nevertheless, he often draws examples from childhood, and at the outset he derives the function of play from animals, children, and archaic cultures (which calls to mind his biogenetic structure of argument): for children, play is "already" something different than it is for animals, and in "passing" over to archaic cultures, "we find that there is more of a mental element 'at play'" (*Homo Ludens: A Study of*

does not focus on destruction, the moment of the creative transformation of the ordinary world as well as the sovereignty of the person(s) at play are nevertheless central for him.[150] Play, for Huizinga, is a "free activity" whose illusory nature is apparent for the player, even though it sometimes proves all-absorbing.[151] Play takes place within certain limits and establishes its own order there.[152] The same holds for poetry (*Dichtung*), which the author likewise bases on play. Compared with religion, science, law, war, and politics, poetic works are granted a higher ontological status. Whereas these other realms have grown distant from their ludic origins, "the function of the poet [...] remains fixed in the play-sphere where it was born."[153] Like other theoreticians of art, Huizinga invokes childhood and 'prehistory.'

> Poetry [...] lies [...] on that more primitive and original level where the child, the animal, the savage and the seer belong, in the region of dream, enchantment, ecstasy, laughter. To understand poetry we must be capable of donning the child's soul like a magic cloak and of forsaking man's wisdom for the child's.[154]

Once again, a genealogical proximity between play, literature, and childhood is affirmed here. Although Huizinga speaks of "enchantment, ecstasy, laughter," in reference to "that more primitive and original level," he simultaneously stresses the poetic tendency toward self-imposed limits, the creation of order, and the regularity of play by focusing on literary *forms*.

The theories discussed in the present chapter hardly agree on what constitutes the artist's creative activity, but they all assert that its mysteries can be unlocked by means of the paradigm of the 'primitive,' whether understood in reference to indigenous peoples, children, or the mentally ill. Regarding the causes of creativity, they fill in the space left blank by earlier notions of genius by seeing archaic, formative forces at work in artists. These forces bring about the expression of a different way of thinking and prompt a dialectic of unconscious drives and conscious acts of will. The theories I am discussing in this chapter thus no longer link artistic creativity to a mysterious 'talent,' but to the 'primitive.' Although 'primitive thinking' is marked by logical fallacies, drives, desires, emotions, and collective ideas, they are convinced that the depths of its existence, origins, and ways of function-

---

the *Play-element in Culture* [London: Routledge & Kegan Paul, 1949], 14). In other words, the play of children represents an embryonic version of cultural development for whole societies.
150 Huizinga, *Homo Ludens*, 19.
151 Huizinga, *Homo Ludens*, 13.
152 Huizinga, *Homo Ludens*, 15–16.
153 Huizinga, *Homo Ludens*, 19.
154 Huizinga, *Homo Ludens*, 119.

ing have been sounded by empirical research in the fields of ethnology and psychology. In brief, they argue that the artist activates the faculty that once defined all mental activity and that still defines the mental activities of children and the mentally ill. Furthermore, they hold it can be accessed even by 'healthy' adult Europeans in altered states of consciousness. At the same time, this explanation also lays the mechanisms of creativity open. Contrary to what Kant had claimed of the work of geniuses, creative acts are now said to follow a discernable pattern and certain procedures – for instance, image-agglutination by means of condensation and displacement. The artistic *outcome* is still original, but the artist's *procedures themselves* lack originality because they are common to a great number of mental acts. Or, in other words, the originality of the work of art is based on recognizable processes that obey certain 'primitive' patterns of thought. The 'genius,' in the Romantic sense of the word, no longer exists.

Chapter 6
'Primitive Language' – Theories of Metaphor

"Do you know what a symbol is? ... Do you want to try to imagine how sacrifice first emerged?"[1] These two questions, posed by Gabriel to Clemens in a dialogue staged in Hugo von Hofmannsthal's "Gespräch über Gedichte" (Conversation on Poetry, 1904), initially seem to Clemens and the reader to have little to do with each other.[2] Yet Gabriel corrects our mistake by explaining that in the act of sacrifice, an animal is substituted for a human victim and that in like manner, the lyrical symbol takes the place of "a state of sensibility" (*ein Zustand des Gemüts*).[3] The experience of sacrifice, he claims, is based on the sacrificer himself dying "for a moment"[4] in the animal, and this momentary co-identity is the precondition for the substitution to function. The same is understood to hold for the lyrical symbol: sensibility dissolves in the symbol in such a way that the recipient understands its meaning immediately without being able to express it in words.

This remarkable theory of symbols is not as speculative as it might first seem. Rather, it is symptomatic of the trend around 1900 of basing theories of language and metaphor on the new disciplines of the human sciences, in particular ethnology and developmental psychology.[5] In these anthropological theories, metaphor was seen to derive from the ways of thinking and speaking exhibited by indigenous communities (who, according to the paradigm of the 'primitive,' were thought to represent early human culture) and by children

---

[1] Hugo von Hofmannsthal, "Das Gespräch über Gedichte," *Gesammelte Werke*, ed. Bernd Schoeller, vol. 7, *Erzählungen, Erfundene Gespräche und Briefe, Reisen* (Frankfurt am Main: Fischer, 1979), 502. The ellipses are Hofmannsthal's.
[2] The German version of this chapter has been published in shortened form: Nicola Gess, "'So ist damit der Blitz zur Schlange geworden.' Anthropologie und Metapherntheorie um 1900," *Deutsche Vierteljahresschrift für Literaturwissenschaft und Geistesgeschichte* 83.4 (2009).
[3] Hofmannsthal, "Das Gespräch über Gedichte," 500.
[4] Hofmannsthal, "Das Gespräch über Gedichte," 502.
[5] Benjamin Specht reaches similar conclusions: "In studies of myth and ethnology," metaphor "also represents the rudiment of a primitive level of culture that is supposed to make it possible to reconstruct the genesis of language and consciousness, [...] 'the Paleontology of the human mind,' as Friedrich Max Müller put it" ("'Verbindung finden wir im Bilde.' Die Metapher in und zwischen wissenschaftlichen Disziplinen im späten 19. Jahrhundert," in *Metaphorologien der Exploration und Dynamik 1800/1900. Historische Wissenschaftsmetaphern und die Möglichkeiten ihrer Historiographie*, ed. Gundhild Berg, Martina King, and Reto Rössler [Hamburg: Meiner, 2018], 44; the author is referring to Max Müller, *Lectures on the Science of Language* [London: Longman, Green, Longman, Roberts, and Green, 1864], 338).

(who, according to Haeckel's law of "biogenetic constitution," were placed in analogy with the latter as well).[6] In this way, the human sciences were elevated to a superior status as supplier of facts, where previously only speculation had reigned. On the other hand, the very same anthropological theories of metaphor also demonstrated that the propositional knowledge of the sciences itself derives from metaphor. That is, it was formed on a foundation traditionally ascribed to rhetoric and literature (a feature evident, for example, in the literary quality of the ethnological and psychological writings examined in Chapters 2 to 4).

In response to ethnological and psychological research and in recognition of the metaphorical basis of all science, three anthropological theories of metaphor emerged around the turn of the century: First, as long as the epistemic ideal of accessing the world-in-itself persisted, this recognition of the metaphorical basis of science could lead to a skeptical attitude toward knowledge per se and to a preference for literature as the realm of conscious illusion. Alternatively, it led to claims of an emphatically other knowledge that is genuinely literary and based on metaphorical thinking. Through such language the world-in-itself becomes accessible – quite in contrast to the operations of scientific knowledge.[7] Finally, a third response involved weakening the boundary separating the "two cultures" of literature and science by rendering the metaphoric and indeed *poietic* dimension of all forms of knowledge more recognizable.[8] In the following I will present these three theories of metaphor. Before that, however, I must address the theories of language that developed around 1900 in ethnology and developmental psychology.

---

6 As one reads in the *Historische Wörterbuch der Philosophie*, the first use of the term *primitivum* in Latin grammar served to distinguish between *verba primitiva* and *verba derivativa*. In this light, modern usage in the 'human sciences' takes up a very early sense of the word ("primitiv," 7: 1316).

7 Here affinity exists with the perspectives of vitalist philosophy, which stresses nonrational modes of relation to the world – e.g., intuition in Bergson, understanding in Dilthey, vision (*Schauung*) in Klages, and fantasy in Jung (who, like the other writers here, classifies it as "primitive").

8 Wolfgang Riedel has devoted a great deal of attention to theories of metaphor in the context of literary primitivism, especially in the essay "Arara ist Bororo." Sabine Schneider discusses Hofmannsthal in light of Riedel's reflections in "Das Leuchten der Bilder in der Sprache," *Hofmannsthal Jahrbuch* 11 (2003).

Constructions of 'Primitive Language': The Cratylist Tradition

From the outset, scientific reflections on 'primitive thinking' were connected with the construction of a 'primitive language.' Four key features stood at the center of deliberations by ethnologists and developmental psychologists on language: The first included the vivid nature of 'primitive languages,' that is, their detailed imagery based on an abundance of metaphors. The second was the naturalness of language, i.e., how it is motivated through its objects. The third was the participation or even co-identity of language with its objects. This latter relationship leads to the fourth characteristic that language was thought to possess: magical power.[9] In the following I will explain these four characteristics in more detail.

Vividness (*Anschaulichkeit*) describes on the one hand 'primitive language' in the sense of *parole*, which ethnologists characterized as lacking in abstraction and having the tendency to describe events in great detail. According to Wilhelm Wundt, for example, a "Bushman" expresses the idea of a "warm welcome from the white man" by means of a series of verbal pictures dramatizing the interaction: "The white man gives him tobacco, he fills his pouch and smokes; the white man gives him meat, he eats this and is happy, etc."[10] Yet, vivid language also describes 'primitive language' in the sense of *langue* and its wealth of grammatic forms and vocabulary that together aim to convey the greatest possible specificity. Lévy-Bruhl, for example, notes an abundance of verbal forms in "Indian languages," which capture shades of meaning entirely unknown to Europeans: "A Ponka Indian in saying that a man killed a rabbit, would have to say: the man, he, one, animate, standing (in the nominative case), purposely killed by shooting an arrow the rabbit, he, the one, animal, sitting (in the objective case)."[11] Similarly, the indigenous lexicon is rich in words for tangible, sensory experience. Instead of classes and kinds, it offers "image-concepts," which always have a particular referent (not "foot" in general but the foot of a certain person, not "fish" but, more specifically, a perch). Lévy-Bruhl explains these "image-concepts" in analogy to highly detailed drawings, and accordingly sees the entire language as a "drawing" bound to the language of signs and gestures: "If verbal language [...] describes and delineates in detail positions, motions, distances, forms, and contours, it is because sign-language uses exactly the same means of expression."[12]

---

[9] On theories of signs and metaphor in the context of primitivism, see also Werkmeister, who reaches similar conclusions in *Kulturen jenseits der Schrift*, 197–247, especially 231–237.
[10] Wundt, *Elements of Folk Psychology*, 72.
[11] Lévy-Bruhl, *How Natives Think*, 119.
[12] Lévy-Bruhl, *How Natives Think*, 140.

However, the vividness of 'primitive language' does not refer only to the utmost specificity of *langue*, but also to the motivated quality of words. In this context, it is important to distinguish between an indexical and iconic relationship between language and world.[13] On the one hand, for ethnologists and developmental psychologists alike, language derives from indexical gestures that they understand as both deictic and expressive. Their phonetic counterpart are demonstratives that trace back to "reflexive vocalizations" accompanying expressive or referential gestures. Clara and William Stern, for example, consider the interjection "there!" a "natural, outwardly directed vocalic gesture."[14] On the other hand, ethnologists and developmental psychologists thought language derived from iconic imitation, gestures that imitate the signified object or trace its outward form.[15] Thus, the Sterns speak, for example, of "sound gestures" made by movements of the mouth but corresponding to certain hand or arm motions that are for their part mimetically motivated.[16]

Another important dimension of the vivid quality of 'primitive languages,' for ethnologists and developmental psychologists alike, was its heavy use of figurative language (*Bildlichkeit*). They noted that, paradoxically, an object is vividly described (*anschaulich abgebildet*) when a word from a *different* context is employed, in other words, when metaphor is used. Thus, E. B. Tylor writes of the "wild and rambling metaphor which represents the habitual expression of savage thought."[17] Along similar lines, James Sully observes

> We may detect a close resemblance between children's language and that of savages. In presence of a new object a savage behaves very much as a child, he shapes a new name out of familiar ones, a name that commonly has much of the metaphorical character.[18]

---

13 On the distinction between deictic and mimetic gestures, cf. Wundt, *Elements of Folk Psychology*, 63–66. Tylor already noted that all languages share "sounds of interjectional or imitative character" (*Primitive Culture*, 1: 145); likewise, Sully considers expression and imitation to be the two sources of human language (*Studies of Childhood*, 147). See also William Stern, *Psychology of Early Childhood*, 90–95; and Clara and William Stern, *Die Kindersprache. Eine psychologische und sprachtheoretische Untersuchung* (Darmstadt: Wissenschaftliche Buchgesellschaft, 1965), 319–320, who distinguish between natural sounds, which provide the first words for affect and desire, and acts of imitation, which represent the first form of objective description.
14 Stern and Stern, *Die Kindersprache*, 368. On expressive motion, cf. Wundt, *Elements of Folk Psychology*, 53–60.
15 On this distinction, cf. Wundt, *Elements of Folk Psychology*, 105–106.
16 Stern and Stern, *Die Kindersprache*, 355.
17 Tylor, *Primitive Culture*, 2: 404.
18 Sully, *Studies of Childhood*, 168.

The reasoning behind this process is supposed to lie as much in the unfamiliarity of the object as in the speaker's need for greater vividness. At any rate, according to these scholars, it is motivated by the similarity between two phenomena or objects, which the metaphor connects.[19] To illustrate the point, Sully notes that "the Aztecs called a boat a water-house."[20]

Ethnologists and developmental psychologists emphasized that such metaphors are viewed as comprising real relations by those who employ them: "given the lexical paucity [*Wortnot*] of the first stages of language, primitive man reaches for the first word that presents itself by chance, in keeping with a vague likeness, [and] uses it as a substantive designation for the object."[21] Indeed, some researchers therefore maintained that "metaphor" is an inaccurate term for these linguistic renderings.[22] Such doubt attests both to a strictly Aristotelian conception of metaphor and to an awareness of the fact that European rhetorical concepts here get transferred onto languages of foreign cultures. One result of this projection was that many scholars ascribed a particular talent to indigenous cultures because of the supposed metaphoricity of their language, even seeing them as the first poets (once again a well-worn topos in the European tradition of the philosophy of language).

---

[19] However, if – like Lévy-Bruhl – one does not assume that 'primitive thinking' is based on association (the working premise of English ethnologists) so much as participation, metaphors do not express a perceived similarity between objects. Instead, the objects are co-present in a single perception, and the name they share indicates their mutual participation.

[20] Sully, *Studies of Childhood*, 168. Richard Thurnwald sums it up as follows: "Characteristically, most languages of peoples in a state of nature derive a new word by synthesizing images commonly in use, for instance: 'spring, well' = 'eye-water.'" He also calls this process a "metaphorical mode of expression" ("Psychologie des Primitiven Menschen," in Gustav Kafka, *Handbuch der vergleichenden Psychologie*, vol. 1, *Die Entwicklungsstufen des Seelenlebens* [Munich: Reinhardt, 1922], 269). Like Lévy-Bruhl, Heinz Werner calls such coinages "concept-images" (Werner, *Einführung in die Entwicklungspsychologie*, 194). Ernst Kretschmer refers to them as image-agglutinations, whereby he makes the distinction (taken from Freud) between processes of condensation and displacement (*Medical Psychology*, 87–88).

[21] Stern and Stern, *Die Kindersprache*, 324.

[22] The Sterns also understand metaphor only as a *consciously* "improper" expression, thus excluding 'primitive' transfers of meaning born of the lack of adequate terminology from being classified in this way. For Heinz Werner, it is only possible to speak of metaphor when "allegorical consciousness" (*Gleichnisbewusstsein*) has emerged (*Die Ursprünge der Metapher*, 28, 34). According to Lévy-Bruhl, it is not a matter of transfer so much as the expression of participation always already in place. Thus, he stresses that signs become necessary for 'primitives' when participation is no longer directly felt but still mythically represented. Hereby, the content of myths is not as important as the mystical atmosphere surrounding words (i.e., their participation with what they designate) (Lévy-Bruhl, *How Natives Think*, 323–327).

Based on its indexical, iconic, and metaphorical features, ethnologists and developmental psychologists deemed 'primitive language' to be 'natural' – that is, directly motivated by its objects.[23] Tylor, for example, contends that "savages possess in a high degree the faculty of uttering their minds directly in emotional tones and interjections, of going straight to nature to furnish themselves with imitative sounds."[24] Thereby, the indexical relationship affirms connection through contiguity, which is understood to be language's physiological motivation: an inner tension is involuntarily discharged in a physical movement that is related to language in a way that can be explained by physiology.[25] In contrast, iconic and metaphorical relationships hold that language is motivated by similarity between word and object and between two objects denoted by the same motivated word, respectively.

The vivid and natural traits of 'primitive language,' as ethnologists and developmental psychologists argued, go hand in hand with the belief that words are directly tied to the objects they designate and therefore possess magical power. This supposed connection signifies more than just the motivation of words. It points to the ontological basis of the claim that signs are not arbitrary, but motivated by their objects. This basis entails the belief that words are either components of the objects to which they refer or have been incompletely separated from them – that is, they still participate in them. The Sterns write,

> [f]or children – as for primitive human beings in general – the word, once acquired, and the object constitute an organically coherent whole [...]. Children and peasants cannot think otherwise, than that the long, dark form baked from flour is not only called, but is "bread." [...] The word is conceived as the quality, the proper intuition of the thing.[26]

As part of its object, the word expresses the object's essence. Knowing the word for an object amounts to recognizing it for what it truly is. Such participation means reversing cause and effect: the object in question counts as the cause of the word that now names it; conversely, the word has the potential to cause events that happen to the object. Thus, Lévy-Bruhl writes of the mystical character of words,

---

23 As already noted, Kretschmer distinguishes between condensation and displacement as two different modes of image agglutination; thus, metaphoricity would be complemented by metonymy (Kretschmer, *Medical Psychology*, 87–88).
24 Tylor, *Primitive Culture*, 1: 147. Tylor explicitly refers to theories of the natural origins of language, e.g., de Brosses (146) but urges caution about indulging in etymological speculation.
25 Cf. Wundt, *Elements of Folk Psychology*, 21–22.
26 Stern and Stern, *Die Kindersprache*, 320.

> [t]he use of words can never be a matter of indifference: the mere fact of uttering them [...] may establish or destroy important and formidable participations. There is magical influence in the word, and therefore precaution is necessary.[27]

Knowing names involves not just recognizing objects, but also gaining power and influence over them.

In their conception of 'primitive language,' representatives of early ethnology and child psychology were obviously working in the Cratylic tradition.[28] Like the philosopher in Plato's dialogue of the same name, turn-of-the-century theorists sought to prove the naturalness of 'primitive language' by pointing to its indexicality, iconicity, and metaphoricity. At the same time, however (and following the admonitions of Socrates), they were forced to acknowledge that their claims often depended on speculative etymology, without which no traces of such natural qualities could be found. In this respect, their theories display an orientation that Gérard Genette would describe as "mimological"[29] by clinging to the always already lost ideal of a seamless correlation between words and things, which they ascribed to a 'primitive language' of their own construction.[30] Thus, once again, turn-of-the-century ethnologists and child psychologists did not develop new concepts so much as they found confirmation for European linguistic tradition in other cultures. 'Primitive culture' was supposed to offer proof of what generations of philosophers had merely speculated or fantasized about: the natural origin of human language.

Malinowski and the Magical Power of Language

To a certain extent, ethnologists and developmental psychologists developed the concept of a 'primitive language' only in passing. They claimed it to be a result of

---

[27] Lévy-Bruhl, *How Natives Think*, 154.
[28] The Sterns explicitly invoke the Platonic dialogue (*Die Kindersprache*, 127, 319).
[29] Gérard Genette, *Mimologics*, trans. Thaïs E. Morgan (Lincoln: University of Nebraska Press, 1995).
[30] Interestingly, this holds both for those who showed sympathy for 'primitive thinking' and its critics. Proponents of rational thought also dreamed of a language that would stand in direct connection with concepts and therefore cause no falsification of them (cf. Charles K. Ogden and I.A. Richards, *The Meaning of Meaning* [San Diego: Harcourt Brace, 1989], who fault the logicians of the late nineteenth and early twentieth centuries for clinging to the ideal of a natural language).

'primitive thinking,' which was the actual focus of their work.[31] The writings of Bronislaw Malinowski, who inaugurated the practice of field research and in so doing founded the modern field of ethnology, provide a contrasting perspective. Not only do they demonstrate the transformation of ethnology and developmental psychology into semiotic theory, but furthermore reverse the relationship that was supposed to hold between thought and language by deriving the former from the latter. To do so, Malinowski starts out with the idea that language possesses magical power. For in this new framework, the issue is not how language *represents* reality, but how language *impacts* reality. In terms of linguistics, the focus is pragmatic. Unlike the authors I have been discussing, Malinowski is not searching for a natural language. Instead, by examining the potential of language to *do* things – its performative dimension (in the sense used by J. L. Austin) – he affirms the validity of indigenous belief systems.

On the basis of linguistic usage among the inhabitants of the Trobriand Islands, Malinowski develops – in *The Problem of Meaning in Primitive Languages*, a supplement to Charles K. Ogden und Ivor A. Richard's *The Meaning of Meaning* (1923) – the concept of the phatic function of language, later taken up by Roman Jakobson.[32] In this capacity, language does not serve as a "means of thinking" so much as a "mode of action."[33] In consequence, Malinowski comes to question Ogden and Richards' principle of "symbolic relativity," that is, the notion that mere convention governs the connection between symbol and referent. In contrast to the authors treated so far, he does not do so in the name of the supposed naturalness of language; instead, he focuses on the analysis of children's language acquistion and use, in which he sees parallels to indigenous peoples. Accordingly, he observes the following:

> To the child, words are [...] not only means of expression but efficient modes of action. The name of a person uttered aloud in a piteous voice possesses the power of materializing this person. Food has to be called for and it appears [...]. Thus infantile experience must leave on the child's mind the deep impression that a name has the power over the person or thing which it signifies.[34]

---

31 This was especially true for ethnologists (see, e. g., Tylor, *Primitive Culture*, 1: 271; Thurnwald, "Psychologie des primitiven Menschen," 266). Among developmental psychologists, a greater interest in language was evident from the outset because it allows the development of thought in children to be observed most fully.
32 Bronislaw Malinowski, "The Problem of Meaning in Primitive Languages," in Ogden and Richards, *The Meaning of Meaning*.
33 Malinowski, "The Problem of Meaning in Primitive Languages," 315.
34 Malinowski, "The Problem of Meaning in Primitive Languages," 320.

Early childhood experience, Malinowski maintains, shapes people for the rest of their lives, and this is therefore where belief in language as a magical force – based on a direct connection between symbol and referent and the power of words bound to that immediacy – begins. Malinowski revisits and expands this thesis in later texts, observing that modern European adults still have experiences over and over again that suggest words possess a magical power: "knowledge of the right words [...] gives man a power over and above his own limited field of personal action."[35] Thus, Malinowski opposes Freud's concept in *Totem and Taboo* of an "omnipotence of thought" with the "omnipotence of words":

> Magic is not a belief in the omnipotence of thought but rather the clear recognition of [...] its impotence. [...] Verbal magic grows out of legitimate uses of speech, and it is only the exaggeration of one aspect of these legitimate uses.[36]

In semiotic terms, the peculiar network of relationships that 'primitive thinking' spins between things is not a result of the naturalness of language, according to Malinowsky, but appears rather as a kind of participation between symbol and speaker experienced during the act of speech. By means of this participation, language acquires magical power and the belief in magic is founded.[37]

Theories of Metaphor around 1900: Nietzsche, Mauthner, Vischer, Biese, Cassirer

At the time, Malinowski's thesis that magical thinking originates in language was a departure from the views of his predecessors in the fields of ethnology and developmental psychology.[38] However, his claim was unexceptional in the

---

35 Malinowski, "An Ethnographic Theory of the Magical Word," in *Coral Gardens and Their Magic*, vol. 2, *The Language of Magic and Gardening* (Bloomington: Indiana University Press, 1965), 235. The text was written in 1935. As Ken Hirschkop has recently observed, for Malinowski, "the force of magic was just a concentrated version of the general pragmatic force of all language, which, in the second half of the century, would become a subfield of linguistics and a live topic in analytic philosophy" (*Linguistic Turns: Writing on Language as Social Theory* [Oxford: Oxford University Press, 2019], 165).
36 Malinowski, "An Ethnographic Theory of the Magical Word," 239.
37 See Robert Stockhammer, *Zaubertexte. Die Wiederkehr der Magie und die Literatur, 1810–1945* (Berlin: Akademie, 2000), 26.
38 With the possible exception of Karl Bühler; on the basis of his study *The Mental Development of the Child*, he had already spent several years investigating the performative nature of language

context of the theories of metaphor elaborated by philosophers and scholars of literature around the turn of the century, who also sought to demonstrate the relevance of metaphor for 'primitive thinking.' Indeed, establishing this conviction was necessary for the subsequent claim that this thinking could be revived in modern literature.[39] Thus, philosophers and literary scholars took up the views of their contemporaries in the human sciences and radicalized them with the help of recent developments in the philosophy of language to claim that 'primitive language' was not the outcome but the starting point of 'primitive thinking' and even formed the basis of contemporary and scientific thought.[40] At the same time, they expanded the definition of metaphor to include all transmission processes involved in the creation of language. Iconic and even indexical relations between object and word were now understood as transfers (from object to gesture or sound) and, in this sense, as metaphors. In this way, 'primitive language' turned out to be completely shaped by metaphors. Thus, for example, Mauthner, whose theory of metaphor likewise drew from findings in anthropology, claims that "the metaphor or the poetic image is the origin and essence of all language," and as such it forms the foundation of modern concepts and thought processes as well.

This tendentious reference to the human sciences enabled theorists of metaphor to go beyond the speculations of Giambattista Vico or Johann Gottfried Herder by determining metaphorical language as original speech in an anthropo-

---

– which he then presented in *Theory of Language: The Representational Function of Language*, trans. Donald Fraser Goodwin (Amsterdam: John Benjamins, 2011). For an overview of affinities, see Stefan Henzler, "Der Handlungscharakter der Sprache bei Karl Bühler und Bronislaw Malinowski," in *Betriebslinguistik und Linguistikbetrieb. Akten des 24. Linguistischen Kolloquiums, Universität Bremen, 4.–6. September 1989*, vol. 1, ed. Eberhard Klein, Françoise Pouradier Duteil, and Karl Heinz Wagner (Tübingen: Niemeyer, 1991).

39 This stands in the context of a widespread critique of civilization: "At the beginning of the twentieth century, myth seems to provide the answer to the negative impression of having lost an original connection to the world in the present. In his 1911 essay, 'Concept and Tragedy of Culture,' [Georg] Simmel gets to the heart of the fundamental critique of modernity." (Anja Schwennsen, "Kunst und Mythos zwischen Präsenz und Repräsentation," in *Zwischen Präsenz und Repräsentation. Formen und Funktionen des Mythos in theoretischen und literarischen Diskursen*, ed. Bent Gebert and Uwe Mayer [Berlin, Boston: De Gruyter, 2014], 207). Schwennsen exempts Cassirer from this "frame of mind, which was paradigmatic for *Lebensphilosophie*": for the latter, art is not "the last refuge" but a "process [...] of giving-form" (209); I will return to this point below.

40 It is important to distinguish between such histories of the development of thought in terms of evolution or simply genealogy and perspectives (which were less widespread) that posted two different types of thought (e.g., Lévy-Bruhl) or multiple, culturally specific types of thought (Boas).

logical and apparently empirically secured way. Subsequently, they were able to come up with a new classification and justification of literature. Against the backdrop of these new disciplines, old questions concerning the nature and purpose of art now received new answers. These theorists of metaphor suggest that the use of tropological language in literary arts relates it to 'primitive thinking,' thus enabling the latter's revival or further development. Opinions differed, however, in terms of the epistemic value attributed to metaphorical language (and therefore literature). As mentioned at the beginning of this chapter, three tendencies may be noted: (1) Because it is regarded as an inauthentic language, metaphor leads to false concepts and prevents any awareness of 'real reality.' If one holds this opinion, one must either lead a futile battle against metaphor and its falsehoods (like the pioneers of analytical philosophy), or one must abandon the quest for knowledge with resignation (like Fritz Mauthner) or with an aesthetic posture in relation to a world of unauthentic illusions (like Nietzsche). (2) As an original and motivated form of language, metaphor represents a privileged access to reality. From this perspective (held, for example, by Alfred Biese and his appreciative reader, the young Hugo von Hofmannsthal), one turns from scientifically based concepts and knowledge to poetry in hopes of gaining immediate access to the world-in-itself. In contrast to Malinowski's concern with the performative nature of language, these first two tendencies look to language's descriptive reference to an extra-linguistic reality. What separates them is the understanding of metaphor as either arbitrary or motivated. In a sense, each view focuses exclusively on just one side of the metaphorical equation. A metaphor declares, "A is B" – which is all that proponents of the second thesis hear. On the basis of the identity posited, they conclude that metaphor is motivated by the world-in-itself and grants privileged access to it. At the same time, saying "A is B" presupposes that the likeness of A and B is not in fact given; proponents of the first thesis stress this implicit non-identity, and they conclude that metaphor is not authentic and misrepresents reality. [41] (3) The third tendency's focus on the positing power (*Setzung*) of language sets it apart from the first two: as a positing language, metaphorical language reveals the *poietic* activities of the human mind, that is, the role of creation in cognition – a view held, for example, by Ernst Cassirer. In the following I will give four examples of the positions mentioned (Nietzsche, Mauthner, Vischer and Biese, and Cassirer), concentrating on

---

[41] See Monika Schmitz-Emans, "Metaphor," formerly in the now-offline *Basislexikon Komparatistik*; it is now accessible via: https://docplayer.org/25246668-Metapher-autorin-monika-schmitz-emans.html.

texts evincing an appreciation for metaphor and poetry (that is, leaving aside theories dismissing the metaphoricity of language).

Nietzsche

Nietzsche's influential "Ueber Wahrheit und Lüge im aussermoralischen Sinne" (1873; "On Truth and Lie in a Nonmoral Sense," 1977) does not offer a theory of metaphor so much as a critique of language and epistemology.[42] However, this critique is based on metaphor in a double sense. Nietzsche simultaneously starts out from both the linguistic dependence of all knowledge as well as the origin of language in a twofold process of transfer, which he calls metaphor: "To transfer a nerve stimulus into an image – first metaphor! The image [is] again copied in a sound – second metaphor! And each time a complete leap [takes place] out of one sphere into an entirely new and different one."[43] Nietzsche does not understand metaphor in the Aristotelian sense but uses this term in the sense of a transfer process from one sensory realm to the other: a nerve stimulus leads to a mental image, which prompts an auditory sensation. At the same time, he denies that any one of these relay-points captures the essence of any 'thing in itself.' In contrast to the scholars and theorists discussed above, Nietzsche considers both stages of transmission to be "arbitrary"[44] since they each necessarily focus on a single aspect of the object to the exclusion of other features.[45] Also,

---

[42] Tylor's study appeared in German translation the same year that Nietzsche wrote this text, and he read it thoroughly. However, Nietzsche first borrowed the translation from the university library in Basel in June 1875, so it is uncertain whether he already knew of it in 1873. On Tylor's influence on Nietzsche, see Hubert Treiber, "Zur 'Logik des Traumes' bei Nietzsche. Anmerkungen zu den Traumaphorismen aus 'Menschliches, Allzumenschliches.'" *Nietzsche-Studien* 23 (1994): 6n16.

[43] Friedrich Nietzsche, "On Truth and Lie in a Nonmoral Sense," in *On Truth and Untruth*, trans. Taylor Carman (New York: Harper, 2010), 26. Gustav Gerber also assumes the metaphorical quality of all languages in *Sprache als Kunst* (Language as Art, 1871) (Hildesheim: Olms, 1961), e.g., 309, 312; Nietzsche borrowed this title from the university library on 28 September 1872 and took up key aspects of it in his own writings (cf. Meijers and Stingelin, "Konkordanz zu den wörtlichen Abschriften und Übernahmen," *Nietzsche-Studien. Internationales Jahrbuch für die Nietzsche-Forschung* 17 (1986). As Benjamin Specht has recently noted apropos of Gerber, Wundt, and Dilthey, Nietzsche "demonstrably drew inspiration from the linguistics of the time [...]. There one finds a corresponding expansion of metaphor as the genetic principle of speech and thought, albeit without the same critique of epistemology" ("'Verbindung finden wir im Bilde,'" 43).

[44] Nietzsche, "On Truth and Lie," 28.

[45] Nietzsche, "On Truth and Lie," 27–28.

the initial stimulus is overly subjective and says more about the individual experiencing it than it does about the thing perceived. Nietzsche concludes that already at its earliest stage, where intuitive metaphors abound, language can convey no knowledge about the world: "We think we know something about the things themselves when we speak of trees, colors, snow, and flowers, yet we possess only metaphors of the things, which in no way correspond to the original essences."[46]

Paradoxically, Nietzsche's insight into the metaphorical basis of all language (that is, the fact that it is *not* motivated) did not lead him to give up on the model of a descriptive language truly depicting reality. He remained attached to the idea, albeit in negative fashion, by stressing the vitiated nature of language and the knowledge it affords; the farther one gets from the "thing in itself," the greater the deficit.[47] In particular, then, Nietzsche's verdict bears on conceptual language, into which individual intuitive metaphors are dissolved during a later stage of language development. The concept no longer fulfills a mnemonic function as the intuitive metaphor does; it does not call to mind a "single, absolutely individualized original experience"[48] but rather serves a systematic purpose, inasmuch as it creates order. The sense of security such order provides is purchased by twofold oblivion: First, one forgets that intuitive metaphors are arbitrary and "takes them for the things themselves."[49] Second, concepts are formed through "forgetting what distinguishes one [thing] from the other"[50] – information that was still given in initial metaphors.

Nietzsche calls those who successfully master this double forgetting and live quietly ever after in their conceptual framework "rational." In contrast, an instinctive and intuitive person will work against these processes of oblivion by shattering traditional concepts with new metaphors and in full cognizance that the latter are simply metaphors. Such individuals relate to the world of objects in a thoroughly "aesthetic" manner insofar as they do not seek to access 'things in themselves.' Instead, they are continuously enacting transfers between registers of meaning (well aware of their arbitrary nature) in order to inhabit a world of metaphor built on "semblance and beauty."[51] This terrain does not be-

---

[46] Nietzsche, "On Truth and Lie," 26–27.
[47] See Klaus Müller-Richter and Arturo Larcati, '*Kampf der Metapher!*' (Vienna: VÖAW, 1986), 225.
[48] Nietzsche, "On Truth and Lie," 27.
[49] Nietzsche, "On Truth and Lie," 35.
[50] Nietzsche, "On Truth and Lie," 28.
[51] Nietzsche, "On Truth and Lie," 47.

long to science, the province of rational beings who believe in objective knowledge but to art, which Nietzsche likens to dreams.

Nietzsche's designation, "intuitive," like other terms he employs, indicates an anthropologically oriented theory of metaphor. Elsewhere in the text, he even more clearly refers to the "*drive [Trieb]* to the formation of metaphor"[52] as a uniquely human trait, an activity that differentiates humans from other animals. At the same time, Nietzsche does not present the forging of metaphors as a cultural feat so much as a matter of raw biology: as an instinctual urge, which is based on a certain use of the intellect induced through the struggle for survival – namely, the dissimulation through which "weaker individuals" continue their existence. Though it had already been in use in the original state of a *bellum omnium contra omnes*, for civilization to emerge (on the basis of a social contract between human beings requiring a peace agreement), it must yield to stable designations, which represent the first step toward conceptual thinking.[53] In brief, metaphor is attributed to humanity's state of nature, while concepts emerge with the beginnings of civilization.

Another of Nietzsche's theses underscores the proximity of metaphor to raw nature: "Everything that distinguishes man from beast hinges on this capacity to dispel intuitive metaphors in a schema, hence to dissolve an image into a concept."[54] In contrast to Nietzsche's declaration above, the metaphorical drive is not what makes human beings human; instead, metaphor remains in the realm of "beasts," from which humankind emerges only through the formation of concepts. In this light, the conduct of "intuitive man" appears regressive. This category of human abandons the achievements of civilization and yields to instinct, which serves individual self-preservation, and in doing so breaks the social contract. Nietzsche speaks of the "primitive world of metaphor" and envisions a new, but also ancient "culture" of conscious and collective self-deception. The world of semblance this yields is treated (i.e., formed and received) as art. Even though art does not provide the means for attaining "true reality," it is invested with the pathos of first beginnings and great originality in descriptions such as "a mass of images that originally flowed forth hot and liquid from the primal power of human imagination."[55]

---

52 Nietzsche, "On Truth and Lie," 42. Emphasis added.
53 Nietzsche, "On Truth and Lie," 26.
54 Nietzsche, "On Truth and Lie," 31.
55 Nietzsche, "On Truth and Lie," 35.

Mauthner

At the heart of Fritz Mauthner's *Beiträge zu einer Kritik der Sprache* (Contributions to a Critique of Language, 1901) is the theory of metaphor developed in its second volume.

> Language [...] grows by transferring a complete word to an incomplete impression, by comparison, that is – through the eternal act of approximation, the eternal paraphrasing and speaking in images, which constitutes the artistic strength and logical weakness of language. [...] Our language grows through metaphors.[56]

Elsewhere in the study, Mauthner makes a distinction between two types of "emphasis of similarity"[57]: The first is analogy, which subsumes a group of things that appear the same (but are in fact only like each other) under the same word. The second type, metaphor, designates a thing with a word, whose meaning had until that point only been similar to the thing. Both are products of the human imagination, its unconscious workings in the case of analogy and conscious operations in that of metaphor. At the same time, Mauthner constructs a hybrid form of the two by invoking the initial metaphor that cannot reach back to preexisting words, which he understands as a "forging of analogies without self-deception."[58] In this way things evincing similarity are designated by the same word, even though the one naming them is conscious of the difference between them. Such underlying awareness, Mauthner continues, vanishes over the course of time – until the term is no longer perceived as a metaphor and enters the "organism of language" as a "proper" word.[59] That said, at still another point in the study, Mauthner suspends this key distinction when he treats the belief of the ancient poets (a product of the unconscious creation of metaphors), the symbolic work of more recent poets (a product of their conscious creation), and knowledge as one and the same: "Thus ends for us the generic distinction between knowing, symbolizing, and believing."[60]

Like Nietzsche, Mauthner holds that metaphor is not a motivated form of language. In fact, he employs the term whenever he wishes to point out the arbitrariness and conventionality of language. For instance, to criticize the claim

---

[56] Fritz Mauthner, *Beiträge zu einer Kritik der Sprache*, vol. 2, *Zur Sprachwissenschaft* (Vienna, Cologne, Weimar: Böhlau, 1999), 451.
[57] Mauthner, *Beiträge zu einer Kritik der Sprache*, 2: 416.
[58] Mauthner, *Beiträge zu einer Kritik der Sprache*, 2: 416.
[59] Mauthner, *Beiträge zu einer Kritik der Sprache*, 2: 451; see also 414.
[60] Mauthner, *Beiträge zu einer Kritik der Sprache*, 2: 469.

that language arose from the imitation of sounds, he draws attention to the metaphorical character of such imitation: "Because the sounds of both dead and living nature in no way equal the articulations of human language, all these new, imitative creations fall under the category of metaphor."[61] In other words, and as he observes repeatedly, language is characterized precisely not by "natural" but rather "conventional" imitation, which in a strict sense is no longer imitation at all. In his eyes, language is constituted precisely by the *difference* from the original sound and the purely conventional connection between sound and representation.

Accordingly, for Mauthner, the decisive question that an onomatopoietic theory of linguistic origin would have to ask is, "How did human beings – in addition to their ability to realistically mimic sounds of nature – come to reshape these same vocalizations by convention?"[62] The answer he ultimately offers is contingency. He concedes that language developed out of necessity, but insists that the figures it shapes in the process of developing are coincidental.[63] Offering – despite himself – an origin scenario, he gives the following example: "Each primal human being, we may presume, associated for some reason (which we must call coincidental) the chosen or involuntary sound with rolling motion."[64] The connection between speech and imagination is initially a coincidence; only habit and usage lead to a given word ultimately coming to be viewed as the only "right" one.[65]

However, this still does not clarify the question of how humanity could emerge from a condition of speechlessness to the formation of the first metaphors. Because of his epistemological doubt, which is based on his insight into the fundamental metaphoricity of language, Mauthner remains cautious here. He hypothesizes that language derives from three "reflexive sounds" with which human beings in "primeval times" expressed three main affects: wonder, pain, and joy.[66] These verbal reflexes came to be used metaphorically, that is, to refer not only to affect but also to its various possible causes.[67] At the same time, vocalizations function as imperatives directed toward a counterpart who is expected to resolve the affect.[68] For Mauthner, the purpose of lan-

---

61 Mauthner, *Beiträge zu einer Kritik der Sprache*, 2: 420.
62 Mauthner, *Beiträge zu einer Kritik der Sprache*, 2: 436.
63 Mauthner, *Beiträge zu einer Kritik der Sprache*, 2: 488.
64 Mauthner, *Beiträge zu einer Kritik der Sprache*, 2: 521.
65 Mauthner, *Beiträge zu einer Kritik der Sprache*, 2: 523.
66 Mauthner, *Beiträge zu einer Kritik der Sprache*, 2: 439.
67 Mauthner, *Beiträge zu einer Kritik der Sprache*, 2: 339.
68 Mauthner, *Beiträge zu einer Kritik der Sprache*, 2: 441.

guage is not communication for its own sake (and certainly not naming for its own sake), but (as with Malinowski) for the sake of a specific goal that the speaker wishes to achieve:

> That first linguistic vocalization [was] neither a noun, nor a verb, nor an adjective, but already an intention: the wish to suggest something to the other who had food – in this case, the mother. [...] Thus, given the purpose of language, we may suppose [...] that in a certain sense the imperative form is older than the concept of "milk."[69]

Later on, Mauthner expands this thesis by attributing language with a mnemonic function: "The original words [sought, with help of a detail taken from an overall image] to recall the image in its entirety; [...] even developed language [affects] nothing more than the evocation of particularly striking [*belichteten*] memory-pictures."[70] These theses can be connected if they are understood as two successive stages of language development. Mauthner suggests as much when he speculatively describes how language develops from poetry to drama to epic.[71] In this picture, affective vocalization corresponds to lyric (which, strictly speaking, would represent a prelinguistic phenomenon), imperative utterances correspond to drama, and acts of recollection to epic.

Unlike Nietzsche, Mauthner – with his insight into the metaphoricity of language – also abandons belief in a world-in-itself independent from language. This is perhaps also why, unlike Nietzsche, he does not arrive at a euphoric affirmation of the world of semblance, since there is no 'actual world' that language belies. Unlike Cassirer, whom I will discuss below, his insights lead him to adopt a resigned attitude.[72] For, as a scholar, he clings to the ideal of objective knowledge, even though he knows language makes it impossible.[73] Accordingly, because standing concepts are not to be trusted, he is unable to further pursue his own hypotheses about the origin of language[74] and declares them to be poetry at best[75] – which, in this scientific context, was understood as devaluing.

Rarely does any alternative to this stance present itself, and when it does it is only when Mauthner turns his attention to the creative power of metaphor. For instance, he states that "the most general form of metaphor" is what gives us

---

69 Mauthner, *Beiträge zu einer Kritik der Sprache*, 2: 445.
70 Mauthner, *Beiträge zu einer Kritik der Sprache*, 2: 524.
71 Mauthner, *Beiträge zu einer Kritik der Sprache*, 2: 441.
72 Mauthner, *Beiträge zu einer Kritik der Sprache*, 2: 440.
73 Mauthner, *Beiträge zu einer Kritik der Sprache*, 2: 454.
74 Mauthner, *Beiträge zu einer Kritik der Sprache*, 2: 440.
75 Mauthner, *Beiträge zu einer Kritik der Sprache*, 2: 437. Hence the richness of Mauthner's metaphors.

"our reality in the first place, in the form of our vocabulary."[76] For this reason, he refuses to compare the 'primitive language' of children with pathological phenomena. It is not (mental) illness but "flourishing poetic force" that has led such language to develop.[77] In contrast to Cassirer's view, the guiding idea here concerns a completely individual language.[78] Mauthner divides the linguistic development of children into two stages: a first phase in which the child forms random words and thereby creates a "personal original language,"[79] and a second one in which the child says goodbye to this original language and learns the language along with the syntax of adults. This transition amounts to a loss of paradise: "For the first time, the child combines two words into a sentence, thereby losing the paradise of youth; whereas an accidental word still harbored a whole world, acquired language is no longer as majestic."[80]

In contrast to Nietzsche, Mauthner does not emphasize the unique and incomparable qualities of the object that get lost with this paradisical language but rather the uniqueness and incomparability of the speaking subject that disappear in the process of submitting to convention. Just as singular is the world that the speaking subject opens up through his or her individual language. Mauthner imagines, for instance, that in this world "the name 'cake'" might represent "a kind of god, who gives [the child] physical contact with the sweet thing"[81] it desires. In this world, the imperative function of language imbues language with a magical force.[82]

---

[76] Mauthner, *Beiträge zu einer Kritik der Sprache*, 2: 472; see also his discussion of Vico, 2: 484, 488.
[77] Mauthner, *Beiträge zu einer Kritik der Sprache*, 2: 411.
[78] As Magnus Klaue shows, this emphasis on radically individual language also has a political dimension: "Mauthner's theory of language and metaphor runs counter to the *völkisch*-nationalistic call for linguistic purity, which [...] was propagated by institutions such as the Allgemeiner Deutscher Sprachverein and popularized by schoolbooks and light entertainment literature" (Klaue, "Aufbauende Zerstörung," *Sprachkunst. Beiträge zur Literaturwissenschaft* 37 [2006]: 46). "Local linguistic practices and dissident forms of discourse such as poetry" he observes, "undermined compulsive homogenization." In this regard, Mauthner understands "linguistic usage [...] as a field of combat" and its "colonization as illusory as [the possibility of] liberation" (48).
[79] Mauthner, *Beiträge zu einer Kritik der Sprache*, 2: 405.
[80] Mauthner, *Beiträge zu einer Kritik der Sprache*, 2: 406.
[81] Mauthner, *Beiträge zu einer Kritik der Sprache*, 2: 389; see also 402.
[82] Klaue points out that in Mauthner's "method of constructive destruction" metaphor represents the "expression of an insoluble aporia: only to the extent that the critic of language constantly calls his basic assumptions into question can he free himself (and, to a certain extent, language itself) from the 'tyranny of language'" ("Aufbauende Zerstörung," 36). Here lies a utopian potential that is only "rarely perceived, since its critique was mostly viewed in terms of the

As I noted above, Mauthner posits an arbitrary connection between the sounds of speech and the objects they signifiy. On a few surprising occasions, however, he expresses an opposing notion, which can eventually be explained by the longing for a primordial language just described (which would be motivated for its speakers). In this spirit he writes,

> [t]he push for such bold metaphors (such as the transfer of space to time, or color to sound) comes from a compulsion lying in the conditions of the real world, which have not yet been revealed. Language is metaphor, but metaphor somehow corresponds to the world.[83]

Here, Mauthner abandons the metaphor as a world-creating force and comes back to the idea of metaphor as a 'true,' motivated representation of the world, passing from the idea of creation to that of discovery. The example he provides to illustrate this idea is how a small space and a large one are imitated by motions of the glottis and mouth, that is, with a narrow aperture for the "i"-sound and a broad one for "o." The German words for "small" and "large" (*klein* and *groß*) are thought to be motivated by their object by means of mimetic gestures. Mauthner goes as far as to speculate that there might be a fundamental "kinship of substance [*Ding-Verwandtschaft*] between the circumstances of reality and sound"[84] that is responsible for the features of spoken language. In Chapter 9 I will return to this notion in the context of Walter Benjamin's writings.

Vischer and Biese

Friedrich Theodor Vischer's influential essay, "Das Symbol" (1887; "The Symbol," 2015), calls upon the second type of anthropological theories of metaphor. Vischer constructs three stages of development for the symbol, which, in his view, correspond to those of humanity as a whole. In the first phase, the symbolic image and its meaning still coincide (or are confused with each other). In the last stage – that of the present day – they stand clearly separated, for it is now clear to the conscious mind that their relationship is the result of a mediation. The second stage represents a peculiar intermediate position between the two

---

'hatred of language.' Only Gustav Landauer recognized the explosive power of the method of innovative destruction" (37).
83 Mauthner, *Beiträge zu einer Kritik der Sprache*, 2: 453.
84 Mauthner, *Beiträge zu einer Kritik der Sprache*, 2: 454.

and is for Vischer the actual home of language.[85] At this point there prevails an "instinctive and nevertheless free, unconscious and yet in a certain sense conscious ensoulment of nature [*Naturbeseelung*],"[86] which he grounds in anthropological terms: it is in the nature of the human soul to project itself and its conditions into other forms of being.[87] Such empathy (*Einfühlung*) is based on "point[s] of comparison," that is, on involuntary moments of perceiving similarities between humans and the natural world (for instance, when natural forms are regarded as expressive faces). Thus, at this level, the symbolic image and its meaning are perceived to stand apart while still interacting in an intimate relation of kinship. Vischer ascribes a "truth in the higher sense" to this empathetic process, which he again explains on an anthropological basis: empathy is an "essential act of the soul" that derives from the fact, and simultaneously proves that "the universe, nature, and spirit [*Geist*] must be one at root."[88] Poetry represents the preservation of such truth in modern times, for its tropological language sustains awareness of the inner relationship of all being.[89]

In response to Vischer, the literary historian Alfred Biese developed an anthropologically-oriented theory of metaphor in *Philosophie des Metaphorischen* (Philosophy of the Metaphorical, 1893). Brigitte Nerlich and David D. Clarke summarize his position as follows:

> Biese agrees with all those who no longer say that metaphor is an abbreviated comparison. He therefore praises Gerber, but also Wilhelm Dilthey who had written around 1880 [...] that figures of speech are not mere decorations of speech but are an integral part of poetic creativity [...]. Biese declared: "Metaphor is not a poetic trope but an original form of cognitive perception."[90]

---

85 Bernhard Buschendorf points out that the second level mediates between the opposing poles (religious versus rational symbolism) and concerns the aesthetic nature of the symbol, whose "animating effect" is especially apparent in language and, more specifically, metaphor ("Zur Begründung der Kulturwissenschaft," in *Edgar Wind. Kunsthistoriker und Philosoph*, ed. Horst Bredekamp, Bernhard Buschendorf, Freia Hartung, and John Michael Krois [Berlin: De Gruyter, 1998], 230, 229).
86 Friedrich Theodor Vischer, "The Symbol," trans. Holly A. Yanacek, *Art in Translation* 7, no. 4 (2015): 428.
87 On Vischer's aesthetics of empathy and its status in the history of science, see Müller-Tamm, *Abstraktion als Einfühlung*, 214–248.
88 Vischer, "The Symbol," 430; see also 446–447.
89 Vischer, "The Symbol," 446.
90 Brigitte Nerlich and David D. Clarke, "Mind, Meaning and Metaphor," *History of the Human Sciences* 14, no. 2 (2001): 49. Specht stresses that Biese is "at the height and vertex of delivering metaphor from rhetoric, not at the beginning" – which started, in his eyes, with Nietzsche's early works ("Verbindung finden wir im Bilde," 43). Nerlich and Clarke do the same: "Biese stands in

Nevertheless, Biese defines "the metaphorical" rather vaguely, as the "reciprocal transfer between inside and outside."[91] This exchange is said to result from the epistemological dilemma that human beings can only make the "foreign" accessible through "what is fully known, i.e., our inner and outer life,"[92] and at the same time have to rely on symbolic forms to give shape to their thoughts and feelings. Accordingly, Biese claims, humans reach for analogy as the "innermost schema of the human psyche," from which the "metaphorical" arises as the "primary form of perception."[93]

Biese also derives language from metaphor in a double sense: language proceeds metaphorically and is itself a metaphor: "Language is metaphorical through and through: it embodies the spiritual, and it spiritualizes the physical; it is an abbreviated image of analogy of all life, which is based on the reciprocal and profound fusion of body and soul."[94] His concept of rhetorical figure thus represents much more than ornament. It reflects the "primary form of perception," as well as language formation and poetic creation.[95]

Invoking Giambattista Vico, Biese concludes that the language of tropes was not invented by writers but instead involved forms of expression necessary to "prehistoric peoples" that were only perceived in our own times as metaphorical transfer. Thus, lyric's tropological manner of expression is the earlier, authentic linguistic form, whereas the prosaic expression of prose discourse is the later, artificial form. The former is so fundamental for all language that even today the analogies it forges "continually proliferate in linguistic creation [*Sprachschöpfung*],"[96] whether in everyday usage or literature in particular. The difference from the practice of "prehistoric peoples" is simply that poetic works now count merely as beautiful semblance. Biese claims this view of literature is false, however. He adduces an array of ontological and epistemological reasons and critiques standing notions of human understanding and conceptual thought in order to demonstrate that an "eternal truth" inhabits the metaphorical language of poetry, namely the "inner harmony" of nature and spirit and ulti-

---

a long line of thinkers on metaphor," e.g., Scherer, Dilthey, Wundt, Brinkmann, Kohfeldt. "All later readers of these major and minor works agreed on the fact that metaphor could no longer be regarded as a shortened comparison" ("Mind, Meaning and Metaphor," 50).
91 Biese, *Die Philosophie des Metaphorischen*, 15.
92 Biese, *Die Philosophie des Metaphorischen*, 3.
93 Biese, *Die Philosophie des Metaphorischen*, 13, 15.
94 Biese, *Die Philosophie des Metaphorischen*, 22.
95 Biese, *Die Philosophie des Metaphorischen*, 220.
96 Biese, *Die Philosophie des Metaphorischen*, 81.

mately the "divine" as the "creative" force in human beings.[97] Although he fails to provide a convincing argument, the sheer quantity of his claims attests unambiguously to his desire to promote belief in the power of metaphor to his contemporary readers.

Nietzsche, Vischer, and Biese entangle themselves in a fundamental contradiction. According to their shared assumption, all language is based on metaphor. In contrast to Mauthner, they do not draw the consequences from this insight and revise their accounts of a descriptive model of language in favor of a constructivist one – that is, they do not give up on the idea of an extra-linguistic reality that language either represents successfully or not. Instead, and despite their insight into the metaphoricity of all language, they all adhere to a descriptive model of language. In Nietzsche's case, this leads to a fundamental skepticism about our (language-based) knowledge of the world. For Biese (and the young Hofmannsthal, who reviewed his work approvingly[98]), the result is that only the metaphor, not the concept, is capable of disclosing the world-in-itself. Yet this view of poetic metaphor is made possible only by ignoring metaphor's positing nature (*Setzung*). Recognizing the metaphoricity of all language, Nietzsche embraces disbelief; Biese, however, adopts what Genette would call a mimological outlook and hopes to gain access to 'true reality' via motivated metaphors.

The difference of Nietzsche, Vischer, and Biese's theses from those advanced by Malinowski and Mauthner is plain. The latters' ethnological and developmental observations lead to their abandonment of the descriptive model in favor of a positing model of language. For Mauthner, this entails resignation or the sentimental longing for a paradisiacal, personal language. For Malinowski, metaphor represents neither failed linguistic representation of reality (that is to be denied, ignored, or skeptically affirmed) nor its ideal realization; instead, metaphor simply demonstrates the characteristic of all language to change discursive reality in the course of its use. According to Malinowski, indigenous peoples understand

---

97 Biese, *Die Philosophie des Metaphorischen*, 224. Hereby, "creation" does not mean the "invention" so much as the "discovery" of the right words (as in Plato's *Cratylus*).
98 In the review, he agrees so much with Biese's view of metaphor as the root of all thinking and speech that he declares it a commonplace. What Biese's book lacks, in his estimation, is consideration of the process by which metaphors emerge. For Hofmannstahl, it is fueled by drives, laden with affect, and belongs – as he illustrates with numerous metaphors of his own – to the realm of the sublime: a "strangely vibrating state, in which metaphors [...] rain down on us amidst terror, lightning, and storms; in this sudden [...] illumination we sense, for a moment, how the whole world fits together" (Hofmannsthal, "Philosophie des Metaphorischen," in *Gesammelte Werke*, ed. Bernd Schoeller, vol. 7, *Erzählungen, Erfundene Gespräche und Briefe, Reisen* [Frankfurt am Main: Fischer, 1979], 47).

this positing force of language as magic. What is mistaken, in his eyes, is not the belief that language can influence reality but the reduction of this power to an ontological unity of sign and referent. This, however, is exactly what Vischer and Biese do. In this regard, their theories of metaphor are committed to a magical conception of language.

Cassirer

Cassirer occupies a position between these earlier theorists of metaphor (Nietzsche, Vischer, Biese) and Malinowski, even though he may not have been familiar with the latter's works. He also engages intensively with the paradigm of 'primitive thinking.' *Sprache und Mythos* (*Language and Myth*, 1946), which appeared in 1925, examines the role of language in what he calls "mythical thinking." The study advances the claim that language did not arise from myth, and myth did not come from language. Instead, both derive – in a phenomenological sense more than an historical one – from one and the same root: "It [the form of spiritual/mental conception (*geistige Auffassung*) in myth and language] is the form which one may denote as *metaphorical thinking;* the nature and meaning of metaphor is what we must start with."[99]

Like Nietzsche and Biese before him, Cassirer does not speak of metaphor in the Aristotelian sense, but of "radical metaphor,"[100] which translates "cognitive or emotive experience" into sounds and mythical forms. This primeval transfer is the condition for all concept formation in language and myth and lives on as the principle of transfer in linguistic and mythical thinking. In either case, "the law of the leveling [...] of specific differences" prevails; "every part of a whole is the whole itself; every specimen is equivalent to the entire species."[101] Thus, for Cassirer, the Aristotelian definition of one of the main types of metaphor (as *pars pro toto*) logically arises from mythical thinking as he defines it but with the key difference that what represents a mere figure in formal rhetoric means "real identification" for both myth and language:[102] "whatever things bear the same appellation [must] appear absolutely similar."[103] Indeed, by means of metaphorical usage, language can have a retroactive effect on myth: "If the visible image of

---

[99] Ernst Cassirer, *Language and Myth*, trans. Susanne K. Langer (New York: Dover, 1953), 84. Emphasis in original.
[100] Cassirer, *Language and Myth*, 87. The term comes from Max Müller ("Metaphor," 377).
[101] Cassirer, *Language and Myth*, 91–92.
[102] Cassirer, *Language and Myth*, 92.
[103] Cassirer, *Language and Myth*, 95.

lightning, as it is fixed by language, is concentrated upon the impression of 'serpentine,' this causes the lightning to *become a snake*."[104]

Up to this point, Cassirer's view still seems compatible with those of Vischer and Biese. Yet he decisively parts ways with them by considering those who relate tropological speech to a longed for world-in-itself as followers of a "naïve realism that regards the reality of objects as something directly and unequivocally given"[105] – a judgment (as my explanations above have shown) that bears on both Vischer and Biese as well as Nietzsche, *ex negativo*. By contrast, Cassirer stresses the positing power of language; one must "see in each of these spiritual forms a spontaneous law of generation; an original way and tendency of expression." From this perspective, language, myth, and art are symbols "in the sense of forces each of which produces and posits a world of its own." At the same time, they perform a descriptive function inasmuch as "in these realms the spirit exhibits itself in that inwardly determined dialectic by virtue of which there is any reality, any organized and definite Being at all."[106]

Not stopping at this insight, Cassirer examines how the first positings (*Setzungen*) and therefore the "genesis" of "primary linguistic concepts" came into being. By his account, they were created in a process that began with an emotional shock to the "primitive" consciousness through an encounter with an existentially significant object or event. Much like what Mauthner describes, this shock, i.e., the sensory content connected to it, was objectified and took the shape of an expressive utterance that continues even after the affect subsided. This first utterance is what Cassirer calls "radical metaphor," the condition for all further language formation.[107]

He furthermore extrapolates the belief in the unity of sign and referent from the moment of shock, an express rejection of claims based on the suggestive or imperative power of speech (whereby he might be referring either to Mauthner or to Malinowski). His point of departure is that, during the moment of shock, "sensory content" forcefully seizes "primitive" consciousness and "reign[s] over practically the whole experiential world." Under such conditions, the radical meta-

---

104 Cassirer, *Language and Myth*, 96.
105 Cassirer, *Language and Myth*, 6.
106 Cassirer, *Language and Myth*, 8.
107 Nor is that all. As Birgit Recki observes in her incisive study, Cassirer's "definitions of transfer into another medium and of pars pro toto make every punctuation mark a semantic fulfillment of sensory experience. With that, the 'form of metaphorical thinking' is described, which lies at the foundation of all symbolic forms. [...] The radical metaphor is the functional principle of culture sought by the question about the unity of symbolic forms" (*Cassirer, Grundwissen Philosophie* [Stuttgart: Reclam, 2013], 40–41).

phor fuses with its content into "an indissoluble unity."[108] In contrast to Vischer and Biese, Cassirer does not accept that there is an analogy-based relation between radical metaphor and the sensory content connected with the shock. Instead he draws on affective theories of 'primitive thinking' that appeal to a physiologically motivated connection between the two that originates in the act of expression. In this way, he defines radical metaphor as a (physiologically and affectively) motivated positing that propagates in linguistic thought and action as the principle of transfer.

Cassirer acknowledges that language and myth drift apart in the course of development, so that in his era language only commands its metaphor-forming force in the literary sphere. But it does not enact this force through the persistence of mythical thinking, as Vischer and Biese would have it. Rather, the word forms itself "into artistic expression."[109] Cassirer, then, does not affirm the revival of mythical identity between sign and referent or the (supposed) truth of myth that this implies.[110] For him, literary works constitute a "world of illusion and fantasy" in which – contra Nietzsche (and Mauthner) – "the realm of pure feeling can find utterance, and can therewith attain its full and concrete actualization" and, most importantly, the "mind [*Geist*]" learns to understand language as "its own self-revelation"[111] (specifically, I should add, a revelation of its affective and physiologically motivated *poietic* activity, which at the same time guarantees human cognition). Understood along Foucauldian lines, this might be taken to mean that in modernity the magical conception of language returns in modified form; it is no longer oriented toward the inner connec-

---

108 Cassirer, *Language and Myth*, 58.
109 See also Schwennsen, who writes, "Through this break, which Cassirer identifies between reflected and unreflected representation, art and myth are separated" ("Kunst und Mythos zwischen Präsenz und Repräsentation," 215). And moreover: "The separation between myth and art begins, according to Cassirer, at the point where aesthetic expression goes beyond a spontaneous outpouring of powerful sensations. Art is not just expressive, but also form-giving and constructive and represents, in its way, the 'dynamic process of life itself' that goes missing with the overcoming of myth" (216).
110 Cassirer, *Language and Myth*, 99. For this reason, I have reservations about Hirschkop's recent claim that – unlike "those who thought that myth was a danger" (e.g., Ogden and Richards, Orwell, Bakhtin, Frege) – Cassirer belongs to a group of thinkers (among others, Viktor Shklovsky, Roman Jacobson, and Walter Benjamin) "who welcome language's mythical inclinations" (*Linguistic Turns*, 162) and for whom "myth remain[s] an ineradicable feature not only of religion, but of every other symbolic form as well – science, art, language – perpetually threatening their progressive achievements" (198).
111 Cassirer, *Language and Myth*, 99.

tion between language and things, but between language and its speakers. As Foucault writes,

> [l]anguage in the nineteenth century [...] was to have an irreducible expressive value [...] for, if language expresses, it does so not in so far as it is an imitation and duplication of things, but in so far as it manifests and translates the fundamental will of those who speak it.[112]

This chapter has shown that around 1900, anthropologically oriented theories of metaphor emerged in domains adjacent to ethnology and developmental psychology that undermined the Aristotelian distinction between proper and improper speech. Metaphor, these theorists were convinced, cannot simply be replaced by the 'proper' word; the content it transports can be expressed only in that one way and none other. Similar views had been voiced a century earlier, but what was now new was the historical-genealogical and especially the supposedly empirical basis of the human sciences, which lent the idea broad currency and persuasive force. Now, metaphor was no longer mere rhetorical ornament. Verified by science, it gained the status of a 'transcendental a priori' anterior to all thinking and speaking.[113] Wolfgang Riedel has described this turn as the beginning of an unprecedented reevaluation of metaphor that continues to this day.[114] Yet, certainly, there is an essential difference between it and today's theories of metaphor. As we saw on the preceding pages, many theorists at the turn of the century – Malinowski and Cassirer excepted – paradoxically still clung to the ideal of language motivated by extralinguistic reality. Either they skeptically viewed metaphor as the most original but nevertheless contingent sign that was capable, at best, of establishing a culture of semblance, which was then to be euphorically embraced or endured with resignation. Or, citing anthropological and epistemological reasons, they elevated metaphor mimologically to the position of a natural sign. This outlook fit with the rather conservative but extraordinarily successful model of *Dichtung*, or poetry and literary arts in general, as a means of disclosing 'true reality' – a quasi-magical way of (re)discovering greater intimacy between human beings and the world they inhabit.

---

112 Foucault, *The Order of Things*, 290.
113 Riedel, "Arara ist Bororo," 238.
114 See Riedel, "Arara ist Bororo," 238–241.

Part Three: **'Primitive Thinking' in German Literary Modernism**

Chapter 7
The "Tropological Nature" of the Poet in Müller and Benn

Robert Müller's novel *Tropen. Der Mythos der Reise. Urkunden eines deutschen Ingenieurs* (Tropics. The Myth of a Voyage. Documents of a German Engineer, 1915) and texts by Gottfried Benn show how closely 'primitive thinking' and metaphor are connected in primitivist discourse of the early twentieth century. Müller's novel, to which the bulk of the following chapter is dedicated, revolves around the homonymic quality of its title, *Tropen*. The text constructs multiple connections between the jungle (the tropics, or in German, *Tropen*) as the home of the 'primitive' and the linguistic figure of transference (tropes, or in German, *Tropen* as well) along the course of its protagonist's quest for both origins and futurity. In the final part of the chapter, I attend to Gottfried Benn's early work and poetological reflections during the early 1930s. In these texts, Benn is also concerned with the polyvalent tropics, whose primeval vegetation, however, he shifts to the human body, which figures as an inscription of the archaic.

A Biological Reverie

Schiller's declaration, "they are what we were," which I discussed at length in the introduction, echoed throughout the cultural history of the nineteenth and early twentieth centuries. This formula took on a new form in the context of the human sciences that developed at that time and that understood man as a historical being. A century earlier, equating non-European peoples, children, or animals with the supposed origins of humanity was nothing more than an analogical operation. But in the context of the human sciences, this analogy soon acquired the status of fact inasmuch as ethnology, developmental psychology, and psychopathology now claimed to have empirical evidence that qualities of the original species were *in fact* present in indigenous cultures, children, and the mentally ill. Such arguments were supported by temporal models based on the notions of survival, the recapitulation of phylogeny in ontogeny, and regression.

The models of survival and recapitulation, in particular, reduced the distance between the modern era and the presumed first beginnings of humanity. Not only did indigenous peoples as a whole represent a survival of the original state, but such survivals abounded even in modern Europe according to the social anthropologist E. B. Tylor. "In our midst," he writes, one still encounters numerous "primaeval monuments of barbaric thought and life."[1] For Tylor, these survivals involve particular collective patterns of behavior, whereas early twentieth-century individual psychology was interested in survivals of an archaic psyche that could affect not only entire cultures – as in Jung's theory of a collective unconscious – but also adult individuals who were seen to retain the patterns and contents of the child's psyche. Moreover, according to the theory of recapitulation, the child's mind corresponded to that of archaic man, and therefore just a few years, not millennia, stood between modern Europeans and the 'barbarous' origins of their culture. Psychoanalysis furthermore held that many components of this childlike, ancient mind were still present in the modern adult and could dominate mature thinking in the process of regression.

Tylor presented an idealist argument inasmuch as he traced the survivals back to passed-down cultural traditions. Psychologists, by contrast, tended to hold the materialist position that an organic memory exists, thus producing a biological version of "they are what we were." As demonstrated in Chapter 3, this perspective is closely related to Ernst Haeckel's popular theories, specifically on biogenetic law, according to which "*ontogenesis is a brief and rapid recapitulation of phylogenesis.*"[2] Haeckel maintains not only that humans, as a species, can look back on an ancestral line reaching back to single-cell organisms but furthermore that the life of each individual, which begins with a single cell, repeats the course of phylogeny as a whole. He suggests that the past lives on in the present in multiple ways. Following Jean-Baptiste de Lamarck, Haeckel posits a mnemonic connection between his contemporaries and their ancestors and is convinced that traces of their experience are archived in the biological achievements that they have inherited from their non-human forebears. From this biogenetic perspective, the "they are" in the phrase "they are what we were" refers not just to unicellular organisms (and even inorganic life) but also to organs and physiological processes in human beings of the present. They originate in earlier stages of development and store the experiences that led to their development in the first place. Thus Haeckel regards his contemporaries as possessing organic memory: not only are "they" what "we were," but rather *we still are* – through our bodies – *what they were.*

---

[1] Tylor, *Primitive Culture*, 1: 19.
[2] Haeckel, *The Riddle of the Universe: At the Close of the Nineteenth Century*, trans. Joseph McCabe (New York: Harper and Brothers, 1900), 81. Emphasis in original.

Examples of this assumption abound in the human sciences of the early twentieth century, each stranger than the next. Thus, the paleoanthropologist Edgar Dacqué writes:

> At first, [humans] must have possessed amphibious and reptilian-looking features. Perhaps he had the sluggish gait of amphibians and webbed fingers and toes [...]. With [...] reptiles he may have shared a partially horny armored body [...]. But with both groups hypothetical prehistoric man probably had [in common] a fully developed parietal organ, i.e., a fully developed eye-like opening on the top of the skull.[3]

According to this anti-Darwinian position, humans did not become humans at the end of phylogenesis but were humans from the very beginning, and the various animals split off from this archetypal form. It follows that the organic memory of humankind reaches back into the most ancient times. Dacqué speaks of this especially when discussing the parietal eye of primordial "man," a feature connected to an "even older stage than that of the primeval amphibian or fish"; its traces, he maintains, may still be found in the human being of the present in the pineal gland, which he identifies as a receded pair of eyes.[4] Dacqué devotes so much attention to the parietal eye's survival because he wishes to rouse the "essentially intellectual man" of his own time to reactivate his "natural vision" (*Natursichtigkeit*), which, in his estimation, represents the "oldest state of mind" and is directly linked to this (supposed) organ.[5] From this reactivation, he proclaims, with prophetic pathos, the coming of a "great world epoch with new mental and physical possibilities that will emerge as we discard the cerebral intellectual state."[6]

The scientific community welcomed and elaborated upon Dacqué's claims that primeval qualities have survived and can be reactivated. Thus, his colleague Eugen Georg asserts that the human organism still contains "about 200 ancient organs," at least in rudimentary form.[7] Inasmuch as every feature of the human body must serve a purpose, he interprets these remnants as sites of possible revival:

---

[3] Edgar Dacqué, *Urwelt, Sage und Menschheit. Eine naturhistorisch-metaphysische Studie* (Munich: Walter de Gruyter, 1924), 70–71.
[4] Dacqué, *Urwelt, Sage und Menschheit*, 73.
[5] Dacqué, *Urwelt, Sage und Menschheit*, 250, 232.
[6] Dacqué, *Urwelt, Sage und Menschheit*, 250.
[7] Eugen Georg, *Verschollene Kulturen. Das Menschheitserlebnis. Ablauf und Deutungsversuch* (Leipzig: Voigtländer, 1930), 149.

> For if all these features (whenever they were acquired) – amphibian-like internal organs, reptilian ornaments [*Reptilienrequisiten*], parietal eye, mammalian rudiments – were constantly dragged along, smuggled through all biological ages in fragments [*Organtorsi*], doesn't it look as if there was some higher intention behind it all, [...] to be activated again one day if necessary?[8]

Reactivation will achieve synthesis in what Georg calls "Quintenary Man," when the "dreamlike-elementary experience of the world" of "Tertiary Man" and the "gnawing intellect" of "Quaternary Man" give way to "prophetic wisdom."[9]

In light of such speculative paleo-anthropological theories, the reason for my reference to "reverie" in the subtitle to this chapter section should be clear. As noted in the introduction, the term comes from Gaston Bachelard, who uses it to point out that scientists are rarely motivated by an objective attitude toward what they examine so much as they are guided by affects, needs, and ideas.[10] (We can leave the question open as to which emotional needs might have motivated these paleoanthropologists. Yet, all of the following are evident in their megalomaniac heralding of a new age and unwillingness to subject speculations to critical examination: the narcissism of viewing humanity as not only the crowning achievement but also the starting point of creation, the possessive need to ascribe special abilities to human beings, and a scientific will-to-power.[11]) Bachelard also uses "scientific poetry" as a synonym for reverie; its forms of expression and procedures, he stresses, are governed by *poèmes*, not *theorèmes*.

The pattern of argument underlying "they are what we were" functions as one such *poème*. In contrast to what scientific writings purport, this formula was not a matter of discovery and subsequent verification; instead, it shaped broad swaths of European cultural historiography (albeit in versions that varied according to their respective epistemic contexts). It should not be understood as a scientific insight so much as a template for speculative knowledge formation: a scheme into which conjectures about human origins are integrated time and again. Furthermore, texts like Dacqué's can also be said to share an affinity with poetic reverie because they take up literary works of the past – especially accounts of myth – and formulate hypotheses on their basis. In other words, these sources are no longer understood as fiction but as historical records

---

8 Georg, *Verschollene Kulturen*, 149–150.
9 Georg, *Verschollene Kulturen*, 150.
10 Bachelard, *The Psychoanalysis of Fire*, 6.
11 Gaston Bachelard, "Scientific Objectivity and Psychoanalysis," in *The Formation of the Scientific Mind*, trans. Mary McAllester Jones (Manchester: Clinamen, 2002).

from the past, "a time," as Dacqué puts it, "when myths were still experiences, i.e., real."¹²

Conversely, speculative science and its *poèmes* correspond to literary texts that make massive use of science and its daydreams in order to pick up on them, reflect on them, and possibly spin them further. This is exemplified, as the next sections will show, by the work of Robert Müller and Gottfried Benn.

The Tropics

In Robert Müller's novel *Tropen*, the plot proper starts in the second chapter with a voyage down a river – a primitivist topos since Joseph Conrad's *Heart of Darkness* (1899). In Conrad's novella, Marlow tells his companions on a ship anchored on the Thames how once, attracted to the last uncharted "blank spaces" on his map of the world, he entered the employ of a Belgian trading company and captained a river boat up the Congo.¹³ He experiences the journey as one into the "heart of darkness" – to an encounter with Kurtz, the head of an ivory trading station, who for Marlow epitomizes the "darkness" that marks the barbarous conduct of European colonializers. Conrad's story has been read as a criticism of the methods and effects of colonization in the Congo Free State in particular and of the hypocrisy of the supposed civilized world more generally, which, under the pretext of philanthropy, indulges a brutal appetite for power and wealth. At the same time, readers have noted the racist portrayal of the native population, who are not only depersonalized (becoming an anonymous mass) and dehumanized (portrayed as suffering creatures), but also used to represent the natural state of "darkness" and "horror" that lurks within all human beings and resurfaces in the colonizers under the effects of the wilderness.¹⁴ It hasn't been unequivocally proven that Robert Müller read *Heart of Darkness*, but he is supposed to have counted Conrad among his favorite au-

---

12 Dacqué, *Urwelt, Sage und Menschheit*, 35. This is also connected with "Henri Bergson's 'commitment' to the intuitive method" (Marcus Hahn, *Gottfried Benn und das Wissen der Moderne* [Göttingen: Wallstein, 2011], 604).
13 Joseph Conrad, *Heart of Darkness*, ed. Owen Knowles and Allan H. Simmons (Cambridge: Cambridge University Press, 2018), 8.
14 First noted by Chinua Achebe, "An Image of Africa: Racism in Conrad's *Heart of Darkness*," in *Heart of Darkness: An Authoritative Text, Backgrounds and Sources, Criticism*, ed. Robert Kimbrough (New York: Norton, 1988).

thors,¹⁵ and as Matthias Lorenz has recently demonstrated in a comparative reading of the two works, it seems very likely that he knew *Heart of Darkness* as well. Thus, the charismatic American Slim, who incites Brandlberger, the first-person narrator of *Tropen*, to search for gold in the jungle, can be read as a "revenant of Conrad's Mr. Kurtz" inasmuch as he "shares several characteristics with him and bears his name in 'diagonal' translation," i.e., from the German word *kurz* (meaning short) to the English "slim."¹⁶

Müller's connection to another pioneer of literary primitivism is even more pronounced. As Monica Wenusch and others have shown, Müller engaged intensively, from at least 1913, with the writings of the Danish author Johannes V. Jensen. The latter's *Skovene* (Forests, 1904; published in German as *Wälder* in 1907) "clearly, deeply, and demonstrably"¹⁷ influenced his novel of the tropics. Jensen's narrative also begins with a river journey that takes a European explorer into the tropical forests of Birubunga – which amounts to a trip back to a primeval time and simultaneously into his own interior: "into a perspective that had been forgotten, but which I knew."¹⁸ The explorer's clichéd intentions in striving for heroic masculinity – by climbing a hitherto unconquered mountain and killing a tiger – fall flat, and scholars have persuasively interpreted them as parody used intentionally by Jensen to ridicule the naïve yearning for adventure among his contemporaries, caught up in their own fears and fantasies, and to tell them to instead "explore their own psyche."¹⁹

In terms of both plot and character, Müller's novel displays many similarities with its two predecessors.²⁰ However, its formal complexity and reflection on language and literary craft, which radicalizes Jensen's insight into the imaginary nature of colonialist journeys into the jungle, go further still. Let us first return to the river journey. What Conrad and Jensen only hint at is made explicit in Müller's novel: the voyage follows the scheme of regression into ontogeny and phylogeny, which ultimately determines the entire plot.²¹ In its beginning, Brandl-

---

15 See Matthias N. Lorenz, *Distant Kinship. Entfernte Verwandtschaft: Joseph Conrads "Heart of Darkness" in der deutschen Literatur von Kafka bis Kracht* (Stuttgart: Metzler, 2017), 182.
16 Lorenz, *Distant Kinship*, 187.
17 Wenusch, Monica, *"… ich bin eben dabei, mir Johannes V. Jensen zu entdecken…" Die Rezeption von Johannes V. Jensen im deutschen Sprachraum* (Vienna: Praesens Verlag, 2016), 275.
18 Johannes V. Jensen, *Wälder*, in *"Die Welt ist Tief." Novellen* (Berlin: Fischer, 1912), 168. See Wenusch, *"… ich bin eben dabei,"* 275–297, for a comparison of *Tropen* and *Skovene*.
19 Volker Zenk, *Innere Forschungsreisen. Literarischer Exotismus in Deutschland zu Beginn des 20. Jahrhunderts* (Oldenburg: Igel, 2003), 78.
20 See also Zenk, *Innere Forschungsreisen*, 103–111.
21 In keeping with the discourse of the day, which associates this mode with insanity, the travelers are later called "madmen." Cf. Thomas Schwarz, "Robert Müllers *Tropen* (1915) als neuras-

berger, dozing on his boat, has the impression "of [having] already [...] experienced all this once before"[22] and arrives suddenly at the following insight:

> In the depths of my consciousness, in the mines of my origins, there slumbered the mood, from prehistoric times, of millions of beings; the maternal suckle and flow of the stream, the incubating warmth of the surroundings, the solicitous calm of idleness had coaxed my primal instincts. How long it had been: ... twenty-three years and nine months ago, my life as one of those gristly cells had reached its peak. My identity with this state was established. Down at the bottom of these viscous fathoms dwelt beings whose dear friend I once had been.[23]

The journey up the jungle is experienced as a return to both an individual and human past. Time has stood still here, so that what has long since passed in Brandlberger's own life and human history is still present. Foreign space is reinterpreted as the survival of the self's origins. Brandlberger finds that the swampy river of the jungle has preserved the initial state of phylogeny, which had been recapitulated in Brandlberger's prenatal existence:

> In ancient times, stem cells settled all over these jungle puddles, scurried greedily along the edges of alien growths, let their pennated antennae flutter under the intermittent gush of waters flowing together and fished with sinewy muscles for other organisms [...]. These forms of life all around [...] once were me.[24]

This passage shows the influence of Haeckel, whose narrative of human development and assertion of the biogenetic law likewise follows a double (onto- and phylogenetic) regression back to the "stem cell" and "a corresponding, unicellular ancestor, a [...] Laurentian protozoon."[25]

---

thenisches Aufschreibesystem," in *Neurasthenie. Die Krankheit der Moderne und die moderne Literatur*, ed. Maximilian Bergengruen, Klaus Müller-Wille, and Caroline Pross (Freiburg: Rombach, 2010).
22 Robert Müller, *Tropen. Der Mythos der Reise. Urkunden eines deutschen Ingenieurs. Herausgegeben von Robert Müller. Anno 1915*, ed. Günter Helmes, 3rd ed. (Hamburg: Igel Verlag Literatur & Wissenschaft, 2010), 17.
23 Müller, *Tropen*, 19.
24 Müller, *Tropen*, 19–20.
25 Haeckel, *The Riddle of the Universe*, 84. Müller was familiar with Haeckel's biogenetic law; cf. his letter dated 4 June 1912, in which he calls his own psyche a "well-preserved record" (*Abschichtungsexemplar*) (in Robert Müller, *Briefe und Verstreutes* [Paderborn: Igel, 1997], 50); see Christian Liederer, *Der Mensch und seine Realität. Anthropologie und Wirklichkeit im poetischen Werk des Expressionisten Robert Müller* (Würzburg: Königshausen + Neumann, 2004), 66; and Thomas Schwarz, *Robert Müllers Tropen. Ein Reiseführer in den imperialen Exotismus* (Heidelberg: Synchron, 2006), 69.

Indeed, Haeckel's considerations also explain why Brandlberger can remember his prenatal and phylogenetic past. For Haeckel, humans' remote unicellular ancestors are still present in multiple ways in modern men. To begin with, the life of the human organism depends on that of the cells constituting it. More still, Haeckel follows Lamarck in positing the existence of a mnemonic connection between man and his ancestors. Even on the most archaic level, "sensations may leave a permanent trace in the psychoplasm, and these may be reproduced by memory."[26] Accordingly, the character portrayed by Müller, a reader of Haeckel, is able to experience the memory, induced by nature in the tropics, of having existed as a "nibbling bundle of cells [...] in the water."[27]

The jungle does not only preserve life on the cellular level. Brandlberger also rediscovers later stages of the evolutionary "scale"[28] – for instance, he sees "his nerves' mode of life" in the panther and the victory of his "democratic nerves" over the "cosmic principle of satiated spiritual calm" (*Weltprinzip der fetten Seelenruhe*) in the butterfly.[29] Elementary characteristics of his own nature are embodied by animals living in this habitat, which he interprets as survivals of corresponding phylogenetic stages of human development. This embodiment provides the basis for Brandlberger's participatory experiences with these animals, which involves an anthropomorphization of animals and conversely the zoomorphization of humans.[30]

The "Indians" of the novel play a particularly important role in Brandlberger's recapitulation of phylo- and ontogenesis. The three jungle travelers in the novel live with them for a spell, and their shaman, Zana, accompanies them until the end of the journey. For Brandlberger, the indigenous people represent an evolutionary stage of life geared entirely toward sensuality and carnal pleasure, in particular. For this reason, he calls them "priests of the senses."[31] The priestess of this priestly people is Zana, who thus embodies her people and

---

26 Haeckel, *The Riddle of the Universe*, 117–118.
27 Müller, *Tropen*, 20.
28 Liederer identifies three developmental stages in Müller's work: vegetative, primitive man; rational and civilized man; and the new man. He finds five corresponding dimensions of perception: being (a line) and stationary space (a plane or surface) for the first stage, depth and time for the second, and absolute consciousness transcending time and space (language, mind, image, paradox) for the third (*Der Mensch und seine Realität*, 135–174).
29 Müller, *Tropen*, 89.
30 For a thorough discussion, see Liederer, *Der Mensch und seine Realität*, 113–123.
31 Müller, *Tropen*, 59. Schwarz demonstrates that the practices Brandlberger observes and adopts are sadomasochistic and situates them in the context of contemporary sexology (*Robert Müllers Tropen*, 175–193).

their principle of life.³² Almost more important than her priesthood, however, is the fact that she is both an "Indian" and a woman. In Brandlberger's eyes, woman – in contrast to man – "has never left the tropics"; in other words, women still inhabit an earlier stage of human history. Zana therefore proves to be doubly 'primitive,' predestined to represent sensuality and sex.³³

As in his encounters with the jungle river and animals, Brandlberger rediscovers a part of himself among the people living there. The "Indians" remind him of his "most audible wishes, physical desire."³⁴ Life in their midst soon transforms the white men into "barbarians" who indulge their instincts: "Barbarian forms of life acted on patented rights; where spiritual hollows stretched wide, dark movements occurred, frothing out from deserts of blood. Primordial forces began to stir."³⁵ At the end of the expedition, in the company of supposed cannibals, Brandlberger finally rids himself of all the remaining inhibitions European culture has imposed on him.

> I [...] became familiar with a stage where the primal drives of man, hunger, and love become to a certain degree identical. My heightened nervousness mobilized all the ancient dispositions harbored within. It overturned inhibitions put in place by millennia of culture, to which a chain of thirty generations had held firm.³⁶

Although Brandlberger fails to notice as much, such behavior on the part of the travelers reveals a difference between the "neo-barbarians" and the tribespeople. For the "Indians" have at their command a particularly elaborate – and therefore "cultivated"³⁷ – understanding of sensuality. Accordingly, the novel speaks of their "physical refinement," their art of extracting "the honey of bodily presence from life,"³⁸ and "physiological enlightenment."³⁹ The rebarbarized Europeans, on the other hand, live out raw, uncultivated drives – which

---

32 See Liederer, *Der Mensch und seine Realität*, 51.
33 Zana represents the pars pro toto not only of her people, but also of the jungle, which Müller codes as feminine – or, more precisely, maternal (with the ambivalence this entails). See Schwarz, *Robert Müllers Tropen*, 68–70. Compared to the female characters in Jensen and Conrad (who belong to the scenery, as it were), Zana plays a much more active role, a variant of the femme fatale. On the role of sexuality in Müller's novel (also in the context of literary primitivism), see Eva Blome, *Reinheit und Vermischung*, 164–189.
34 Müller, *Tropen*, 89.
35 Müller, *Tropen*, 51.
36 Müller, *Tropen*, 280.
37 Müller, *Tropen*, 72.
38 Müller, *Tropen*, 58.
39 Müller, *Tropen*, 72; Müller-Tamm draws attention to the "organized social system" of the jungle inhabitants (*Abstraktion als Einfühlung*, 351).

lead to three mysterious sexual murders.⁴⁰ Nonetheless, Brandlberger equates this behavior of the "rebarbarized" with the ways of the "Indians." He identifies the "Indians" (or their stand-in, Zana) with the "sensuality of the [jungle] nature" and the "terrible, confusing drive" also raging in the "white man" when he states, "I think of *drive,* the tropics in the nature [*Gemüt*] of the white man."⁴¹ Because of this misidentification and the resulting misunderstanding of the culture of the jungle inhabitants, the travelers' expedition can only fail. They violate the customs of their hosts when one of their number gives free rein to his lust, interrupts a ritual dance between Zana and the chief, and challenges the latter to a duel – in which he dies. Here (and in contrast to Conrad's novella), the text is wiser than its homodiegetic narrator in that it provides the reader with insight into the culture of the jungle people and into the Europeans' mistakes that it denies Brandlberger.

Other passages also make it clear that the novel adopts a critical stance toward its narrator, Brandlberger. This includes the failure of the three Europeans' hunt for treasure and of Brandlberger's project to bring forth a new human race. As the putative editor, "Robert Müller," writes in the foreword, Brandlberger was unable to found the "Freeland colony" that he intended to establish and was killed in an "Indian uprising."⁴² Moreover, the editor's prefatory remarks identify the explorer as an outdated – and not particularly likeable – "type":

> Hans Brandlberger was a young man of the dawning twentieth century and quite like all other young people of that ancient time, [...] without any real talent or character, indeed hardly a spiritual person [...], too lax and troublesome [...], petty, [...] amoral – always a little angry and irritated with himself.⁴³

---

40 Accordingly, Schwarz's diagnosis of sadomasochistic practices (*Robert Müllers Tropen*, 175–193) should be qualified: the "Indians" have cultivated them, whereas the travelers – who are inexperienced in this regard and have repressed such desires until now – lose control when they succumb to them; in consequence, their actions lead to murder-rape.
41 Müller, *Tropen*, 26. Emphasis in the original.
42 Werkmeister therefore speaks of the novel's postcolonial perspective (*Kulturen jenseits der Schrift*, 370). This argument finds support in the parody of typical gestures of conquest, which Schwarz observes by reading Theodor Koch-Grünberg's Amazon ethnographies alongside the novel (*Robert Müllers Tropen*, 107–115).
43 Müller, *Tropen*, 7–8.

Thus, from the outset readers are encouraged to adopt a skeptical attitude toward the narrator of the supposed travelogue, in keeping with the editor's own "suspicion."[44]

The relationship between Brandlberger and the story he tells of the jungle inhabitants is characterized by othering and nostrification, features also found in the works of Conrad and Jensen.[45] On the one hand, the "Indians" – especially at the beginning of the encounter – are denigrated as "animals" and rigorously set in opposition to the white men, who put on the airs of infinitely superior "masters." At the fore stands their distancing from the natives, with whom as few common traits as possible are acknowledged: "It would have been embarrassing to find our equal [einen Duzbruder] among these beasts."[46] Brandlberger retains this bearing to the very end, seeing a wildcat in Zana and fancying himself a "new man" facing this "primeval woman" (Urweib).[47] At the same time, however, the "Indians" are also nostrified to the European narrator inasmuch as he understands them as the survival of an earlier stage of human development and as representatives of the lost characteristics of modern-day men. Thus, it is not the case that the civilized Europeans discover the sophistication of tribesmen; instead, they discover their own latent barbarism.

But othering and nostrification do not contradict so much as complement each other, for the latter concerns strangers who are already othered. They are animals from Brandlberger's perspective, but Europeans once were animals, too – and still are, beneath the surface. Nostrification does not involve approaching or relating to the tribespeople, then; instead, the Europeans project an estranged version of themselves onto them, a tendency that Brandlberger occasionally recognizes in moments of reflection. The process of nostrification culminates in identification. Indeed, Brandlberger asserts in the final sentence of the novel, "I am the tropics." [48] This returns to themes already introduced at the start of the novel, when he reflects on his "identity with" the "condition" of the jungle river.[49]

Identification in the novel follows the pattern of appropriation by continuing the imperialist gesture of the "master's" superiority. It is connected with the assertion that "northern man" is the proper carrier of the tropics: "He, the north-

---

44 Müller, Tropen, 8. For discussion of the foreword, see Dietrich, Poetik der Paradoxie. Zu Robert Müllers fiktionaler Prosa (Siegen: Carl Böschen, 1997), 17–22.
45 On the double strategy of othering nostrification, see Michael C. Frank, "Überlebsel," 160.
46 Müller, Tropen, 48. On animalization, see Liederer, Der Mensch und seine Realität, 113–119.
47 Müller, Tropen, 276.
48 Müller, Tropen, 283.
49 Müller, Tropen, 19.

erner [*Nordländer*], is much more southern [...] in his instincts than the southernmost race."[50] At the same time, identification is accompanied by a distancing gesture: the journey into the jungle was supposedly only made for the purpose of study, to observe the "primordial existence" of humankind. For the European traveler seeks to appropriate this origin only so that primeval sensuality and civilized rationality may achieve synthesis in a "new man [*neuer Mensch*]."[51] As Brandlberger declares: "We are conquering the savage [...]. Now, we're taking back for ourselves what we had traded for our brains, but without giving anything back. We're holding on to our possessions."[52]

Similar to Brandlberger's position, Müller vehemently criticizes exoticism in this novel and elsewhere. Yet at the same time, or at least in the early 1910s, he had been a passionate supporter of imperialism. The idea that the "new man" is supposed to be a hybrid of the inhabitants of the jungle and Europeans does not mean – either for Brandlberger or the author – that the foreign is accepted on its own terms, or that the categories of self and other are deconstructed (as has been claimed[53]). On the contrary, this concept of the "new man" stands under the sign of imperial ambitions. As Thomas Schwarz has demonstrated, Müller understands "hybridization as an imperialist project."[54] The task is to incorporate the foreign in order to balance out deficits of one's own, thereby ensuring continued dominance. In this spirit, Müller's essay "Was erwartet Österreich von seinem jungen Thronfolger?" (What does Austria expect of its heir to the throne? 1915) claims that the "circulation and metabolism of a civilized nation [*Kulturstaat*]" are "dependent on the fodder" it gets its hands on: a "filthy but imposing process of digestion" comprises the "healthiest way to prepare a highly evolved brain."[55] In *Tropen*, this process takes the form of Brandlberger seeking

---

50 Müller, *Tropen*, 282.
51 Müller, *Tropen*, 276.
52 Müller, *Tropen*, 129.
53 See Riedel, "'What's the difference?' Robert Müllers Tropen (1915)," in *Schwellen. Germanistische Erkundungen einer Metapher*, ed. Nicholas Saul, Daniel Steuer, Frank Möbus, and Birgit Illner (Würzburg: Königshausen + Neumann, 1999), 69; Müller-Tamm, *Abstraktion als Einfühlung*, 353.
54 Schwarz, *Robert Müllers Tropen*, 221. See also 221–276, 305–320. Transferred to the register of sexuality, this fits with the sadism of the colonists (Schwarz, *Robert Müllers Tropen*, 175–193).
55 Robert Müller, "Was erwartet Österreich von seinem jungen Thronfolger?" in *Gesammelte Essays*, ed. Michael M. Schardt (Paderborn: Igel, 1995), 63. Also quoted by Schwarz, *Robert Müllers Tropen*, 76. Lucas Gisi, drawing attention to the same passage, points to the "founding of a new race as a brutal fantasy of colonization and subjugation" ("Die Biologisierung der Utopie als Apokalypse. Der neue Mensch in Robert Müllers Tropen," in *Utopie und Apokalypse in der Moderne*, ed. Reto Sorg and Stefan Bodo Würffel [Munich: Fink, 2010], 223).

to use Zana to breed a new humanity that, by incorporating tropical sensuality, would be immune to signs of European degeneration ("syphilis, consumption, and mobs") and thus ensure the superiority of the colonizer-race in the future. That the novel, at least in its frame narrative, also expresses reservations about such plans may reflect the doubts Müller would later entertain about his "imperialist megalomania."[56]

Accordingly, to follow Stephen Greenblatt's lead, Brandlberger's relationship with the inhabitants of the jungle can be read as metonymic.[57] It is not characterized by perceived similarity (that is, by identity and difference subsisting side by side) or impartiality (renouncing appropriation). Rather, the protagonist-narrator perceives the other as a part of what is actually already his own, something to be instrumentalized in order to fortify a position of strength and maintain supremacy. Correspondingly, the relationship between (northern) man and the tropics is also shaped in the novel as a metonymic one: at some points, man mirrors the tropics (he duplicates them on a small scale), and at others the tropics are only a mirror (in the sense of a product) of man.

Tropological Language

The novel posits a connection between the tropics (the jungle) and tropes (figures of speech) in several ways. First, it stresses that the jungle is only a linguistic image into which the European transfers a part of himself: "Why am I talking about the tropics so much? The savage doesn't know them, only the northerner does – they're a figure for his ardor and the burning fever in his nerves. They're his invention, a metaphor he creates."[58] The very beginning of the book already confirms this inasmuch as the jungle tropics admit representation only by means of an array of linguistic tropes. For instance, the descriptions of the river's course accumulate figures of image and sound in a striking manner: "linguistic realization itself *is* the jungle."[59]

---

56 See Schwarz, *Robert Müllers Tropen*, 73–82. In "Robert Müllers *Tropen* (1915) als neurasthenisches Aufschreibesystem" (155), Schwarz emphasizes that after World War I, Müller increasingly distanced himself from his "imperialist megalomania," and especially from the martial ideal of toughness associated with it, which also defines Brandlberger's fantasy of breeding a new race.
57 Greenblatt, *Marvelous Possessions*, chapter 3. On the two different attitudes toward the foreign, see page 135 of the same book as well as chapter 2 of the study at hand.
58 Müller, *Tropen*, 214.
59 Dietrich, *Poetik der Paradoxie*, 49–50.

Brandlberger's claim of the tropological nature of the tropics finds further confirmation in numerous references to the jungle as a fictional and literary environment. By way of the title, *Tropen* refers to a book, and its inhabitants are the novel's characters – specifically, the letters of the alphabet in the manuscript.[60] Brandlberger and Slim both announce their intentions to write a novel called *Tropen*;[61] and the latter even wants to have the "whole story told by someone who never has been in the tropics."[62] Such passages emphasize not only the tropics' fictionality[63] but also the writing process and the materiality of writing. The actual author, Robert Müller, who represents himself in the novel's subtitle as the editor of Brandlberger's work, claims to have found Brandlberger's "typewritten manuscript"[64] in his desk. In this same document, Brandlberger writes that he "is going upstream on a desk" (*als Schreibtisch einen Strom hinauf*) and will "write the story he has yet to experience" (*das Buch, das er erst erleben wird*).[65]

The tropics are the product of writing, then, and many of the people who appear on its pages are described as written characters: "He [Checho] was tall and thin, like a letter";[66] "[Meme's] calves stretched above flat, wide soles; between them, like the letter *M*, the upper part of his body hung suspended, a pliant pyramid of delicate bones, muscles and nerves."[67] Werkmeister has observed that the letters' pictoriality, not their symbolic function, occupies the foreground: writing itself "grows primitive."[68] At the same time, personages in the novel are identified quite literally as figures of speech, revitalizing a dead metaphor in the process. This reanimation of metaphor and its literal interpretation also point to a primitivistic use of language, passing from concept back to vivid description – both defining features of the 'primitive' use of language in contemporary theories of metaphor.

Second, the opposite claim is advanced (as noted above): "northern man" would be a mere trope (etymologically, a "turn") of the jungle:

> He [bears] the tropics within himself. [...] He's the means by which nature preserves the tropics, which are dying out. The tropics are the foundation of his organism and vital

---

60 See Werkmeister, *Kulturen jenseits der Schrift*, 373.
61 Müller, *Tropen*, 213, 234.
62 Müller, *Tropen*, 234.
63 For a detailed discussion, see Dietrich, *Poetik der Paradoxie*, 58–68.
64 Müller, *Tropen*, 6.
65 Müller, *Tropen*, 27.
66 Müller, *Tropen*, 36.
67 Müller, *Tropen*, 68.
68 Werkmeister, *Kulturen jenseits der Schrift*, 372.

power; he is built following their principle, and everything is repeated within him in miniature. He, the human being, might be said to be a trope of the tropics [im Verhältnis zu den Tropen ein Tropus].[69]

Here, the connection is not linguistic so much as biological: the concept of the trope provides a metaphor for the metonymic relationship between the macroscopic tropics (of the primeval forest) and the microscopic tropics (of the human organism). Nature, not man, thus appears as the agent. Brandlberger finds that the original state of all life has been preserved in the jungle river, the source of human life from which physical existence – "nibbling bundles of cells"[70] – derives. The observation concerns the brain in particular, which has evolved from teeming cells: "It turns out that he, the northerner, has the tropics/tropes [Tropen] in him. [...] His brain, filled with a lush vegetation of tropics and metaphors [Gleichnisse], can be explained by the residues of his ancestry."[71] The tropics (Tropen) of the brain allude on the one hand to its organic composition, its development from rampant cells that still exist.

> My head on fire, I saw myself plunging into the unrealized possibilities of immature conditions, wild, primeval states and elemental battles, into the swamps of my blood and vegetal contentment. My brain cycled through the whole process through which the world had come to be and brought it forth anew.[72]

On the other hand, the tropes (Tropen) of the brain allude to figures of speech – that is, to the brain's creative force.

Thirdly, in this way the novel also establishes a semantic connection between tropes and the tropics, so that in German the word Tropen changes from a homonym into a polyseme. Linguistic acts of creation performed by the human brain obey the same principles as natural acts of creation in the jungle, for the latter has achieved biological immortality in the former. The same proliferation and crossbreeding are at work. The figures of speech are characterized as tropical tropes because they in fact originate in the processes of the tropics of nature.

Describing the tropics as a European trope and human beings as a trope of the jungle forges a causal chain that renders it impossible to distinguish between cause and effect: the primeval forest is a trope of man, who is a trope of the primeval forest, which is a trope of man, and so on. Man turns out to be an image of

---

69 Müller, *Tropen*, 282.
70 Müller, *Tropen*, 20.
71 Müller, *Tropen*, 234.
72 Müller, *Tropen*, 134.

an origin that is itself a metaphor made by man.[73] The identification of a first cause is replaced by the process of transference itself: all that exists emerges from this process. Both metaphorically and literally, the trope ascends to the position of the creative principle determining the world of the novel, *Tropen*.[74]

Brandlberger discovers this principle at the very outset of his journey, when he first encounters the jungle river: "*Tatwamasi:* it is you!" However, the way he handles the matter makes it clear that it is necessary for us to differentiate between specific tropes. Whereas metaphor is based on similarity and therefore the simultaneous perception of identity and difference, the novel's protagonist follows the principle of radical de-differentiation and identification.[75] As I noted above, this results in a *metonymic* construction or in a primitivistic, literal reading of metaphors. For Brandlberger, the tropes of language indicate an actual identity or at least an actual connection:

> So there I sat and felt that the equator really is a glowing hoop passing through the intestines. [...] I entertain relationships with a natural world that is female through and through. Sexuality floats over the waters, and I combine hymns of blood into a chorus. The forest is nature's immense heart, and the brown water of the river the holiest blood of my own.[76]

In light of his biogenetic convictions, we may be sure that Brandlberger means these words literally. However, taking things literally also leads to misunderstanding. For instance, when Slim speaks of streams extending from one person to another, Brandlberger thinks that he means the water where they have pitched camp, which he wants to follow in order to return to civilization.[77]

Brandlberger's de-differentiating attitude is clearly expressed in his habit of considering things that stand in relation to each other as "the same." For instance, in the following quotes:

> Whatever one experiences, it's always the same adventure; it doesn't matter if you fall into the clutches of a panther or under a bus, and what matters least of all is whether her name is Zana or Miss So-and-So.[78]

---

73 Müller-Tamm aptly sums it up: "the subject is a mere metaphor of its irretrievable origins, at once the performer and the effect of historical projection" (*Abstraktion als Einfühlung*, 357).
74 Contra Werkmeister, this "pure mediality" has nothing to do with the "primitive coincidence of signs and things" (Werkmeister, *Kulturen jenseits der Schrift*, 377); in the first case, emphasis falls on the material nature of the sign, and in the other, on the materiality of the object.
75 See Werkmeister, *Kulturen jenseits der Schrift*, 371.
76 Müller, *Tropen*, 24–25.
77 Müller, *Tropen*, 236.
78 Müller, *Tropen*, 26.

> Where one reality exists, there can also be another. [...] So analysis is the same as synthesis.[79]
>
> Isn't everything always the symbol of one and the same thing: the human being?[80]
>
> Everything takes on a meaning; I'm moved to see how facts and symbols come together and yield the same thing.[81]

These passages make it clear that for Brandlberger similarity always flows into identity. The way he understands tropological phenomena slips toward metonymy, taking figurative language literally and ignoring the interplay of identity and difference.

Symbolic transfer, which usurps the principle of a first cause in the novel, to Brandlberger amounts to a projection mechanism for creating tropes of oneself. A case in point is the scene illustrating Brandlberger's "tropical delirium." In keeping with the phrase's double meaning,[82] the madness to which the Europeans succumb is not just a physical fever but also a fit of tropological identificatory transference. When Brandlberger – affect-driven and without apparent motivation – shoots a pair of storks, he reasons as follows:

> You're aiming at something outside yourself, a beautiful, red fetish – a red ideal – and ultimately you mean yourself. But if, one day, you make the formal decision to do yourself harm, then absent-mindedness will take care of that and you'll do it to your neighbor. You execute yourself in a doll – man, you're suspicious; it looks to me like you're an incurable poet.[83]

The fit leads to the blurring, if not the complete collapse, of differences between self and other, whether human or animal. Brandlberger *is* the stork; he kills himself in it, as he later does in his companions. His affective participation does not involve exchange with the other, but blindness to the other in a delusive self-reflection: the stork and his companions are mere "dolls" for a (mis)identifying transference of himself.

---

[79] Müller, *Tropen*, 235.
[80] Müller, *Tropen*, 244.
[81] Müller, *Tropen*, 276.
[82] On the discourse of *Tropenkoller*, see Schwarz, *Robert Müllers Tropen*, 159–163.
[83] Müller, *Tropen*, 215–216.

Poets

As I noted in the previous chapter, Hugo von Hofmannsthal derived the poetic symbol from the sacrificial ritual. Brandlberger's reflections on killing the pair of storks follows the same logic: one who executes himself in an animal is an "incurable poet" because he transfers himself to another.[84] The poetic self that was killed in the stork concerns a "type" that gives itself over "to its atavism," that is, whose body and behavior still attest to the animal(s) he once was at earlier stages of evolution. Brandlberger accepts "only things and creatures similar to himself, or that have been drilled into him."[85] By way of the sacrificial animal, Brandlberger, the poet, defines himself as a type lying in wait for similarity, language providing the decisive hints: "Haven't you noticed that madmen are fatalists about language? Watch out, man. Coincidences in language are the destinies of thought."[86]

At the same time, however, Brandlberger acts counter to this type when he kills the storks because the act demanded a "further education," a search for something different. This search he immediately attributes to "the conquerors, the colonizers": "They go for it and hold fast to life."[87] Brandlberger also calls people like this poets: "a kind of poet, with healthy [...] digestion, at least. When they have cramps and vomit, they're at the peak of contentment. Above all other states, this is poetry, and they thrive on it."[88] Turning away from similarity and seeking out the unknown therefore does not occur for the sake of recognizing the unfamiliar, but in order to appropriate it. As in Hofmannsthal's rite of sacrifice, a primitivist motif is used here to justify the poet: anthropophagy. However, it does not derive from substitution (the act of sacrifice) so much as incorporation/digestion (cannibalism). Here, Müller takes up the metaphorical imagery from the *Thronfolger* essay, which discusses imperialist states "eating up" colonized territories and "digesting" them as well as "sucking in the globe" and "pumping the marrow of the earth" into their "brains."[89] Müller's return to

---

[84] Müller, *Tropen*, 216.
[85] Müller, *Tropen*, 217.
[86] Müller, *Tropen*, 264.
[87] Müller, *Tropen*, 217.
[88] Müller, *Tropen*, 217–218.
[89] Müller, "Was erwartet Österreich," 64. Also quoted in Schwarz, *Robert Müllers Tropen*, 76. Schwarz connects Müller's reflections with the *Will to Power*, where the power of a people is said to depend on incorporating "foreign" material and "transforming it into blood, so to speak" (Schwarz, quoting Nietzsche, *Robert Müllers Tropen*, 221).

the metaphor of cannibalism in *Tropen* points to the imperialist ambitions of the poet developed there.

The novel also describes the new type of poet as the forerunner of a "new man," who Brandlberger sees emerging in himself. This ideal is characterized by its synthesis of "primitive sensuality" and civilized intellect – a topical notion in primitivist discourse contemporary to the book. He states, "I climbed down the ladder of evolution, and now I'm climbing back up. Soon I'll be with the man of the future again, having been among the beings of prehistory."[90] The journey into the jungle amounts to a regression into the future, a sensual, post-rational mode of existence (which, as the reader already knows thanks to the novel's preface, Brandlberger will not achieve[91]). The rationalism of Europe is not replaced by "dream logic" (evident in Slim's tendency toward pan-signification) so much as by insight into the relativity of waking thought and nighttime visions: "Both experiences are real, only the accent has changed. [...] After all, what's logical interpretation but something illogical – a mere interpretation, poetry."[92]

At issue stands an inversion effect, much discussed in the psychology of perception of the time, which Brandlberger discovers at the outset of his journey[93] – for instance, when looking down from the ship into the world "upside down" in the water:

> I practiced a little, and before long I could snap back and forth like a thin sheet of metal. This sensory illusion worked perfectly. It was just the accent moving around – that's it, the

---

[90] Müller, *Tropen*, 89. On the three stages, see Liederer, *Der Mensch und seine Realität*, 135–174.
[91] Müller-Tamm observes that the journey only appears to be going backwards because (supposed) prehistory is in fact a "culturally specific self-projection defined by the biological and cultural-theoretical thought patterns of evolutionism" (*Abstraktion als Einfühlung*, 356). In other words, the voyager never gets away from his own time and culture. Gisi also points out that this goal is never achieved and therefore speaks of a "paradoxical utopia" ("Die Biologisierung der Utopie als Apokalypse," 223). Liederer resolves the contradiction by claiming that for Müller humankind "always [stands] on the threshold of the next anthropological stage of evolution: in the process of becoming, because the reader represents the 'last rung' of its realization" (*Der Mensch und seine Realität*, 142).
[92] Müller, *Tropen*, 226.
[93] According to Werkmeister, the figure of inversion (which he traces back to Erich Moritz von Hornbostel's experimental psychology, among other sources) plays a "key role in primitivistic discourse" (*Kulturen jenseits der Schrift*, 354) to the extent that it represents upside-down ethnology – the "primitivization" of thinking and literature (355); Müller's "project of inversion" makes it impossible to separate normal and altered states (356). Before Werkmeister, Müller-Tamm had already pointed to the relevant intertext in experimental psychology (*Abstraktion als Einfühlung*, 364–365).

accent! I had it. The accent reflects whole perspectives, whole realities rest on it. By so-called sensory illusion, I could turn the world upside-down and make another one. So who can say which one's right and which one's wrong?[94]

Rational thinking and dream logic are but one of many examples of this shift in accentuation in the novel. The accents fall differently in the dream, inverting the images presented by rational thought. As Brandlberger notes in the same passage, linguistic tropes produce the same effect: "Does it mean nothing when we speak in symbols and parables – is the refreshment of a fertile lie nothing at all?"[95] "Symbols," for him, enact "accentuated reflections" that make the world appear "in reverse."[96]

However, Brandlberger's reaction to the epistemological doubt caused by the inversion-effect is not to affirm the higher truth of the inverted world. Instead, he recognizes the relativity of perception and how it is processed. He preaches a paradoxical way of thinking that can think image and counter-image at one and the same time. The sensual, post-rational man is characterized by such thinking; and at the the same time he is the product of such a paradox: "One age [i.e., that of sensuality, NG] is the paradox of the other [i.e., that of rationality, NG]."[97] Insofar as it sublates rationality and dream, thinking-in-paradox resembles dreaming reason or analytical dreaming: seeking knowledge, but with the help of imagination, intuition, and creative combinatorics. Liederer therefore calls it "somnambulistic-intuitive thinking" that enables "insight according to the principle of creative synthesis."[98]

Unmasking logical interpretations to be a form of literary art,[99] Brandlberger formulates an epistemological critique: the world of rational thought is as fictional as the world of dreams. At the same time, he performs a constructivist turn.[100] Both worlds are the creation of a perceiving and interpreting subject: "It's just a matter of our preference, our creative will for change [...]. Learn to scan the meter of reality [*Lernet die Wirklichkeit skandieren*]!"[101] "We were the first to find out there's no reality, and we're also the first to invent new ones

---

94 Müller, *Tropen*, 38–39.
95 Müller, *Tropen*, 39.
96 Müller, *Tropen*, 40.
97 Müller, *Tropen*, 39.
98 Liederer, *Der Mensch und seine Realität*, 199.
99 Müller, *Tropen*, 226.
100 See Müller-Tamm, *Abstraktion als Einfühlung*, 365–366, as well as 368–380 (on "phantoplasm"); see Liederer, *Der Mensch und seine Realität*, 187–203.
101 Müller, *Tropen*, 39.

over and over!"¹⁰² Scholars often deem this perspective a legacy of Nietzsche. In fact, it radicalizes the latter's position: instead of depicting creative activity as *failing* to capture the world-in-itself, it affirms – and celebrates – the insight that what has been produced is *the only* true reality. Consequently, then, Brandlberger declares, "Observing and learning something from nature means creating something new in it. Seeing and producing are one and the same."¹⁰³

The poet's task, in Müller's novel, is to disclose the world-as-poetry and thereby create conditions favorable for the emergence of the new human race. In his capacity as poet, Brandlberger stands at the dawn of a new age. The novel he wishes to write – or has already written – is meant to fulfill this objective: "I proclaim the mirror, the world upside-down, paradox! This will be my other great contribution to humanity."¹⁰⁴ Brandlberger does not only sermonize this insight; the novel *Tropen* also already realizes it, confronting the reader with an inverted and paradoxical world that leads to unfamiliar thought processes and gaps that encourage further creative thought.¹⁰⁵

For Brandlberger, the jungle represents the survival of an early stage of humankind, and it functions as a "paradox of another [epoch]"¹⁰⁶ from which the European travelers and the book's readers descend. As such, they encounter here an "upside-down" world where it is not intellect but the senses, not individuated life but communal participation, not logic but "the wildest thinking"¹⁰⁷ that determine life for the "Indians'" and eventually for the foreigners. Their life is governed by habits of mind deriving from language, madness, and dream and that are defined by the principle of de-differentiating transference described above.

This way of thinking is not only depicted when Brandlberger falls victim to tropical fever. Since he is the narrator of the novel, it shapes the novel's characterization and narrative style as a whole. According to him, he and his companions indulge in the "wildest thinking":¹⁰⁸ "That's how brains work [*das ist das System der Gehirne*]. They stand under each other's spell";¹⁰⁹ "we've all become

---

102 Müller, *Tropen*, 224.
103 Müller, *Tropen*, 235.
104 Müller, *Tropen*, 40.
105 Dietrich's analyses fall under the heading of the "poetics of paradox," which, in his eyes, shapes Müller's fictional prose (*Poetik der Paradoxie*, 72–80). At the same time, the affirmation (or normalization) of paradox coincides with the "logic of myth" (75). With Liederer, I would assign them to different levels of evolution, however.
106 Müller, *Tropen*, 39.
107 Müller, *Tropen*, 224.
108 Müller, *Tropen*, 224.
109 Müller, *Tropen*, 236.

the same person since we've had to live together like this."[110] The novel puts this dissolution into effect when, toward the end, the characters merge with each other more and more. Clear characterization and demarcation of characters yield to their condensation and displacement.[111] As a result, it is often ambiguous who is who, or who has done what. For instance, "The man uttered that deep rutting cry, then I saw him sitting quietly in the boat. There I sat myself. The boat was gliding off, over two worlds."[112] Or: "Indeed, [van den Dusen] now looked a bit like Slim."[113] And: "You [Brandlberger] really look like Slim! If only you knew how much you're like him!"[114] Such displacements make it impossible to solve Slim's crime-novel-like murder: "How did it all happen? You know? No. And you? Me neither."[115] The same effect is brought about inasmuch as the narrative style increasingly lacks logical order.[116] Linear progression is replaced by a network of scenes whose chronological sequence and causal connections necessarily remain opaque to the reader, in some cases because they are recounted in different versions (e.g., the deaths of Rulc, Slim, and van den Dusen).

The novel does not only present characters who think "wildly"[117]; it is itself determined by this "wildest thinking."[118] Thus, it presents an inverted world to the European reader. Together, the reader's world and the world of the book form the paradox that is thematized as the trigger for insight into the *poietic* nature of everything encountered. In addition, the novel stages further paradoxes (e.g., different accounts of the same death noted above).[119] Through these internal events and the manifest contradiction between the novel and real life, the

---

110 Müller, *Tropen*, 266. Liederer provides further examples (*Der Mensch und seine Realität*, 77–86); see also Müller-Tamm, *Abstraktion als Einfühlung*, 360; as well as Riedel, "'What's the difference?'" 72–76, who speaks of "mythical thought" in this context.
111 See Werkmeister, *Kulturen jenseits der Schrift*, 371.
112 Müller, *Tropen*, 269.
113 Müller, *Tropen*, 258.
114 Müller, *Tropen*, 266.
115 Müller, *Tropen*, 249.
116 On the novel's narrative technique, see Müller-Tamm, *Abstraktion als Einfühlung*, 348–350; Liederer, *Der Mensch und seine Realität*, 224–233; Dietrich, *Poetik der Paradoxie*, 30–36, 48–68; Gardian, *Sprachvisionen*, especially 154–169.
117 Brandlberger is not distinguished by paradoxical reasoning – as Liederer claims – so much as by wild thinking, in relation to which Schwarz discerns an affinity to the paranoid "writing-down-system" of Daniel Paul Schreber ("Robert Müllers *Tropen* [1915] als neurasthenisches Aufschreibesystem," 147).
118 Müller, *Tropen*, 224.
119 See Liederer, *Der Mensch und seine Realität*, 198–201, for discussion of the murders as examples of "paradoxical parallel- or alternative realities."

reader confronts mounting perplexity – the true jungle of *Tropen* ("I think this jungle [i.e., the novel] is inaccessible to the reader"[120]). Both the novel and the world it depicts demand another way of thinking, one that does not resolve paradox but aims to integrate it into one's own view of the world. This way of thinking is aware of its *poietic* force, the ability to perceive and/or create other (and contradictory) worlds through interpretation. Accordingly, as Christian Liederer has shown, the poet Robert Müller induces those who "enter" the tropics of *Tropen* to modify their thinking and complete the events described in creative fashion.[121] Thereby, the reader comes to do what Brandlberger envisions: to contribute to the development of the "new man" whom he sees already announced in himself.

What is the relationship between the constructivist project of this 'new poet,' for whom Brandlberger wishes to clear the way by means of what he writes, to the imperial ambitions of this poetic ideal within the novel? It would seem that a more refined version of incorporating the foreign is at work, insofar as the foreign world is here reduced to a mere construction. Although the same holds for the world of the familiar, this constructivist turn does not delight the "Indians" so much as the Europeans. The familiar, not the foreign, profits from reality being constructed (in keeping with certain historical, sociological, and material preconditions the foreigners do not share).[122] And it is not the foreigner, but the European who, on the basis of this insight, wants to create a new race and to instrumentalize the foreigner without letting him in on his plans. In this constructivist imperialism, it is not European civilization that disappears from the world stage (instead it turns into a civilization of colonizers) but that of the so-called 'primitives,' who are destined to serve as breeding stock. It is much easier to justify brutal colonial practices if the worlds they destroy are nothing more than constructions of reality anyway, not the (only) reality of other human beings. The creative freedom of seeing the world only as one wants combined with political power amounts to such willful ignorance of others' 'perspectives' (which, for them, are realities) that destroying these same people is seen as acceptable – and is often even carried out. Against this background, the idea that the world is nothing more than a trope does not mean that the differences between the self and other are valued. Instead it signifies that one has the license to turn the other into a metonymy of the self where the jungle is European poetry

---

120 Müller, *Tropen*, 234–235.
121 See Liederer, *Der Mensch und seine Realität*, 355–369.
122 "Phantoplasm" might also be understood to point to how fantasies rest on certain material preconditions and to this extent are not arbitrary (or equally available to all); see the next footnote.

and can be 'reworked' at any time. Needless to say, this reworking has material consequences for other people and the milieu they inhabit: namely, brutal repression or even extinction. This may be the sinister deeper meaning of Brandlberger's polyvalent concept of "phanto-plasma":[123] the biopolitical consequences of Europeans' imperial fantasies.[124]

Returning to Primordial Slime

Müller stages Schiller's "they are" through an exotic locale and its foreign inhabitants, both of which represent survivals of the phylo- and ontogenetic past. Gottfried Benn relocates this topological and chronological scheme to the body. While Müller merely hints that primordial, tropical vegetation also determines the cerebral physiology of modern Europeans, this relationship is Benn's central focus.[125] In "Unter der Großhirnrinde. Briefe vom Meer" (Below the Cerebral Cortex: Letters from the Sea, 1911), one of his first literary efforts, his early scientific interests in evolutionary biology, neurophysiology, and psychiatry converge. In this fictional letter written during a trip to the seaside, the first-person narrator

---

[123] The term has elicited any number of belabored interpretations, which Liederer discusses in detail (*Der Mensch und seine Realität*, 264–369). For his own part, Liederer considers the "'nature' of phantoplasm" to be "what is mutable, floating, and preliminary," that is, the *variability of form*" (266; emphasis in original); worlds created by different principles are therefore different phantoplasms (356). Müller-Tamm has offered a substantive corrective by pointing out that the terms mark "the physiological quality of all transmitted and projected perceptions constituting the world" – with an emphasis on *physiological*. She points out that *plasma* is a biological concept circulating broadly because of Haeckel; for Müller, however (and in contrast to Haeckel), the term "no longer designates the materiality of psychic functions, but instead refers to reality as the effect of psychic functions" (*Abstraktion als Einfühlung*, 370).

[124] Yet Brandlberger only appears to have the last word. In fact, his theories break down to the extent, as Schwarz explains, that the novel identifies them as "formations of delusional systems" and showcases their "megalomania." Brandlberger's delusions of grandeur are deconstructed by the affinity of his "wild narration" to a "neurasthenic writing system" ("Robert Müllers *Tropen* [1915] als neurasthenisches Aufschreibesystem," 154).

[125] For readings of Benn in the context of the history of science, see the following monographs: Regine Anacker, *Aspekte einer Anthropologie der Kunst in Gottfried Benns Werk* (Würzburg: Königshausen + Neumann, 2004); Ursula Kirchdörfer-Boßmann, *"Eine Pranke in den Nacken der Erkenntnis." Zur Beziehung von Dichtung und Naturwissenschaft im Frühwerk Gottfried Benns* (St. Ingberg: Röhrig, 2003); Gerlinde Miller, *Die Bedeutung des Entwicklungsbegriffs für Menschenbild und Dichtungstheorie bei Gottfried Benn* (New York, Bern: Peter Lang, 1990); and, most recently, Hahn, *Gottfried Benn und das Wissen der Moderne* (which incorporates earlier essays by the author).

– a former doctor and researcher (like Benn himself who had given up a career in psychology in favor of medical practice) – formulates a pointed critique of science, taking aim, in particular, at the paradigm of psychophysics, the associated theory of localization (which reductively traces all self-determined actions and abilities to specific centers in the brain), as well as epistemological constructivism (which deprives the self of an independently existing external world). The only scientific undertaking to escape criticism is paleoanthropology; indeed, it is used to construct a utopian return to archaic conditions.

At the beginning, the first-person narrator describes a regression that takes place along paleoanthropological and neurophysiological lines, both chronologically and in terms of the brain's structure. The journey involves going "back to the past" and simultaneously "sinking from the surface"; it leads from thinking that lies like "lichen [*Flechte*] on the brain," from nausea "above," on the "cerebral cortex," down toward a space "deep below in the mud," "in cracks, crevices, and under the foliage," to the "lower centers" "under the cerebral cortex."[126] The writing subject ("I") identifies with his "forefathers" – by which he means not just human ancestors but also primordial organisms of uncertain identity: "Maybe it wasn't a jellyfish, but just a pile of slime from a plant [...], from which everything else started."[127]

This retrograde fantasy is connected to an unconscious, vegetative state. The letter's writer relativizes thought as only one of various possible "cycles of the psychic process [...] which are just as lawful and regular."[128] One such (potentially "happier") process involves the "softening of the brain" (*Gehirnerweichung*), which brings about the same condition that the neuroanatomist Paul Emil Flechsig associated with instinctual and potentially criminal actions. But in contrast to Müller's Brandlberger, Benn's narrator does not want to act on these instincts; instead, the goal is a specific state of unconsciousness. What the first-person narrator develops in his regression fantasies is reminiscent of the "calm [...] dreamless sleep" that Flechsig describes as enveloping the body whose drives have been satisfied and cerebral function suspended.[129] The aim is to achieve a feeling of mute security in slimy caves, calling to mind a return to the uterus, though without explicit reference to ontogenetic regression. Correspondingly, the sea

---

126 Gottfried Benn, "Unter der Großhirnrinde. Briefe vom Meer," in *Sämtliche Werke. Stuttgarter Ausgabe*, vol. 7.1, *Szenen und andere Schriften*, ed. Holger Hof (Stuttgart: Klett-Cotta, 2003), 355–356.
127 Benn, "Unter der Großhirnrinde," 7.1: 356.
128 Benn, "Unter der Großhirnrinde," 7.1: 358.
129 Paul Flechsig, *Gehirn und Seele. Rede, gehalten am 31. Oktober 1894 in der Universitätskirche zu Leipzig*, 2nd ed. (Leipzig: Veit & Comp., 1896), 17.

to which the narrator has gone is declared the "Cambrian Sea" of prehistoric times. In writing the latter, the narrator performs the transition to the desired state through a shift in verbal tense. The letter begins in the simple past, establishing a distance between the narrator and the narrated events, and transforms into the present tense, thus lending the archaic world an imaginary reality: "I sank into ages past [...]. Huge, greenish dragonflies with heads as wide as a child's skull shoot through the air and deliver treacherous stings."[130]

At the same time, this work, but also other early writings by Benn, question such a descent inasmuch as it is repeatedly interrupted by thought or deferred to the future. The self-reflexivity of these texts, which pushes plot and/or dialogue into the background, can be traced back to a sentimental signature, to use Schiller's term.[131] Benn's oeuvre as a whole deals with the findings of natural science in a paradoxical manner. "Unter der Großhirnrinde" makes that particularly clear: On the one hand, the narrative "I" rejects the methods and worldview of natural science, dismissing its claims as relative or even false. On the other hand, the insights of natural science provide the basis for what the letter elaborates.[132] This takes place in a dual sense: Benn moves the narrator into the sentimental position, namely seeing the state of naivety, but being unable to return to that state due to the knowledge he has acquired in the meantime – regardless of whether or not it is false.[133] Also, Benn enlists bioscientific insights to sketch

---

130 Benn, "Unter der Großhirnrinde," 7.1: 356. In addition to meditation and sleep, the end of the letter presents a third possibility of return: the escapist cliché of vacation in Italy (horse-drawn carriages in Naples, warm sun on one's back, the roar of the ocean in one's ears, mountaintop vistas). The only original feature and point of interest in this passage is the infusion of archaic energy into the impressionist idyll: the ocean, where the narrator would find "Cambrian seas" again, is described as a "pool of cornflower blood" (363). The metaphor is vexing because it combines imagistic spheres with opposing connotations. In characteristic fashion, Benn stands commonplace utopias on their heads. A benign scene of nature and clichéd attitude ("I want to encounter things in a pure and brotherly way; [...] just look at them, contemplate them, smile at them, rejoice in them. Let the world grow around me like a meadow [of corn flowers]") turns into identification with "lower" life forms, both physiologically and "geologically," right down to plants and the sea, the starting point of all life.
131 On the sentimental in Benn, see Riedel, "Endogene Bilder. Anthropologie und Poetik bei Gottfried Benn," in *Poetik der Evidenz. Die Herausforderung der Bilder in der Literatur um 1900*, ed. Helmut Pfotenhauer, Wolfgang Riedel, and Sabine Schneider (Würzburg: Königshausen + Neumann, 2005), 196; and Riedel, "Wandlungen und Symbole des Todestriebs," 110–113; on Benn and Schiller, see Antje Büssgen, *Glaubensverlust und Kunstautonomie. Über die ästhetische Erziehung des Menschen bei Friedrich Schiller und Gottfried Benn* (Heidelberg: Winter, 2006).
132 See Hahn, *Gottfried Benn und das Wissen der Moderne*, 81–84.
133 See Riedel, "Wandlungen und Symbole des Todestriebs."

the very archaic utopia to which the letter-writing ego wishes to return. In other words, there's no getting around the natural sciences. On the contrary, the longing of the letter-writer for an original state of naivety is doubly shaped by precisely these same fields, in that it is caused by them and in that they provide the template for his imagination of an original state.

The Body as Hieroglyph of the Archaic

Some twenty years later, Benn was no longer a candidate for an advanced degree in psychiatry. After a self-diagnosed depersonalization disorder, he had obtained accreditation in dermatological and sexually transmitted diseases and was operating, with some frustration, his own practice in Berlin-Kreuzberg. As his essays of the early 1930s demonstrate, the thought-figure, "they are what we were," still preoccupied him. But in contrast to what he had written in the early 1910s, his focus was no longer on the regression into primordial slime so much as revitalizing the memory of archaic ages stored in and activated by means of the body, which *Dichtung* (poetry/literature) is supposed to give voice to. In so doing, Benn followed the lead of speculative paleoanthropology.[134] In his 1930 essay, "Der Aufbau der Persönlichkeit" (The Development of Personality), he writes – like Eugen Georg before him – that the body harbors some two hundred rudiments dating back to the prehistoric emergence of human beings.[135] The lower parts of the brain and such bodily fluids as blood and pus – which earlier works describe as physical carriers of the archaic – are joined in this essay by the vegetative nervous system and endocrine system (among other organs). Time and again, Benn seeks out traces of ancient memories materialized here, which are supposed to still shape the personality of modern day individuals and can even be directly perceived in certain physiological processes (e.g., orgasm and intoxication).

However, Benn's claims about the immediate, physical experience of the archaic conflict with his repeated references to the body as a hieroglyph. Contemporary psychoanalytic theories on organic memory as well as on symbolization

---

[134] For a thorough account of the theories of Dacqué and Georg, see Hahn, *Gottfried Benn und das Wissen der Moderne*, 151–173. On the author's "turn to primordial times" (125) in light of his reading of Jung and Erich Unger, see Dieter Wellershoff, *Gottfried Benn, Phänotyp der Stunde. Eine Studie über den Problemgehalt seines Werkes* (Cologne: Kiepenheuer & Witsch, 1986), 125–152; Kirchdörfer-Boßmann, *"Eine Pranke in den Nacken der Erkenntnis,"* 269–273n125.
[135] Benn, "Der Aufbau der Persönlichkeit. Grundriss einer Geologie des Ich," in *Sämtliche Werke*, vol. 3, *Prosa 1* (1910–1932), ed. Gerhard Schuster (Stuttgart: Klett-Cotta, 1987), 272.

shed light on the logic at work here (see Chapter 4). In the second part of *Thalassa: Versuch einer Genitaltheorie* (1924; *A Theory of Genitality*, 1938), Sándor Ferenczi speculated about a phylogenetic parallel between individual birth and the expulsion of humankind's distant ancestors from the water, noting

> the extraordinary frequency with which, in the most varied creations of the mind, both normal and pathological, in the products of the individual and the collective psyche, both the sexual act and the interuterine situation are expressed by the symbol of the *fish*, that is, the depiction of a fish moving or swimming in the water.

For Ferenczi, this symbol offers the opportunity to speculate about human origins in water in literal terms. The fish swimming is not a metaphor for the similar state of the embryo in the womb so much as its "primal scene." Prehistoric humanity emerged from these depths. In other words, "a bit of phylogenetic recognition of our descent" has been stored in this symbol.[136] Ultimately, Ferenczi's reflections result in a reverse symbolism. According to Ferenczi's theory, the fish in the sea cannot be understood as the symbol for "uterine existence." On the contrary, life in the womb amounts to a physical symbol of the "maritime existence" of humankind's "animal ancestor": "In accordance with the 'reversed symbolism' already met with several times, the mother would, properly, be the symbol of and partial substitute for the sea, not the other way about."[137] In sum, Ferenczi understood organs such as the uterus as physical symbols formed by mnemonically charged germ plasma for repeating phylogenetic catastrophes such as the expulsion of primitive humanity from water in an attenuated and modified manner, thereby relieving ancient trauma over the ages. From these ideas, he formed his theory of "bioanalysis," which applied psychoanalytic insights to the body itself, offering a hermeneutic approach to organic life.

Benn most likely read Ferenczi as his poem "Regressiv" (written in 1927 or before; "Thalassal Regression," 1953) cites the latter's concept of "thalassal regression."[138] This proximity to the psychoanalyst's reflections illuminates the poet's concept of the body recorded in his essays from the early 1930s. Like Ferenczi, Benn discerns both a natural and a symbolic side of the body. Thus, "Zur Problematik des Dichterischen" (On the Problematic of the Poetic, 1930) speaks of it as a "transcendence of non-metaphorical race [*Geschlecht*], [...] reality with

---

136 Ferenczi, *Thalassa*, 44–45.
137 Ferenczi, *Thalassa*, 54.
138 See Riedel, "Endogene Bilder," 186, and "Wandlungen und Symbole des Todestriebs," 106. The English translation of this poem was published as "Thalassal Regression," trans. Edgar Lohner and Cid Corman, *Quarterly Review of Literature*, 7 (1953): 290–297.

mad symbols, canon of the natural and hieroglyph formed from phantasms, matter without idea, yet the medium from which to drink magic."[139]

This figuration of the body as hieroglyph calls to mind the Romantic idea of nature as a hieroglyphic system. The hieroglyph is understood as a paradoxical sign that sublates its own signification, which promises unity between the signifier and signified. Indeed this sign is barely legible, i.e., the unity is as difficult to grasp as it is to communicate. In Romanticism, this task falls to artists and their work. Benn's conception of the body-as-hieroglyph realizes the Romantic hope for unity between signifier and signified on the basis of biology. For Benn as for Ferenczi, the modern human body does not stand in a relationship of similarity so much as one of identity with the archaic. It not only represents the archaic but *is* the archaic inasmuch as it has emerged from it. Deciphering its hieroglyphs also occurs along two lines. Not only can the archaic be read in organs and physiological processes, but it is simultanously experienced directly through the same body: the representation momentarily turns into the signified's presence. The "they" and "we" collapse, and the "are" and "were" coincide: We – the bodies of the present – are thus in this moment what they once were.

Reading the hieroglyphic body comprises a compulsive repetition of archaic experiences; it leads to a regression that simultaneously creates something new: "Everything takes shape out of [the body's] hieroglyph: style and knowledge"; "the body, suddenly, is the creative force; physical being [*der Leib*] transcends the soul."[140] In states of intoxication, "creative desire and pleasure" (*schöpferische Lust*) reemerge, which, for Benn, comprise the biological "law of the productive" (*Gesetz des Produktiven*): a constant alternation between giving form and destroying it.[141] However, Benn locates the decisive difference between the biologically driven creative productivity of supposed primordial humans (whose thinking Benn places in a "sphere of organic interests") and that of modern humans in the notion that the productivity of the latter is not simply determined by unconscious experience and bodily drives, but rather involves the *memory* of the archaic. This point enhances the distance already inherent in the temporal notion of memory and also in the symbolic character of the hieroglyph, where reading, i.e., a conscious, analytical, and thus more distanced approach, is indicated. In the penultimate paragraph of "Zur Problematik des Dichterischen," Benn

---

[139] Benn, "Problematik des Dichterischen," in *Sämtliche Werke*, 3: 246.
[140] Benn, "Problematik des Dichterischen," 3: 246; Benn, "Akademie-Rede," in *Gesammelte Werke in vier Bänden*, vol. 1, *Essays, Reden, Vorträge*, ed. Dieter Wellershof (Wiesbaden: Limes Verlag, 1959), 437.
[141] Benn, "Der Aufbau der Persönlichkeit," 3: 437–438. Cf. Miller, *Die Bedeutung des Entwicklungsbegriffs*, 211–225.

underscores this distance: "Mystical participation is over." But *memory* of mystical participation endures: "memory of its totalization is forever."[142] At issue here is not a sentimental longing for regression any longer, but a memory that is inscribed into the body. Accordingly, the "poet" has the task of tracking down such memories by yielding to the repetition compulsion, reading, and experiencing the archaic in the body's hieroglyphs in order to bring about a poetry that expresses the biological law of creation.

The Way to Fascism

Benn's bodily version of "they are what we were" no longer pretends to be interested in what or who is truly other. Instead, from the outset, it attends only to what is foreign within the self, in one's own body. Compared to the fantasies of Müller (or rather Brandlberger), this attitude avoids appropriating 'real-world' others (human beings).

At the same time, however, this does not occur in the name of acknowledging them so much as to affirm the self's superiority: everything already lies within and merely awaits reactivation. In the works discussed here, Benn does not yet make a turn toward *völkisch* thinking, but they are certainly compatible. Ernst Bloch notes as much in *Heritage of Our Times*, where he identifies the murmurs and flashes of primordial types in Benn's work as a language "only [...] of escape, of self-enjoyed frenzy, [...] of purely antithetical and hence insubstantial demonism" pointing to the nihilism at the heart of fascism.[143] Nor did the sentimental signature and reflection of memory prevent Benn from the lure of National Socialist ideology. Until at least the mid-1930s, he shared this ideology and supported it in the form a poetic vision that used a fascination with the archaic to bind collective conformity in a strict, martial form, and he deployed artistic support to cultivate, discipline, and "breed" (*züchten*) such submission.[144]

In the essay "Dorische Welt" (Doric World, 1934), for instance, an imaginary Sparta takes the place of the primal human community, the "strong, mighty, beautiful body of breeding and discipline" replaces the atavistic and ecstatic

---

142 Benn, "Problematik des Dichterischen," 3: 247.
143 Bloch, "Songs of Remoteness," in *Heritage of Our Times*, 182.
144 Gottfried Benn, "Dorische Welt," in *Sämtliche Werke*, vol. 4, *Prosa 2 (1933–1945)*, ed. Gerhard Schuster (Stuttgart: Klett-Cotta, 1989). Cf. Bernhard Fischer, "'Stil' und 'Züchtung' – Gottfried Benns Kunsttheorie und das Jahr 1933," *Internationales Archiv für Sozialgeschichte der deutschen Literatur* 12.1 (2009).

body of 'primitive man,'[145] and the stone column, materializing the spirit of power, pushes aside states of intoxication and amorphousness.[146] The content changes, but the figure of thought remains the same: a remote "they" is identified with the "we" of the present, who are destined to inherit the past and complete it: "One can't say it's far off, the ancient world. Not at all! Antiquity is very close, completely within us; the cycle of culture [*Kulturkreis*] is not yet complete."[147] In carrying out this inheritance, the "poet" no longer exercises the function of re-experiencing the archaic physically; instead, he is to follow the "law" the Doric world imposes on the present: "a law for heroes alone, only for one who works in marble and casts heads with helmets."[148] Disciplined and strictly stylized poetry is meant to cultivate the modern warrior-race in a perversion of the humanistic ideal of education (*Bildung*) – analogous to the Spartan column (*Bildsäule*): "das bildet."[149]

---

145 Benn, "Dorische Welt," 4: 137–138.
146 "Die Macht reinigt das Individuum, [...] macht es kunstfähig" (Benn, "Dorische Welt," 4: 150).
147 Benn, "Dorische Welt," 4: 147.
148 Benn, "Dorische Welt," 4: 153.
149 "Human being, that's race with style" (*Der Mensch, das ist die Rasse mit Stil*) (Benn, "Dorische Welt," 4: 152). For Paul de Man, this perversion is inscribed in the very program of aesthetic education: "The aesthetic, as is clear from Schiller's formulation, is primarily a social and political model [...]. The 'state' that is here being advocated is not just a state of mind or of soul, but a principle of political value and authority that has its own claims on the shape and limits of our freedom." ("Aesthetic Formalization in Kleist," in *The Rhetoric of Romanticism* [New York: Columbia University Press, 1984], 264).

Chapter 8
A Sister in Madness: Figures of 'Primitive Thinking' in Robert Musil

Robert Musil knew and admired Robert Müller. In the obituary he wrote for him in 1924, he calls *Tropen* "one of the best novels of modern literature."[1] The two authors share not only an interest in ethnological and psychological literature on the 'primitive,' but also a fondness for expedition stories. The chapter at hand treats Musil's readings in ethnological research and demonstrates its relevance for his literary works. Examining the character Clarisse in *Der Mann ohne Eigenschaften* (1930–1944; *The Man Without Qualities*, 1953) and the motifs of madness, music, and language associated with her, I argue that the novel contains a complex engagement with the paradigm of the 'primitive' and that the movement from expedition to self-experiment that I trace in the author's earlier narratives emerges as the principle of construction behind his major novel as well. The book's protagonist and its author share equally in the primitivisms the novel contemplates, as a mimetic concept of primitivist narration is overwritten with a reflective one, thus realizing the potential of primitivist discourse for a genuinely modern understanding of literature.[2]

Musil's Ethnological Readings

Musil's interest in ethnology has so far received relatively little attention from scholars. Even though there has been discussion of his engagement with the works of Lévy-Bruhl in the essay, "Ansätze zu neuer Ästhetik" (1925, "Toward a New Aesthetic," 1990), this interest has only rarely been examined as part of his larger, in-depth study of additional ethnological and ethnopsychological

---

[1] Robert Musil, "Robert Müller," in *Klagenfurter Ausgabe. Kommentierte Edition sämtlicher Werke, Briefe und nachgelassener Schriften. Mit Transkriptionen und Faksimiles aller Handschriften*, ed. Walter Fanta, Klaus Amann, and Carl Corino, *Lesetexte*, vol. 12 (Klagenfurt: Drava, 2009), 9. Hereafter *KA*.
[2] An earlier and shorter version of this chapter appeared as Nicola Gess, "Expeditionen im Mann ohne Eigenschaften. Zum Primitivismus bei Robert Musil," in *Robert Musil und die Fremdheit der Kultur. Musil-Studien 2010*, ed. Norbert Christian Wolf and Rosmarie Zeller (Munich: Fink, 2011).

Open Access. © 2022 the author(s), published by De Gruyter. This work is licensed under the Creative Commons Attribution-NonCommercial-NoDerivatives 4.0 International License.
https://doi.org/10.1515/9783110695090-008

works,[3] which took place primarily during the early 1920s and continued to a lesser extent into the 1930s.[4] Apart from Lévy-Bruhl's *How Natives Think* (translated into German in 1923), Musil took particular interest in Erich Rudolf Jaensch's "Die Völkerkunde und der eidetische Tatsachenkreis" ("Ethnology and the Eidetic Ring of Facts," 1923),[5] Ernst Kretschmer's chapter "Entwicklungs-

---

3 For a thorough discussion of Lévy-Bruhl, see Renate von Heydebrand, *Die Reflexionen Ulrichs in Robert Musils Roman 'Der Mann ohne Eigenschaften'* (Münster: Aschendorff, 1966), 103–111. Likewise, Roger Willemsen devotes a chapter to Musil's interest in ethnology and contextualizes his reading of Lévy-Bruhl with conceptions of language deriving from Giambattista Vico and Erich Rudolf Jaensch's *Das Existenzrecht der Dichtung* (Munich: Fink, 1984), 286–297. A thorough analysis of Lévy-Bruhl is also found in Ritchie Robertson, "Musil and the 'Primitive' Mentality," in *Robert Musil and the Literary Landscape of His Time*, ed. Hannah Hickman (Salford: University of Salford Press, 1991). Wolfgang Schraml has offered a more comprehensive engagement with Musil's ethnological readings: besides Lévy-Bruhl, Müller-Lyer, Jaensch, and Groos, Schraml takes particular interest in the "appetitive" depiction of the "archaic" in Musil's works (*Relativismus und Anthropologie. Studien zum Werk Robert Musils und zur Literatur der zwanziger Jahre* [Munich: Eberhard, 1994], 127–139). Florence Vatan (*Robert Musil et la question anthropologique* [Paris: PUF, 2000], 73–88) points to the affinities between Agathe, Moosbrugger, and Clarisse and the thinking of 'primitives' in Lévy-Bruhl, as well as to collective rituals in which Musil sees the society of his own day connected to "the childhood of civilization" (80). As the original version of the book at hand was being written, Sven Werkmeister published *Kulturen jenseits der Schrift*, which discusses Musil's interest in ethnology (338–346), but not *The Man Without Qualities*. Other relevant works include Brigitte Weingart, "Verbindungen, Vorverbindungen. Zur Poetik der 'Partizipation' (Lévy-Bruhl) bei Musil," in Ulrich Johannes Beil, Michael Gamper, and Karl Wagner, eds., *Medien, Technik, Wissenschaft: Wissensübertragung bei Robert Musil und in seiner Zeit* (Zurich: Chronos, 2011), and Marcus Hahn, "Zusammenfließende Eichhörnchen. Über Lucien Lévy-Bruhl und die Ethnologie-Rezeption Robert Musils," in the same collection. Confirming my own findings (Gess, "Expeditionen im Mann ohne Eigenschaften"), cf. Norbert Christian Wolf, "Das wilde Denken und die Kunst: Hofmannsthal, Musil, Bachelard," in *Poetik des Wilden. Wolfgang Riedel zum 60. Geburtstag*, ed. Jörg Robert and Friederike F. Günther (Würzburg: Königshausen + Neumann, 2012); and Florian Kappeler, *Situiertes Geschlecht. Organisation, Psychiatrie und Anthropologie in Robert Musils Roman "Der Mann ohne Eigenschaften"* (Munich: Fink, 2012), who, especially in chapter 3.1–3.2 and in reference to Lévy-Bruhl, elaborates in detail the role played by the theorem of "pre-logical thinking" in Musil's critique of modernity.
4 Wolfgang Schraml points out that Musil developed an interest in anthropology as early as 1913–1914 on a visit to Rome, where he visited the institute for anthropology and ethnology, the insane asylum, and the "monkey island" at Villa Borghese (among other places) (*Relativismus und Anthropologie*, 89).
5 Musil refers to Jaensch in the context of his review, "Aus der Begabungs- und Vererbungsforschung" (Musil, *KA*, *Lesetexte*, vol. 10, n.p.), and also in Musil, *KA*, *Transkriptionen*, Mappe II/8/22.

geschichte der Seele" ("Evolution of the Psyche") from his *Text-book of Medical Psychology*,[6] and Franz Carl Müller-Lyer's *Phasen der Kultur* (Phases of Culture, 1915/1908),[7] from which Musil copied out excerpts into his 1922–1923 notebook entries. Other texts include Erich von Hornbostel's reflections on poetry and music[8] and Ernst Cassirer's discussion of the "languages of peoples living in a state of nature,"[9] which Musil read in the the early 1930s. Numerous related titles that the author made note of in 1923 also warrant mention, even though there is no record of him having written out excerpts from them. These include studies by Konrad Theodor Preuss (e. g., *Die geistige Kultur der Naturvölker* [The psychic culture of primitive people, 1914]), Richard Thurnwald's *Forschungen auf den Salomo Inseln und dem Bismarck Archipel* (Research on the Solomon Islands and the Bismarck Archipelago, 1912), Alfred Vierkandt's *Naturvölker und Kulturvölker* (Primitive Peoples and Civilized Peoples, 1896), and a series of publications on early petroglyphs.[10]

In his readings, Musil proceeded selectively, picking out particular aspects of often voluminous studies. A common thread runs through his choices: First, his interest concerns the perception, imagination, and thinking of indigenous peoples – or, more precisely, the phenomenon of participation, their eidetic faculties, and affect-driven thought. Second, he concentrates on the peculiarities of the languages these peoples speak and their relations to them, especially the vividness of such languages, literal understandings of figurative expressions, and the participation of words with their objects. Finally, he takes note of the magical function of indigenous art, which is oriented on producing, not representing, the desired object. These three points will each be briefly elaborated in the following sections.

---

6 Musil refers again and again to Kretschmer (e. g., Mappe II/9/166; IV/3/299, 300, 301, 305; Mappe V/4/19, 108, 109; Heft 21/59). In a footnote to "Toward a New Aesthetic" (in *Precision and Soul: Essays and Addresses*, trans. Burton Pike and David S. Luft [Chicago: University of Chicago Press, 1994], 197), he makes reference to Kretschmer's *Text-book of Medical Psychology*.
7 Musil, *KA*, *Transkriptionen*, Heft 21/55.
8 Musil, *KA*, *Transkriptionen*, Mappe VI/3/6.
9 Musil, *KA*, *Transkriptionen*, Mappe II/9/144.
10 For the complete list, see Musil, *KA*, *Transkriptionen*, Heft 21/114–115. The preceding page presents another bibliographical list taken from Ludwig Klages's *Zum kosmogonischen Eros* (Munich: G. Müller, 1922) on mythology, especially ancient cults and mysteries. There are also further texts that appear neither in bibliographies nor in excerpted form but are mentioned briefly. For instance, in the same notebook, Alexander von Humboldt's *Reise in die Äquinoktial-Gegenden* (Heft 21/73) and "Frobenius's Africa book" (*KA*, *Transkriptionen*, Mappe VII/11/36) are noted.

Red Parrots or 'Primitive Thinking'

In his 1923 review of Jaensch's "Die Völkerkunde und der eidetische Tatsachenkreis,"[11] Musil focuses on a case treated extensively by ethnologists: the Bororo tribe in Brazil, whose members claim to be red parrots, a phenomenon that Lévy-Bruhl explains as "mystical participation" with a totem animal, appealing to collective representations that shape perception. Musil takes up Lévy-Bruhl's notion in "Toward a New Aesthetic" in order to describe the "extraconceptual correspondence of the human being with the world along with abnormal or correlative moments" in "ancient cultural conditions," whose "late form of development" is to be found in the "experience of art."[12] However, he does not take up Lévy-Bruhl's sociological explanation of participation as a form of collective representation, and indeed he contradicts it by tying participation to the *rupturing* of "preformed stable representations"[13] and affirming its status as a radically subjective experience.

In contrast, "Toward a New Aesthetic" displays a closer proximity to Kretschmer's alternative explanation of the same phenomenon. Kretschmer traces the "magical thinking" of the Bororo back to affective projection and catathymia (that is, affect-driven thought): "If scientific thinking (in terms of causality) classifies objects in accordance with the principle of coincidence, magical thinking relates things on the principle of affective identity."[14] For Musil, in order to establish the opposite of "the normal condition of our relationship to the world,"[15] art takes "pre-civilized" measures that obey catathymic laws: "images stimulated by the same affect are [condensed] in masses, to which the sum of the affect attaches itself"; alternatively, a single image becomes "laden with the inexplicably high affective value of the whole" through the process of displacement.[16] It follows that the conception of affect-driven thinking is central for the diverse manifestations of "the other condition" in *The Man Without Qualities* and their recourse to figurative language, whether those be mysticism, delusion, or incestous sibling-love.

Finally, Jaensch's essay offers Musil yet another explanation for the Bororo's parrot phenomenon. Jaensch traces the peculiar "identification of primitives" back to "[eidetic] images" and groupings of seemingly heterogeneous elements

---

11 Musil, "Aus der Begabungs- und Vererbungsforschung," in *KA*, *Lesetexte*, vol. 10, n.p.
12 Musil, "Toward a New Aesthetic," 196–197.
13 Musil, "Toward a New Aesthetic," 201.
14 Kretschmer, *A Text-book of Medical Psychology*, 96.
15 Musil, "Toward a New Aesthetic," 198.
16 Musil, "Toward a New Aesthetic," 195.

into "inner images" that precede optical perception in time (evolutionarily and phenomenologically).[17] Musil's detailed and largely appreciative review affirms his interest in this theory of eidetic intuition among children and indigenous peoples. Jaensch's theories also influence Musil's literary works. For example, in *The Man Without Qualities*, intuitive modes of cognition shape the characters of Moosbrugger and Clarisse. The resulting blurring of the boundaries between perception and cognition culminates in illusions and hallucinations, and the lines separating image from object and self from other vanish. In such cases, the novel presents what Jaensch would call an "immediate" and "emotional" experience of the "inner essence" of things,[18] which Musil in his review of Jaensch's essay associates with "mystic vision,"[19] also featured prominently in *The Man Without Qualities*.

In particular, however, Musil embraces Jaensch's nonjudgmental handling of eidetics, his warning against drawing the wrong conclusion from the atrophied condition of the eidetic faculty, and his aim to return to "primitive stages of development."[20] He follows Jaensch's enjoinder, however, that one should "not return to one's first home without at the same time building a higher and worthier storey"[21] over it. Jaensch's scientific treatise fulfills this demand by representing such a superstructure: it provides rational insights into eidetic phenomena without evaluating them. Musil does the same by calling for a higher level – what he calls "*supra*rationalism" – that examines the "logic of the analogical and the irrational."[22] Musil's essays, as a genre straddling science and literature and aspiring to investigate supposedly irrational phenomena with both logic and sensitivity, thus attempt to answer Jaensch's call.

Given his many points of overlap with the latter, it is not surprising that in "Toward a New Aesthetic" – though it includes no mention of him – Musil's explanation of the "extraconceptual correspondence of man with the world" closely resembles Jaensch's theory of eidetics. What is new with Musil, however, is the moment of rupture (*Sprengung*), which always only acknowledges such a correspondence momentarily and only against the backdrop of a long-established world of schemata that is in need of disruption.

---

[17] Erich Rudolf Jaensch, "Die Völkerkunde und der eidetische Tatsachenkreis," *Zeitschrift für Psychologie* 91 (1923): 106–107.
[18] Jaensch, "Die Völkerkunde und der eidetische Tatsachenkreis," 107.
[19] Musil, "Aus der Begabungs- und Vererbungsforschung," in *KA, Lesetexte*, vol. 10, n.p.
[20] Jaensch, "Die Völkerkunde und der eidetische Tatsachenkreis," 97.
[21] Jaensch, "Die Völkerkunde und der eidetische Tatsachenkreis," 111.
[22] Musil, "Mind and Experience," in *Precision and Soul*, 142. Emphasis added.

On 'Primitive Language' and Its Magic

Musil's notes reveal his particular interest in the languages of indigenous peoples and their understandings of language. Vivid or pictorial language is central for him. Thus, he appreciatively remarks on Franz Carl Müller-Lyer's word choice in his distinction between the "paper money of words" and the "hard currency of mental representation" (*Scheidemünze der Vorstellung*).[23] Even though Müller-Lyer wishes to express the superiority of conceptual thinking ("paper money"), Musil makes use of those same metaphors to reach the opposite conclusion. He is drawn more to the imagery employed by language (ascribed by Müller-Lyer to "peoples in a state of nature") than to its conceptual content.

For Lévy-Bruhl, the vividness of indigenous languages results from their precise and detailed representation of objects. Kretschmer, on the other hand, focuses on their pictorial or figurative character by examining the laws of "image agglutination" and stylization at work in them. Musil took notes on the latter's analysis,[24] and his later notes on Cassirer demonstrate interest in the similar argument that the first "image-concepts" are formed according to pre-existing similarities or analogies.[25] Hereby, Musil distinguishes between sensory and emotional associations. The latter hold particular significance; his notes on Kretschmer emphasize this point, and in *The Man Without Qualities* characters associated with the "other condition" think along lines determined by affect and make use of figurative language.

According to Musil's notes on Lévy-Bruhl and Jaensch, this dimension and understanding of language went missing in the course of phylogenetic development. "Advance in conceptual and abstract thought is accompanied by a diminution in the descriptive material which served to express the thought when it was more concrete,"[26] Lévy-Bruhl observes. Or, in Musil's own words (apropos of Jaensch), "precisely this transition from intuition to non-intuition is [...] connected with the acquisition of conceptual thought."[27] The notes on Lévy-Bruhl extend this thesis from individual words to sentences, which, in Musil's view, are subject to the same "process of typification."[28] This is one reason why he laments that

---

[23] Franz Carl Müller-Lyer, *Phasen der Kultur und Richtlinien des Fortschritts. Soziologische Überblicke* (Munich: Lehmann, 1915), 34; Musil, *KA*, *Transkriptionen*, Heft 21/59.
[24] Musil, *KA*, *Transkriptionen*, Mappe I/5/171.
[25] Musil, *KA*, *Transkriptionen*, Mappe II/9/144.
[26] Lévy-Bruhl, *How Natives Think*, 152.
[27] Musil, "Aus der Begabungs- und Vererbungsforschung," in *KA*, *Lesetexte*, vol. 10, n.p. Jaensch, "Die Völkerkunde und der eidetische Tatsachenkreis," 105.
[28] Musil, *KA*, *Transkriptionen*, Heft 21/115.

the modern world lacks a formal language for expressing "the non-ratioid," his coinage for what lies beyond the grasp of reason.

Unlike many of his contemporaries, this insight does not prompt Musil to demand a "redemption from the conceptual" (for which he criticizes Béla Balázs in his notes on Kretschmer).[29] Instead, taking up Kretschmer's reflections, he credits "primitives" with having concepts – "primitive art is conceptual"[30] – albeit ones that follow different rules or still retain pictorial traits. To employ Lévy-Bruhl's and Kretschmer's shared terminology, at issue are image-concepts whose meanings still refer to singular entities but are also already inching toward classification. This is why Musil has no wish to do without the concept, as it would lead to amorphous "chaos."[31] Instead he seeks to revitalize the pictorial or figurative potential of the concept in order to articulate the non-ratioid.

In addition to the vividness of language treated thus far, Musil is interested in the identification of figurative language with its object. This aspect of language is equally emphasized by Lévy-Bruhl, Jaensch, and Kretschmer and plays an important role in Musil's notes on them. It involves taking figures of speech literally as well as the idea that an actual, ontological connection exists between objects joined by figurative language (e.g., the lightning *is* a snake). At the same time, the operation involves the idea that the word participates in the object it designates. Thus, in a notation to excerpts he had copied from Kretschmer, Musil observes, "these 'schizophrenic symbols are .. products of incomplete thinking, imagistic forerunners of concepts that .. do not .. get formed' [...]: burning becomes real fire, etc."[32]

According to Lévy-Bruhl, figurative language's double identification with its object derives from the participatory thinking of indigenous peoples: "To their minds [...] there is no perception unaccompanied by a mystic complex, no phenomenon which is simply a phenomenon, no sign that is not more than a sign: how, then, could a word be merely a word?" Indeed, because the vividness of the language creates a relationship of likeness, the word displays the same mystical qualities as the object. Hence the participations it can produce may be equally significant and "frightening": "There is magic influence in the word, and therefore precaution is necessary. Special languages for certain occasions, languages reserved for certain classes of persons, begin to take shape."[33] This linguistic magic stands at the center of Musil's interest in ethnology in the early 1930s.

---

29 Musil, *KA, Transkriptionen*, Mappe IV/3/303.
30 Musil, *KA, Transkriptionen*, Mappe IV/3/307, 303.
31 Musil, "Toward a New Aesthetic," 204.
32 Musil, *KA, Transkriptionen*, Mappe I/5/171.
33 Lévy-Bruhl, *How Natives Think*, 154.

In the context of the essay "Literat und Literatur" (1931; "Literati and Literature," 1990), he took detailed notes on a lecture by Hornbostel on the music/poetry of ancient civilizations. Two aspects of the discussion command his attention: First, poetry is ritual song that does not represent an event but produces it. Second, to this end, a highly specific content is necessary, which says "what must be done," as is a highly specific form that indicates "how it must be done." At the same time, "the form [is] given in the course of the event, which is its content."[34] This complex structure amounts to the same performative logic of ritual theorized by Durkheim as well as Mauss and Hubert addressed in Chapter 2. Accordingly, Musil also talks of magic words that express nothing other than the performative force of speech itself.[35]

In sum, Musil's theory of 'primitive language' concentrates on its vividness, guided by an affective logic, as well as on its performative force, where representation and creation collapse as one. As we will see below, Musil adapted these notions for his own writing.

Animal-Humans: From Expedition to (Self-)Experimentation

Musil's fiction of the early 1920s also reflects his fascination with ethnology. On the model of works such as Müller's *Tropen* and Jensen's *Skovene* (1907), which he held in high regard,[36] Musil wrote expedition stories including *Grigia* (1921; in *Five Women*, 1999) and the unfinished *Land über dem Südpol* (Land over the South Pole, 1911–1929). Both works describe a shift from the foreign to the familiar (accompanied by a turn from expedition to self-experiment) and reflect on the animalistic nature of humans.

The beginnings of *Land über dem Südpol* date back to 1911, and the author's final notes on the project are from 1929. The outlines for its plot call to mind

---

34 Musil, *KA, Transkriptionen*, Mappe VI/3/6.
35 Musil, *KA, Transkriptionen*, Mappe VI/3/8.
36 Musil mentions both texts in his obituary for Müller. Müller also wrote other texts of this genre, e.g., the novella *Das Inselmädchen* (The Island Girl, 1919), which Musil also mentions, though without going into detail. In a letter Musil asks Arne Laurin to send a number of books to a reviewer, including several titles of the same kind, e.g., Douglas Mawson's *The Home of the Blizzard* (1915), Paul Gauguin's *Letters from the South Seas*, Count Vay de Vayas's account of emigration to America (1908), J.V. Jensen's *Das verlorene Land* (1920; The Lost Land [of *The Long Journey* series]; Danish 1908–1922) and Ejnar Mikkelsen's *Frozen Justice: A Story from Alaska* (1922; Danish 1920) (Musil to Arne Laurin, 12 March 1921, in *KA, Lesetexte*, 1921, n.p.). This is further indication that Musil was familiar with the genre (or at least interested in it).

other travel narratives of the day such as Alfred Kubin's account of a fantastic journey, *Die Andere Seite* (1909; The Other Side, 1973), and Müller's *Tropen*. As in the latter work, Musil's story is framed by the remarks of a fictitious editor, who claims to have been given a manuscript by the actual traveler-protagonist. Like Müller's narrator-protagonist, the main character and narrator of the manuscript is a scientist (a mathematician). Grappling with the moral and intellectual limits of his day, he embarks on an expedition to a distant country that is both promising and unknown to him. As in Kubin's novel (whose protagonist is a draftsman), the fictitious editor questions the narrator's mental health from the outset. Musil's narrator is obsessed with the idea that a planet exists above the South Pole and believes he can prove its existence mathematically. The manuscript reports his experiences on this other planet, much like Kubin's explorer records an undiscovered empire deep in Siberia. In each case, the utopian otherness of an alien society fascinates its visitors. Kubin depicts a place where the laws of dreams prevail and where people live somewhere between the realm of "fairytales" and that of "mass hypnosis," both protected by and under the thumb of a dictator whose metamorphoses make him present and absent at once. Musil, in turn, devises a culture centered on human experiments occurring on the planet Ed. Kubin's novel culminates in an apocalyptic scenario triggered by the explorer's eventual resistance to the way of life of the inhabitants of the foreign society, whose entire culture is then wiped out. Musil never finished *Land über dem Südpol*; the fate of Ed remains unknown, but the protagonist winds up in an insane asylum.

The notebook in which Musil collected notes on Müller-Lyer, Lévy-Bruhl, and other ethnologists also contains a number of entries that loosen the boundary between animals and humans.[37] In one regard, this is found in attributions of human (i.e., cultivated) behaviors to animals – for instance (apropos of a work by Müller-Lyer), chimpanzees build huts similar to those of indigenous peoples.[38] Yet, in the other direction, humans are ascribed behaviors that seem animalistic, even brutish. Thus (again relying on Müller-Lyer) Musil writes, "the inferior hunter" will gladly eat "ant eggs, worms, frogs, larvae [...], snakes, lice," and other things that elicit disgust in the civilized European;[39] he also records that an "intelligent Indian" with whom Alexander von Humboldt interacted was a cannibal.[40] Along similar lines, Musil notes how simple fishermen on

---

37 Musil, *KA, Transkriptionen*, Heft 21.
38 Musil, *KA, Transkriptionen*, Heft 21/59.
39 Musil, *KA, Transkriptionen*, Heft 21/59.
40 Musil, *KA, Transkriptionen*, Heft 21/73.

the island of Usedom tear apart worms and impale them with hooks with perfect peace of mind.[41] One page earlier, he mentions the cruelty of children at play – which he explains as a means of acting out "the mightiest instincts of the genus," thereby also likening them to animals (the same occurs in Karl Groos's *Die Spiele der Tiere* [1896; *Play of Animals*, 1898], to which Musil is referring).[42]

This convergence is also evident in Musil's most fully outlined human experiment in his drafts for *Land über dem Südpol*. Here, "human beings are made to run through all [cycles] of animals. By biological means. They sate themselves and work off [animal impulses]."[43] Specifically, Musil took keen interest in the reproductive habits of amphibians and took copious notes on them from Alfred Edmund Brehm's *Tierleben* (1864, 1869; *Brehm's Life of Animals*, 1896), particularly on rites involving the killing of a mate and passages finding parallels between those and human behaviors (such as sexual murder, exhibitionism):

> Mating with newts and salamanders: They swim past each other several times, then the male deposits the sperm on the ground, the female fetches it and introduces it to herself. – Utter cessation of the pleasure of coitus, the stealth of the female, like an exhibitionist at the streetlight.[44] The praying mantis begins eating the male already during the sexual act, which neither one minds. Male toad carried around for days by the female. Sometimes jumps on passing fish, rides clinging to eyes and doesn't let go until it has killed it. (Sex murder). [...] Men whose whole frame is enlarged in erection. Grows four times as large in length and width.[45]

Musil's notes reflect the stereotypical assumption that every human harbors a "beast" within that must be "worked off" and that such abreaction can take the form of sexual violence.[46] This may sound like a culture of perverse desires, but in fact it is devoted to experimentation:

> a world ruled by few who live in a kind of monastery. A world divided into experimental fields. [...] People who live out all intellectually possible constellations [...], people, whose energy lies entirely in the spiritual realm.[47] Their moral experiments cannot be verified on earth, or only with difficulty, and are therefore restricted to their astral laboratory.[48]

---

41 Musil, *KA*, *Transkriptionen*, Heft 21/54, 56.
42 Musil, *KA*, *Transkriptionen*, Heft 21/53.
43 Musil, *KA*, *Lesetexte*, vol. 16, IV. Wien/Berlin, 21: *Die zwanzig Werke* III, 210.
44 Musil, *KA*, *Transkriptionen*, Mappe IV/2/521.
45 Musil, *KA*, *Lesetexte*, vol. 16, IV. Wien/Berlin, 8: *Die zwanzig Werke* I, 210–211.
46 Musil, *KA*, *Lesetexte*, vol. 16, IV. Wien/Berlin, 21: *Die zwanzig Werke* III, 210.
47 Musil, *KA*, *Lesetexte*, vol. 15, *Erzählerische Fragmente*, *Die zwanzig Werke*, *Das Land über dem Südpol* (Planet Ed), 1–2.
48 Musil, *KA*, *Lesetexte*, vol. 16, III. Erster Weltkrieg, II: Klein Grau, 79.

As both experimenters and test objects, these people lead separate mental and physical lives, but they identify only with the first state of existence. The events that affect them corporeally concern them but little. Their only true concern is their research.

> They themselves are not subject to passion [...] While they do have sympathies and the like, they know what they derive from (the animal kingdom) and don't take them seriously. Their life might be monotonous, but an immense pressure to work makes it complete. Ataraxia, or the contemplation of God, might represent something higher, but this and other things are precisely what they're trying to discover. They don't have any solutions yet – there's not enough time for that – but they're conducting interesting experiments.[49]

These researchers display an attitude that may be described as selective self-estrangement: they observe the animal inside, but they do not identify with its appetites or actions. Here, the exploration into alien space leads to an experiment on the alienated element within the explorer-protagonist. Tellingly, the 'aliens' living on the planet turn out to be alienated forms of the self, Germans who had emigrated long ago. The visitor is thus not conducting ethnology so much as inverse-ethnology.[50]

The novella, *Grigia*, where Musil processes his experiences of the First World War, tells of another expedition. Homo, a scientist who has grown weary of civilization, sets out for foreign terrain, but, like Müller's *Tropen*, this work is set in a pre-civilized society rather than a futuristic setting.[51] The protagonist arrives at a "pre-historic lake village built on piles" where the ways of "bygone centuries" live on, and the narrative compares the practices of the villagers to those of "Negroes": they speak "magic words," and the explorer's time among them is likened to "living among savages."[52] As on the planet Ed, the female residents, who alone retain the third-person limited omniscient narrator's interest, display a marked proximity to animal existence characterized not by bizarre sexual practices but because the women exhibit a gentle bovine nature. Their instincts are intact (e. g., the animalistic will to survive), yet they are expressed through an un-

---

49 Musil, *KA*, *Lesetexte*, vol. 16, III. Erster Weltkrieg, II: Klein Grau, 79–80.
50 According to Vatan (*Robert Musil et la question anthropologique*, 75), Musil adopts the attitude of an ethnologist in order to distance himself from the world with which he is familiar and obtain a new perspective. See also Werkmeister, *Kulturen jenseits der Schrift*, 341–343.
51 This is convincingly demonstrated by Schraml (*Relativismus und Anthropologie*, 140–142). See Werkmeister, *Kulturen jenseits der Schrift*, 343–353, for a reading of Musil's ethnological interest in light of media theory.
52 Robert Musil, "Grigia," in *Five Women*, trans. Eithne Wilkins and Ernst Kaiser (Boston: Godine, 1999), 21, 27, 50.

reflecting oneness with nature, involving a placid fulfillment of basic natural needs, including sexual ones. Accordingly, the traveler calls his local love interest by the name of her cow, Grigia.

As in *Land über dem Südpol*, the novella soon reveals that the women of this faraway village actually represent an alienated form of the self – they descend from German emigrants – and the expedition into the foreign turns out to be inverse-ethnology instead. Moreover, and as occurs in Müller's *Tropen*, the travelers themselves quickly regress to a state of savagery. However, the post-civilized, masculine savagery takes the form of uninhibited indulgence in egocentric desires – for instance, sexual assaults on the women, cruel punishments of workers, and the sadistic slaughter of animals. In the course of evening drinking bouts, members of the expedition even lose the ability to speak, interacting with each other only in an "animal language"; meanwhile, the women communicate in a language that belongs to the past but is still human.[53]

The bestiality displayed by once-civilized men – which serves Musil's critique of civilization itself – contrasts with the experience of participation that Homo makes in the course of the same regression:

> Here, amid the secrets of Nature, their belonging together was only one secret more. [...] Among the trees with their arsenic-green beards he sank down on one knee and spread out his arms, a thing he had never done before in all his life, and it was as though in this moment someone lifted him out of his own embrace [...] as though he were being cast in the mould of some other body.[54]

This "belonging together" involves Homo's relationships to his absent wife and to nature, which both merge in Grigia. Yet in contrast to Grigia, such belonging leads Homo to an anticipation of death, which distinguishes him from her as a former member of modern civilization. For him participation is incompatible with the civilized view of life, defined as an agonistic relationship between the self and other, where no integration between the two is possible outside of death.

Thus, Musil's ethnological interest in the phenomenon of participation reemerges in *Grigia*. Once again, this phenomenon is relocated from an ethnological expedition to a confrontation with the alienated self. This estranged self has two faces: The first is the post-civilizational face of savage masculinity that conceives of sexuality in relation to cruelty and crime. The second is the face of a pre-civilizational, feminine, and creaturely devotion to the laws of nature. Homo – whose very name dictates his anthropological significance to the reader

---

53 Musil, "Grigia," 37–38.
54 Musil, "Grigia," 32.

– wears both countenances. He stalks the women of the village and calmly observes the agonizing death of a fly while musing about how he can "feel the presence of God and yet kill."[55] But in other instances, he experiences moments of participation like the one I have cited above. On this score, he resembles the inhabitants of Ed, who experiment with cruel mating rituals while at the same time aiming to see God.

Clarisse – A Sister in Madness (*The Man Without Qualities*)

Musil's great novel, *The Man Without Qualities*, also contains a complex treatment of the paradigm of the 'primitive.' Here, too, the movement from expedition to self-experimentation acts as the novel's constructive principle. By this I do not refer only to the laboratory gaze of the narrator, Ulrich, and the author himself (in his comments on the novel),[56] but also to how the thematic treatment of the 'primitive' in his earlier stories returns in the structure of this highly reflective work. Musil's early works expose the 'primitive' as an obvious construction – after all, no land or planet exists above the South Pole. Rather, the 'primitive' refers to a defamiliarized version of the familiar and an alternative relationship to the world. Therefore, in his novel, and in contrast to Müller and his protagonist Brandlberger, Musil and his alter ego, Ulrich, are not interested in examining the ethnographic other so much as inverting this ethnological perspective to discern the foreign within their own culture: "The primitives here at home are more foreign to us than those of the South Seas."[57] Thus, psychology replaces ethnology; however, it remains inspired by an ethnological perspective. Among other things, this means that foreign aspects of one's own culture remain encoded as 'primitive.' Such coding is tied to a genealogy (whereby alien elements of the self have ancient roots), to an anthropological thesis (that an authentically human faculty is at work in such elements), and to a negative evaluation of re-

---

55 Musil, "Grigia," 40.
56 Cf. Thomas Hake, *'Gefühlserkenntnisse und Denkerschütterungen.' Robert Musils 'Nachlaß zu Lebzeiten'* (Bielefeld: Aisthesis, 1998), 122–161; see also Andrea Pelmter, *'Experimentierfeld des Seinkönnens' – Dichtung als "Versuchsstätte." Zur Rolle des Experiments im Werk Robert Musils* (Würzburg: Königshausen + Neumann, 2008). On experimentation in literature more broadly, cf. Marcus Krause and Nicolas Pethes, eds., *Literarische Experimentalkulturen. Poetologien des Experiments im 19. Jahrhundert* (Würzburg: Königshausen + Neumann, 2005).
57 Robert Musil, "Bücher und Literatur," in *Gesammelte Werke in neun Bänden*, ed. Adolf Frisé (Reinbek: Rowohlt, 1978) 8: 1171. Werkmeister also speaks of "inverted ethnology" in reference to the "possibility of a foreign culture calling into question European forms of knowledge and thinking" (*Kulturen jenseits der Schrift*, 341–342).

gressive movements. The self-experimenter's gaze that forms over the course of the expedition furthermore differs from a typical laboratory situation in that the subject conducting the experiment is himself involved in it – be that in the sense of a change in thinking, as seen in Ulrich, or in writing, as was the case for Musil himself. Both participate in the primitivisms the novel reflects.

The following develops the first part of my thesis (that the novel can be read as an expedition into the 'primitive') by examining the figure of Clarisse. (The second part of my thesis – that the novel moves from expedition to self-experiment – will be treated in the second half of the chapter.) Earlier drafts of *The Man Without Qualities* contain numerous indications that the mentally ill Clarisse has been shaped by the paradigm of the 'primitive.'[58] In a draft from the "Siamese Twins" stage of the novel (1923–1926), when Clarisse is hospitalized for the first time, she writes that in earlier times "religiously awakened" individuals such as Francis of Assisi had the opportunity to "live, teach, and lead their contemporaries" – whereas now they get locked up in psychiatric institutions for mania.[59] In a draft from the "Spy" stage of the novel (1918–1921), Musil still brings up these thoughts in his reflections on his novel, noting, "What

---

[58] The novel's genesis is complex and took place over multiple writing phases. Musil worked on it between 1918 and 1921 under the title "Der Spion" (The Spy), during the period from 1921–1922 under the title "Der Erlöser" (The Savior), between 1923–1926 under the title "Die Zwillingsschwester" (The Siamese Twins), and finally during the period from 1927–1830 under the title *Der Mann ohne Eigenschaften* (*The Man without Qualities*). In 1930, the first volume of the novel was published in two parts (cited here as Kap. I/1–I/19; I/20–I/123). From 1930 to 1932 Musil worked on the second volume and published its first part in 1932 (Kap. II/1–II/38). Countless sketches, notes, and drafts dating back as far as 1918 that relate to both volumes are found in Musil's posthumously published papers. After 1932 Musil worked on various strands for the continuation of the second volume, including the chapter complex "Clarisse" (1933–1936), which would eventually consist of six chapters (including "Besuch" [Visit] und "Insel" [Island]), but these were only published posthumously. The so-called "Druckfahnen-Kapitel" (Galley-proof chapters), which also include the chapters on the psychology of emotion, are dated to 1937–1938 and include 20 chapters that Musil put into print during these years, but then continued to work on and withdrew from publication. These chapters should have been the sequel to Part 1 of the novel's second volume, but were in fact never published during Musil's lifetime. The "Mappen" (files) und "Hefte" (notebooks) cited in the following are part of Musil's posthumously published papers, as are the s-Drafts (1924–1925) and the C-Drafts (1918–1921). These papers are cited from the Klagenfurter Edition (*KA*), which is organized into four parts: *Lesetexte* (Readings), *Transkriptionen* (Transcriptions), *Faksimiles* (Facsimiles), and *Kommentare und Apparate* (Commentary and Critical Apparatus).

[59] Whenever possible, quotes are taken from Robert Musil, *The Man Without Qualities*, trans. Sophie Wilkins, 2 vols. (New York: Vintage, 1996). Otherwise, they follow Robert Musil, *Der Mann ohne Eigenschaften*, ed. Adolf Frisé (Reinbek: Rowohlt, 1995). Here: Musil, *Der Mann ohne Eigenschaften*, 1734.

today is still possible only in a mountain village (with consequences corresponding to this situation), then [could] occur at a center of culture."[60] In other words, Clarisse is identified as a relative of the primitivist figure of Grigia. Other indications of the relationship between Clarisse and figurations of the 'primitive' are the "discoveries" she makes in the s- and C-drafts of the "Island" chapter (1924–1925, 1918–1921, respectively) – for instance, that "vanished forests of the carboniferous era [...] are being freed again today [...] as psychic forces." Accordingly, the narrator compares the nighttime noises on the island to the excitement of an "African village starting a ritual dance,"[61] and credits Clarisse with a native understanding for the secret "dance rhythms of primal peoples."[62] The drafts for *The Man Without Qualities* thus provide some cues that the Clarisse storyline involves yet another variation on Musil's expedition narratives. In 1932, Musil in fact called the book "a mental expedition and research trip."[63] In the following, the novel's references to Clarisse as a figuration of the 'primitive' will be explored in light of three complexes: primitivism in mania and schizophrenia, in music, and in language and poetry.

Primitivism in Mania and Schizophrenia

Early twentieth-century psychopathology understood certain mental disorders, especially schizophrenia, in terms of regression to the phylogenetic stage of 'primitive thinking' (see Chapter 4). Ernst Kretschmer expresses a view held by many others: "In schizophrenic thinking, [...] large cohesive features of the primitive world-pictures are made to live again before our eyes."[64] Freud, for instance, expected psychoanalysis to shed light not only on mental illness but also on the mental operations of prehistoric humankind.

> Neuroses seem to have preserved more mental antiquities than we could have imagined possible; so that psycho-analysis may claim a high place among the sciences which are concerned with the reconstruction of the earliest and most obscure periods of the beginnings of the human race.[65]

---

60 Musil, *Der Mann ohne Eigenschaften*, 1801.
61 Musil, *The Man Without Qualities*, 1566.
62 Musil, *Der Mann ohne Eigenschaften*, 1787.
63 Musil, *KA, Transkriptionen*, Mappe II/1/65.
64 Kretschmer, *A Text-book of Medical Psychology*, 134.
65 Freud, *The Interpretation of Dreams*, 550; cf. 566.

According to this reasoning, the appropriate science for researching the 'primitive' would not necessarily be ethnology, but rather psychology, which uses psychological disorders of its present day to explore the secrets of 'primitive thinking.' In this sense, medical psychopathology and depth psychology of the early twentieth century establish a primitivism of mental illness. Musil takes up this discourse by characterizing Clarisse as a manic-schizophrenic and therefore 'primitive figure.'[66]

Three studies in particular informed his work on the novel in the early 1930s, when Clarisse's character was taking shape in the chapters "Early-morning Walk," "Armistice," and "Hermaphrodite," where her illness (Musil speaks of "incipient manic activity"[67]) erupts and becomes a chronic condition.[68] In 1933–1934 he engaged intensively with Eugen Bleuler's *Lehrbuch der Psychiatrie* (1904; *Text-book of Psychiatry*, 1924) (whose chapter on affectivity he had already taken copious notes on while working on another of the novel's characters, Moosbrugger, in 1927). Other works the author consulted at this time include Kretschmer's *Text-book on Medical Psychology* (whose chapter on folk psychology [*Völkerpsychologie*] he had already read carefully while writing "Toward a New Aesthetic" in 1924) and Traugott Konstantin Oesterreich's *Die Phänomenologie des Ich in ihren Grundproblemen* (The Phenomenology of Ego in its Fundamental Problems, 1910).

During this writing period, many of Clarisse's idiosyncrasies, which had already been developed in previous chapters and chapter drafts, are reworked into manic symptoms and receive diagnosis. The latter, however, is only made explicit in the notes. And while the narrator still draws attention to the symptoms in early drafts, these markers are cut from later versions. These revisions point to the difference between the Clarisse narrative and the case history genre. Case histories usually serve to prove a certain theory or diagnosis.[69] Yet here the case comprises the theory, or rather the latter is not to be had without the former.

---

66 For the cultural and historical context, cf. Anz, "Schizophrenie als epochale Symptomatik."
67 Musil, *KA*, Transkriptionen, Mappe V/4/205.
68 In contrast to discussions of Moosbrugger, scholars have devoted little attention to Clarisse. See especially, Silvia Bonacchi, *Die Gestalt der Dichtung. Der Einfluss der Gestalttheorie auf das Werk Robert Musils* (Bern: Peter Lang, 1998), 249–259; Gislind Erna Pietsch Pentecost, "Clarisse. Analyse der Gestalt in Robert Musils Roman 'Der Mann ohne Eigenschaften'" (PhD thesis: Purdue University, 1990), 12–44; at the same time as the German version of the study at hand appeared, so did Norbert Christian Wolf, *Kakanien als Gesellschaftskonstruktion. Robert Musils Sozioanalyse des 20. Jahrhunderts* (Vienna: Böhlau, 2011), cf. 684–694.
69 On the relation of case history and literature in general, cf. Pethes, "'Vom Einzelfall zur Menschheit'"; and Pethes, "Versuchsobjekt Mensch. Gedankenexperimente und Fallgeschichten als Erzählformen des Menschenversuchs," in *Experiment und Literatur. Themen, Methoden, Theorien*, ed. Michael Gamper (Göttingen: Wallstein, 2010).

I will return below to Musil's criticism of the procedures of psychology implicit in this arrangement.

According to Bleuler and Kretschmer, a key feature of mania is the flight of ideas:

> The *thinking* of the manic is flighty. He jumps by by-paths from one subject to another and cannot adhere to anything. With this the ideas run along very easily and involuntarily, even so freely that it may be felt as unpleasant by the patient. [...] The thinking is incomplete and flighty but not "unclear" in the sense of psychopathology. Up to advanced stages of the disease one can converse with the manic.[70]

Clarisse displays precisely this way of thinking even in the early stages of her illness:

> In fluttering mists, images sprang up, overlapped, fused, faded – that was Clarisse's thinking. She had her own way of thinking; sometimes several ideas were intertwined simultaneously, sometimes none at all, but then one could feel the thoughts lurking like demons behind the stage. The temporal sequence of events that gives such real support to most people became in Clarisse a veil that threw its folds one over the other, only to dissolve them into a barely visible puff of air.[71]

In turn, a flighty state of mind becomes her permanent condition. As Musil puts it in a draft,

> [w]hen she gave herself over to reflection, a thousand and one things occurred to her. For example, just as she could see herself addressed as a man [...] and feel herself a man and a woman at once, and therefore really [...] a double being, [...] she could feel herself also to be related to the [mentally] ill, for they are double beings, too [...]. But relations also extended in many other directions. [...] The reciprocal relations yielded a whole, and new points of departure emerged in almost unlimited number.[72]

Indeed, "a thousand and one things" (*vom Hundertsten ins Tausendste*, literally, "from the hundreth to the thousandth," meaning to ramble and get carried away) is a turn of phrase taken from Bleuler, who uses it to describe thoughts that jump erratically from one thing to the next. Such activity is defined by lightness and speed and evidently leads to deception as "the patient spends much less than the normal time on the individual idea."[73] As Kretschmer elaborates, "in the case of the manic very many more images are rushed through the

---

70 Eugen Bleuler, *Textbook of Psychiatry*, trans. A. A. Brill (New York: Macmillan, 1924), 466.
71 Musil, *KA, Transkriptionen*, Mappe I/5/107; quoted from *The Man Without Qualities*, 152.
72 Musil, *KA, Transkriptionen*, Mappe II/1/12.
73 Bleuler, *Textbook of Psychiatry*, 72.

focal zone of consciousness,"[74] but none of them stays put, and nothing is thought through. In Bleuler's words, "flighty thinking is not aimless in content, although its aim is forever changing."[75] The rapidity with which mental representations change implies, as Kretschmer writes, that the manic individual "feels a pressure of thought."[76] This pressure is exhibited in Clarisse's sense that her thoughts are not her own, that she is their object and not the other way around: "Her thoughts went now one way – as if she were only an instrument on which a strange and higher being were playing."[77] This impression corresponds to Bleuler's remark that "patients feel that 'it' thinks in them."[78]

Musil accounts for the correlations between the various thoughts racing through Clarisse's mind in different ways. In the narrator's estimation, they follow a logically accidental but nonetheless organic sequence:

> [w]hat of such possibilities became reality in Clarisse's mind and what did not, could not be predicted in detail, and it may be called coincidence. She felt it herself. There was something in her that could not be expressed in ordinary terms, and from it arose, as naturally as a tree brings forth a thousand leaves, manifold thoughts, indeed an unlimited multiplicity of them.[79]

The coincidental impression made by such thoughts is due to them having no discernible bracket or heading grouping them together. The appearance of organic continuity prevails, however, because individual connections between elements can be identified. In Kretschmer's words, there is "no trace of a dominant concept; on the contrary, each idea is recognizably linked to the next." Because the dominant concept is not clear, he also describes such flights of ideas as "picture-strip thinking,"[80] whose proximity to the "asyntactical series of images" that play a key role in concepts of 'primitive thinking' is unmistakeable.[81] Whereas, according to Kretschmer, 'primitive thinking' works its way up to abstraction by agglutinating images that ultimately yield logical categories, thinking directed by flights of ideas takes the opposite course; here, sensory images, which are given with language, "become once more detached from the abstractions."[82]

---

[74] Kretschmer, *A Text-book of Medical Psychology*, 148.
[75] Bleuler, *Textbook of Psychiatry*, 72.
[76] Kretschmer, *A Text-book of Medical Psychology*, 149.
[77] Musil, *The Man Without Qualities*, 1611.
[78] Bleuler, *Textbook of Psychiatry*, 152.
[79] Musil, *KA, Transkriptionen*, Mappe II/1/12.
[80] Kretschmer, *A Text-book of Medical Psychology*, 147.
[81] Kretschmer, *A Text-book of Medical Psychology*, 84.
[82] Kretschmer, *A Text-book of Medical Psychology*, 143.

In brief, such thought becomes eidetic again – without super- or subordination – structured only by juxtaposition.

The principle governing juxtaposition, according to the narrator, is association or analogy: "It was analogies in which [...] Clarisse thought [...]. For what Clarisse considered a special way of thinking were apparently analogies."[83] This view corresponds to Kretschmer's position when he observes that elements of thought entertain a relationship of contiguity or similarity. Likewise, Bleuler stresses secondary associations, that is, "external" features that do not demonstrate logical coherence, as when he notes that "in place of inner associations there may be accidental connections, [...] which do not even emanate from the sense of the word but from its sound."[84] This dynamic characterizes Clarisse's thinking throughout, for example in chapter I/38:

> "Snakes!" Clarisse thought. "Snakes!" These events entangled her, trapped her, kept her from getting where she wanted to go, were slippery, and made her aim at a target she did not want. Snakes, snares, slippery; that was life's way. Her thoughts began to race like life.[85]

Since such thought lacks a dominant concept, it can be led astray not only by secondary associations, but also – and just as readily – by sensory impressions. For both Bleuler and Kretschmer, manic thinking is especially susceptible to distraction.[86] The passage quoted above illustrates this point. Playing the piano, Clarisse's thoughts form out of the image of the black and white keys moving before her.

> She saw it all before her like swarms of black birds fluttering around a little girl standing in the snow. But somewhat later she saw a black wall with white spots in it; black stood for – she didn't know, and while the white ran together to form little, and sometimes larger, islands, the black remained unchangingly infinite. This blackness emitted fear and agitation. "Is this the devil?" she thought. "Has the devil turned into Moosbrugger?" Between the white spots she now noticed thin gray tracks; on these she had moved from one thing to the next in her life.[87]

The scene also shows yet another feature of "flights of ideas" at work. When a perceptible contiguity or similarity between successive thoughts is missing, intel-

---

83 Musil, *KA, Transkriptionen*, Mappe II/1/13.
84 Bleuler, *Textbook of Psychiatry*, 72.
85 Musil, *The Man Without Qualities*, 154.
86 Kretschmer, *A Text-book of Medical Psychology*, 145; Bleuler, *Textbook of Psychiatry*, 72, 237.
87 Musil, *The Man Without Qualities*, 154.

lectual activity proceeds along the lines of affect or "catathymia." Bleuler writes that affect conditions thinking in two ways: by favoring associations that correspond to an actual emotion and by increasing the value attached to ideas associated with it. In consequence, "logic becomes falsified." Bleuler speaks of a catathymic impact of unconscious affects on thought processes.[88] By contrast, Kretschmer applies the term catathymia to every "transformation of the psychic content by affective influences."[89] Catathymia also plays a central role in Musil's characterization of Clarisse's mental world, as displayed in his notes on her character: "some catathymic tendencies of feeling produce a much larger, indeed an unlimited, multiplicity of thoughts."[90]

Considerations of affect-driven thinking shaped the character of Clarisse from the outset. Thus, in one of the s-drafts from the early 1920s, one reads,

> [w]hat was taking place might have been causal, necessary, mechanical, and psychological, but aside from that it was moved by a secret driving force; it

might have happened precisely that way the day before, but today, in

> some indescribable and fortunate way, it was different. – Oh – Clarisse immediately said to herself – I am freed from the law of necessity, where every thing depends on some other thing. [...] Clarisse discovered that what she was acting from was a veil of emotions, with things on the other side.[91]

At this point in the novel's composition, her symptoms still entertain a relationship with Musil's apperceptor theory of the early 1910s, inspired by Oesterreich's *Entfremdung der Wahrnehmungswelt* (Alienation of the World of Perception, 1907). In the fifth notebook of his journal, Musil notes that the theory applies directly to Alice, whose name would later change to Clarisse. The central idea here is that the balance between the individual and the world is regulated by the apperceptor, an organ mediating emotional adaptation between the individual and the world.[92] Under normal circumstances, equilibrium prevails, but in manic phases, the emotional adaptation of the world gets the upper hand ("all things change in harmony with this; one might say they remain the same but now find themselves in some other space, or that everything is tinged with another

---

88 Bleuler, *Textbook of Psychiatry*, 33.
89 Kretschmer, *A Text-book of Medical Psychology*, 96.
90 Musil, *KA*, *Transkriptionen*, Mappe I/5/107.
91 Musil, *The Man Without Qualities*, 1560–1561.
92 For a detailed discussion, see Bonacchi, *Die Gestalt der Dichtung*, 94–126.

sense"[93]). In the s-drafts of the novel, Clarisse relates this shift to genius, and the narrator follows her lead as his alter ego does the same. The difference between the mentally ill and geniuses (among whom Clarisse numbers herself) is that, for the latter, instability is in fact "full of strength"; while "constantly disturbed," it is "constantly inventing new forms of equilibrium."[94] Later drafts split this affect-driven thinking into a pathological form attributed to Clarisse, and a utopian one explored by Ulrich in his reflections on emotional psychology.

Affect-driven thinking also defines Clarisse's character in the published novel (1930, 1932). In the scene at the piano (quoted above), her thoughts shift from the color black to the devil and then to Moosbrugger, mediated in turn by the affect each mental representation triggers. In Chapter II/14, the narrator observes, "it was not so much that her ideas were confused as that they left out connections, or that they were saturated with affect in many places where other people have no such inner wellspring."[95] Finally, in late (unpublished) chapters, Clarisse's catathymic tendencies are omnipresent. Thus, when she fails to see Moosbrugger at the prison, she claims that she is being prevented from doing so so that he will disappear. The narrator comments, "Surely other, much more probable explanations could have been found, but this one also agreed with the uncanniness of [the asylum] through which she had wandered [...], bringing forth an uncannily clear kind of deep certainty."[96]

Catathymic mental connections are much more difficult for others to understand than contiguities or similarities perceptible to the senses. According to Bleuler, comprehensibility is precisely what distinguishes mania from schizophrenia.[97] Kretschmer hints at the same point of difference. In his view, the flighty thoughts of the manic individual are linked with each other "on the principle of the simple laws of association."[98] In contrast, he considers schizophrenic thinking to be "almost entirely catathymic, even to the extent of the loss of all contact with the realities of the moment."[99] Inasmuch as catathymia is in evidence, Clarisse's condition approaches schizophrenia. Not only is "a peg on which the whole hangs"[100] missing; the connection between thought elements

---

[93] Musil, *The Man Without Qualities*, 1561.
[94] Musil, *The Man Without Qualities*, 1562.
[95] Musil, *The Man Without Qualities*, 857. The author also took note of this passage in reference to Clarisse's thinking: Musil, *KA*, Transkriptionen, Mappe I/5/107.
[96] Musil, *KA*, *Transkriptionen*, Mappe II/1/11–12.
[97] Bleuler, *Textbook of Psychiatry*, 79.
[98] Kretschmer, *A Text-book of Medical Psychology*, 148.
[99] Kretschmer, *A Text-book of Medical Psychology*, 134.
[100] Kretschmer, *A Text-book of Medical Psychology*, 145.

is imperceptible for her interlocutors because it is completely detached from any sensory experience they would be able to comprehend. What Clarisse says prompts confusion. For example: "Ulrich observed her, trying to understand. He must have missed something – an analogy, or some 'as if' that might have given a meaning to what she was saying"[101]; "I can never follow your leaps from one point to another, or see how it all hangs together."[102] As Bleuler observes, all connection disappears in extreme cases, leaving only individual words; the "lyricism" Clarisse displays in the C-drafts of the "Island" chapter tends in this direction: "Swallow. Arrow. / Flows green into God. / Steeply rising alb easily / Greened in god" (*Schwalbe. Pfeil. / Fährt grün in Gott. / Steil steigende Albe leicht / Vergrünt in Gott*).[103] Achilles, Ulrich's predecessor, admires this utterance for its re-concretization of words and syntactical revitalization.

According to Bleuler, "the separation of associations from experience" facilitates so-called dereistic thinking, whereby "the faintest wishes and fears are endowed with the subjective reality of the delusion."[104] Such thought obeys only the "logic of feeling."[105] The examples he cites include not only the connections among ideas in schizophrenia and dreams, but also the dual beings of mythology. In the "Armistice" chapter, Clarisse invokes this notion when speaking to Walter":

> "The insane are just double beings."
> "Well, you said that before. But what does it mean?" [...]
> Clarisse reflected. "In many depictions, Apollo is man and woman. On the other hand, the Apollo with the arrow was not the Apollo with the lyre, and the Diana of Ephesus wasn't the Diana of Athens. The Greek gods were double beings, and we've forgotten that, but we're double beings too."[106]

A pillar of Clarisse's delusion is the theory of dual beings. The fact that she has a system – fixed points of reference to which she returns again and again – also points to her schizophrenic tendencies. By Bleuler's account, certain thoughts tend to recur persistently in schizophrenia. At the same time, the existence of a delusional system is typical for the condition, with affective connections taking the place of logical ones.[107] The more common elements Bleuler mentions – meg-

---

101 Musil, *The Man Without Qualities*, 233.
102 Musil, *The Man Without Qualities*, 719.
103 Musil, *Der Mann ohne Eigenschaften*, 1796.
104 Bleuler, *Textbook of Psychiatry*, 79.
105 Bleuler, *Textbook of Psychiatry*, 45.
106 Musil, *The Man Without Qualities*, 1378.
107 Bleuler, *Textbook of Psychiatry*, 390.

alomania, a perceived change of sex, and split personality – are all evident in Clarisse.

Musil also drew heavily on Kretschmer, and even more so on Oesterreich's *Phänomenologie des Ich*, to represent Clarisse's condition.[108] In the former's works he found the phenomenon of "double consciousness,"[109] which allows the schizophrenic to be two people at once without perceiving a problem. For instance, " [a] female patient perceives her doctor as her doctor but at the same time as her former lover, possibly even as a third person, her father perhaps or an elderly neighbour."[110] Along these lines, Clarisse believes Nietzsche has been reborn in her doctor.[111] The same double perspective also bears on her own person. In notes to the passage above, Musil wrote, "Cl. experiences herself in the actions of others a[nd] feels the actions and feelings of others in herself."[112] Indeed, this is one of the first symptoms she presents with. Chapter I/97 speaks of Clarisse being

> so beside herself that she can't tell where she is, except that she is definitely not absent. On the contrary, she could be said to be more inwardly present than ever, inside some deep inner space somehow contained inside the space her body occupies in the world.[113]

Similarly, later passages in the published novel connect the experience of "going outside oneself" with the sense of being connected with everything[114] or, alternatively, making room for another person within oneself: "When do you understand another human being? When you feel with him. [...] You have to be like him: not by putting yourself *into him* but by taking him *out* into yourself! We redeem *outward*."[115] The unequaled importance of this passage for Clarisse's character is made plain by the fact that Musil works it into the novel at least four times (in various forms) before featuring it front-and-center in her letter in Chapter II/7.

---

[108] Regarding Oesterreich's importance for Musil's emotion-based conception of ego psychology as it bears on Clarisse (and incorporating discussion of Oesterreich's account of mystical experience), see Bonacchi, *Die Gestalt der Dichtung*, 249–258.
[109] Musil, *KA, Transkriptionen*, Mappe I/5/171.
[110] Kretschmer, *A Text-book of Medical Psychology*, 135.
[111] Musil, *The Man Without Qualities*, 1578.
[112] Musil, *KA, Transkriptionen*, Mappe V/4/19; Kretschmer, *A Text-book of Medical Psychology*, 135.
[113] Musil, *The Man Without Qualities*, 482.
[114] Musil, *The Man Without Qualities*, 719.
[115] Musil, *The Man Without Qualities*, 775.

Clarisse's proximity to participation as the "extraconceptual correspondence of the human being with the world along with abnormal or correlative moments" (as Musil puts it in "Toward a New Aesthetic"[116]) is particularly pronounced here, and with that her proximity to the "other condition" that Ulrich orbits and whose characterization is at this point shaped by Musil's reading of Lévy-Bruhl's work on "primitive mentality." Thus she declares,

> [t]here are days when I can slip out of myself. [...] one feels connected by the air with everything there is, like a Siamese twin. [...] But you know all about that. That's what you meant when you said that there's something impossible about reality.[117]

Her willingness to detach herself from herself goes along with the tendency toward a split personality. The core delusion in all these splintered identities is Clarisse's belief that she has changed sex, becoming at different times "the seventh son of our Emperor,"[118] a boy, or a doctor.

> She had [...] been transformed into a doctor [...]. A wonderful feeling opened [...] her eyes. A deep aperture of her whole being – similar to the [...] repeated address of royal majesty – made her feel, with inexpressible pleasure, the spiritual condition of doctorhood [*Arztseligkeit*], that she [...], in mysterious fashion, was a man.[119]

The narrator also describes this transformation as the "voluptuous protrusion of another being out of the root of her own,"[120] which underscores the affective logic and psychic heteronomy (i.e., the dependency on an alter ego) at work in this event.

Oesterreich's survey of split personality also offered Musil a treasure trove of case studies, from which he made extensive excerpts. Under the heading "Cl." in his notebook, he groups numerous examples of depersonalization and split personality in cases of conversion,[121] acute schizophrenia,[122] experiences of reincarnation (which Oesterreich describes as a "well-known form of som-

---

[116] Musil, "Toward a New Aesthetic," 196–197.
[117] Musil, *The Man Without Qualities*, 719.
[118] Musil, *The Man Without Qualities*, 1072.
[119] Musil, KA, *Transkriptionen*, Mappe II/1/8.
[120] Musil, KA, *Transkriptionen*, Mappe V/4/200.
[121] Musil, KA, *Transkriptionen*, Mappe II/1/79; Konstantin Traugott Oesterreich, *Die Phänomenologie des Ich in ihren Grundproblemen* (Leipzig: J.A. Barth, 1910), 346–349.
[122] Musil, KA, *Transkriptionen*, Mappe II/1/79; Oesterreich, *Die Phänomenologie des Ich*, 359.

nambulism"¹²³), and devilish whisperings, among others.¹²⁴ All of these symptoms appear in connection with Clarisse in draft or published chapters. To take just two of many examples: she recognizes that the "ecstatic thirst for love" is "nothing other than an incarnation, [...] a manifestation in the flesh of something not of the flesh: a meaning, a mission, a destiny, such as is written in the stars for the elect"¹²⁵; later, she identifies Ulrich as "a great devil" who "knows what's good"¹²⁶ and could redeem the world with her help. In a note to the "Early-morning Walk" chapter, Musil stakes out the space extending from mere change in Clarisse to her full-scale transformation.¹²⁷ The model he employs is "alternation from manic to depressive,"¹²⁸ which is accompanied by a comprehensive shift in self-perception and sense of the world explained above in reference to catathymia. The motif bracketing all of Clarisses's metamorphoses is her passage from "a figure of sin to a figure of light."¹²⁹ In this context, the splits in her character appear as precursors to the final metamorphosis into one being that transcends and thus unifies the division within herself.

The figure of light is the "double being" at the utopian center of her system of madness. It takes concrete form first of all in the Nietzschean satyr (that is, the fusion of "goat" and "god," which Clarisse envisions as a sublimation of the "great forces of desire in people": "the goat would become the god!"¹³⁰). And secondly in the related idea of the "hermaphrodite" Clarisse believes herself to be: "I'm not a woman! Clarisse exclaimed, and jumped up. (Didn't you call me 'little fellow' when I was fifteen years old?) [...] I'm no woman, Meingast! *I am the hermaphrodite!*"¹³¹ As such, Clarisse reconciles herself with her double existence consisting of a passive, feminine existence that flows into the world, and an active, masculine life that takes in and shapes the world. As "the hermaphrodite," she can also redeem other people who suffer division from the other halves of themselves – for instance, Nietzsche and Christ ("Both the hostile kings, Nietzsche and Christ, met up in her. [...] Each died in his halfness. Yet they were one, the two great enemies, together a Whole! Double-man [*Der Doppel-*

---

123 Oesterreich, *Die Phänomenologie des Ich*, 365; Musil, *KA, Transkriptionen*, Mappe II/1/79.
124 Oesterreich, *Die Phänomenologie des Ich*, 422–424; Musil, *KA, Transkriptionen*, Mappe II/1/79.
125 Musil, *The Man Without Qualities*, 481.
126 Musil, *The Man Without Qualities*, 718.
127 Musil, *KA, Transkriptionen*, Mappe V/4/204.
128 Musil, *KA, Transkriptionen*, Mappe V/4/204.
129 Musil, *KA, Transkriptionen*, Mappe V/4/204.
130 Musil, *The Man Without Qualities*, 1380.
131 Musil, *The Man Without Qualities*, 1596. Emphasis in original.

mensch]! The body of a woman overcame them by unifying them"[132]). Invoking Greek mythology, Clarisse extends the notion of a double being to everyone around her and affirms the belief (which Walter identifies as totemism) that "every person has an animal in which he can recognize his fate."[133]

The construction of dual existence in *The Man Without Qualities* follows the principle of condensation outlined by Freud in *Interpretation of Dreams*, which served as a key reference for Bleuler and for Kretschmer especially, who elaborates the idea first in relation to 'primitive thinking' and then to schizophrenia. The examples Kretschmer cites include not only the doctor, who is at once also a lover, neighbor and father, but also the "hybrid forms of man and animal"[134] often encountered, according to Kretschmer, in 'primitive thinking.' Such entities do not split so much as accumulate the capacities of both identities so that they acquire a superhuman, quasi-divine status. Kretschmer describes these combinations of man and animal as "symbols" but maintains that "the primitive man" is unaware of the symbolic nature of such ideas.[135] Bleuler, whom Kretschmer cites, identifies this peculiarity as a symptom of schizophrenia and observes that many schizophrenic delusions are meant "entirely symbolically" – for instance, when "a female patient 'is' the cranes of Ibycus because she is 'free from blame and error,' and 'free,' that is, she should not be confined."[136] However, the symbol also usurps the "original concept," and thus the same patient "hallucinates" symbols "as realities"[137] and truly believes she is a bird.[138]

Clarisse, in contrast, demonstrates occasional awareness of the symbolic quality of her double being. In the "Early-morning Walk" chapter, she whispers to Walter (who is trying to understand her delusional system) "that the goat *signified* sensuality, which had everywhere separated itself from the rest of man-

---

132 Musil, *Der Mann ohne Eigenschaften*, 1737.
133 Musil, *The Man Without Qualities*, 1379.
134 Kretschmer, *A Text-book of Medical Psychology*, 86. Caption to Figure 9.
135 Kretschmer, *A Text-book of Medical Psychology*, 87.
136 Bleuler, *Textbook of Psychiatry*, 390.
137 Bleuler, *Textbook of Psychiatry*, 375.
138 The same holds for Bleuler and Kretschmer with regard to symbols stemming from the process of displacement. For Clarisse, many more arise from condensation. In particular, the image of the cross warrants mention, which derives its power from its original religious matrix: "She took a new sheet of paper and drew a cross through the middle. It hardly surprised her any more that this naked, broken piece of paper sprang to life right away. One needed only take a look to find a wealth of strange confirmations. [...] 'There are no more men today!' Cl. said to herself. When this had been confirmed by the magic cross, she looked for new facts. [...] Didn't the cross tell her everything she had experienced in her weeks-long struggle against pity for W.?!" (Musil, *Der Mann ohne Eigenschaften*, 1736).

kind."[139] This statement identifies her as having a manic personality, which, unlike schizophrenics, supposedly remains tied not only to 'primitive' but also – through the process of decomposition – to 'developed' culture. Consequently, Clarisse accrues a creative sovereignty that (as Kretschmer observes of an unusual case of schizophrenia[140]) brings her close to being perceived as an artist.

With these double human/animal identities, Musil returns to another aspect of his earlier ethnologically-inspired expedition narratives. As I have noted, animals play both a symbolic and participatory role in Clarisse's madness,[141] and they are also concurrently invested with the bestial function of Musil's earlier works. This feature is revealed in the figures of the bear and the goat (both of which she associates with Walter and Ulrich, among others). The beasts stand for a split-off (and thus barbarized) form of sensuality appearing especially as male sexuality. As Walter exclaims to Clarisse, "You seem to associate all men with 'goat'!"[142] The figurative meaning she attaches to the goat is also made evident by her association of it with the exhibitionist she sees ("She was telling him that the man under the window had been sent by the goat"[143]). Her perception of male sexuality as violent and perverse has also a biographical basis. As a girl she was abused by her father, and as an adult she was raped by her husband. By creating the goat figure, she mythologizes and excuses these acts of sexual violence. At the same time, she takes on the role of redeemer who unites goat and God and thus takes the guilt of man's weakness upon herself.

In this case, this perspective on male sexuality comes from the mouth of a mentally ill individual. Yet male sexuality is cast in the same dubious light elsewhere in *The Man Without Qualities*. Besides the three sex crimes perpetrated by men already noted, the novel also features Moosbrugger, the murderer of girls, who fascinates all of the novel's main characters. What is more, Ulrich's sexual encounters are characterized more by aggression and emotional coldness than devotion or tenderness. Consistent with this, transcending sexual desire forms part of the experiment in the sibling-love he undertakes with his sister Agathe, which is why the experiment fails when abstinence is abandoned. In this manner, Musil's novel itself carries out a dissociation from male sexuality and the resulting bestialization of it that shape Clarisse's delusional ideas.

---

139 Musil, *The Man Without Qualities*, 1380. Emphasis added.
140 Kretschmer provides the rare example of a young schizophrenic and connects the disintegration of the abstract concept into an asyntactic series of images to Expressionism (*A Text-book of Medical Psychology*, 137–138).
141 Musil, *The Man Without Qualities*, 1380.
142 Musil, *The Man Without Qualities*, 1379.
143 Musil, *The Man Without Qualities*, 1380.

Bearing these stereotypes in mind, one finds that Clarisse occupies a special position in regard to gender. Musil does not describe her with the typically feminine attributes of sexual devotion and passivity, but in terms of the conventionally masculine trait of sexual aggression. This is especially evident when she tries to force Ulrich to have sex with her: "Clarisse suddenly made a physical assault on him. She flung an arm around his neck and pressed her lips to his so quickly that it took him completely by surprise and he had no time to resist."[144] In this context, it is not Clarisse's delusion so much as the novel itself which identifies her as a hermaphrodite. At once masculine and feminine (in psychological terms), she has identified so completely with the perpetrator that she now must bear "the figure of the goat" within herself as an aggressor.[145] A critical perspective on this aspect of Clarisse's delusion thus emerges. It expresses a repressive and stereotyped treatment of sexuality and gender that shaped both Musil's novel and early twentieth-century bourgeois society and was essentialized in primitivist fantasies about the beast within man.

The above sub-chapter demonstrates the extent to which Musil drew on psychological theory to characterize Clarisse as a manic figure with schizophrenic tendencies. Up until this point, I have shown the similarities between *The Man Without Qualities* and contemporary psychological research. (The distance that the author and his protagonist, Ulrich, take from psychology will be thoroughly discussed below.) Furthermore, I have made clear that the manic-schizophrenic thinking constructed by Musil for the Clarisse character is closely related in its structure and content to notions of the 'primitive' outlined in the ethnological and psychological sources he so carefully consulted. This relationship involves especially the recurrence of the asyntactic, eidetic images typical of mania, which obey the principle of association on the basis of external similarities or contiguities. At the same time, Clarisse's mania follows a catathymic logic, which is yet another feature it shares with so-called 'primitive thinking.' Finally, with desire-driven, dereistic thinking, Bleuler, Musil, and ultimately also Clarisse herself refer to the relationship between schizophrenic and mythical thinking. Bleuler writes, "Apollo is split into several personalities [...] indeed he may even be a woman although he is ordinarily a man."[146] As we saw above, Musil takes up this observation in Clarisse's theory of double beings. Through this theory, Clarisse's insanity is not only structurally related to the participatory worlds of 'primitive thinking,' but also by virtue of its content: Clarisse takes

---

144 Musil, *The Man Without Qualities*, 720.
145 Musil, *The Man Without Qualities*, 1379.
146 Bleuler, *Textbook of Psychiatry*, 45–46.

the symbolic construction of double beings literally, understands them as metamorphoses in her own person, and views human beings as animals on the way to divine status. From the associative links of 'primitive thinking,' Kretscher also derives the belief in magical practices. Such belief is likewise shared by Clarisse when she reads patterns formed by animals and colors to discern omens of and instructions for the future.[147] Indeed, she practices magic by onomastic means: "Wherever his name fell, the earth melted. When she uttered it her tongue was like a wisp of sun in a mild rain."[148] At several points she even calls herself a sorceress: "'I deem myself a Thessalian witch!' she screamed into the uproar that now broke loose from all sides" of a thunderstorm (believing that she has caused it).[149]

Music and 'Primitive Thinking'

In an early draft, Ulrich displays an ambivalent attitude toward music. The making of music, primitivity, and mental illness are brought into direct relation because of the emotional volatility they supposedly share.

> You're primitive, you musicians. What kind of subtle, unheard-of motivation does it take to produce a raging outburst after sinking into oneself in silence! You do it with five notes!
>
> – It's something you don't understand, Uli. Clarisse laughed. [...] – You were never sick.[150]

If one pursues this connection, a genealogical and symptomatic proximity between Clarisse's manic 'primitive thinking' and music emerges.[151] The earliest indications of her flight of ideas appear when she plays the piano. The passage quoted above concerns the way her mind works under such circumstances: "In fluttering mists, images sprang up, overlapped, fused, faded – that was Clarisse's thinking."[152] Her flitting thoughts follow musical laws: "the music did not stop for a second."[153] The shift from sequential to synchronous mental representations ("several thoughts were often present in one another at the same time")

---

147 Musil, *The Man Without Qualities*, 1005.
148 Musil, *The Man Without Qualities*, 1626.
149 Musil, *The Man Without Qualities*, 1573.
150 Musil, *The Man Without Qualities*, 1546.
151 For discussion of the role of music, see Pietsch Pentecost, "Clarisse," 73–96, who does not, however, remark on the connection to Clarisse's primitivistic, manic-schizophrenic traits.
152 Musil, *The Man Without Qualities*, 152.
153 Musil, *The Man Without Qualities*, 155.

matches the layering of motifs in polyphonic compositions, as does the impression of constant thinking and images blending into one another, which corresponds to the modifications of musical motifs carried in development and fortspinnung.[154] By the same token, flights of ideas resemble instrumental music inasmuch as propositional content tends to be absent in both. Even though any number of manic thoughts appear in alternation, their substance plays a role secondary to the process through which they succeed each other. Clarisse's mood swings, which are typical of mania, also find a model in music:

> The transitions from charming, gentle, and soft to gloomy, heroic, and tumultuous, which the music went through several times within the space of a quarter hour [...] seemed [...] [like] the carryings-on of a company of drunks that alternates periodically between sentimentality and fistfights.[155]

Ulrich, who makes this observation, resists music's emotional influence. In contrast, Clarisse gives herself over to it: "When I hear music I'd like to either laugh or cry or run away."[156] The process offers a model for the split personality. Her experience of music leads to her claim that one can only understand another person by incarnating their personality ("You have to be like him: not by putting yourself *into him* but by taking him *out* into yourself!"). Clarisse proceeds with other people in the same way she does as a performer or listener of music: she *"plays it inside [herself]."*[157]

The musical genre whose conventions come closest to the principle at work in flights of ideas is free fantasia.[158] This kind of music, which emerged at the end of the eighteenth century in works for the piano, gives composers license to indulge their imaginations in a manner bordering on improvisation. It is distinguished by a "loosening of bars, free choice of theme, rapid shifts between parts and in tone, freedom from distinct periods, harmonic norms, and formal order oriented on other genres."[159] Its earliest and best-known exponent was Carl Philipp Immanuel Bach. At first the lack of determinate content and absence

---

[154] It is not said what kind of piece Clarisse and Walter are playing. However, in light of music identified elsewhere, one may assume that it is a Romantic or late-Romantic composition.
[155] Musil, *The Man Without Qualities*, 1615.
[156] Musil, *The Man Without Qualities*, 1546.
[157] Musil, *The Man Without Qualities*, 775.
[158] On the proximity of free fantasia to inner monologue, see Nicola Gess, "Intermedialität reconsidered. Vom Paragone bei Hoffmann bis zum Inneren Monolog bei Schnitzler," *Poetica. Zeitschrift für Sprach- und Literaturwissenschaft* 42, no. 1–2 (2010).
[159] Peter Schleuning, *Die Freie Fantasie. Ein Beitrag zur Erforschung der klassischen Klaviermusik* (Göttingen: A. Kümmerle, 1973), 36, 104.

of recognizable forms posed a problem for contemporary critics, who considered such music the expression of insanity – and a potential cause of madness for those who heard it. Advocates of free fantasia, however, thought these very qualities gave voice to the imagination, but not so much to its contents as to its motions as such. In other words, they did not approach this music semantically, but understood it as a reflection of particular mental processes. Their affinities with the principle of flights of ideas are plain: rapid changes of theme and affect, connections between individual elements without regard for overall coherence, and hints of madness tempered by thought devoted to the drive of imagination alone. Comparable to the thesis advocated by free fantasia's defenders is the thesis held by Ulrich's predecessors Anders and Achilles that Clarisse's so-called madness exemplifies the pathologization of an artistic norm-breaking creativity.

The musical origins of Clarisse's delusions are also evident in their synaesthetic quality. When experiencing her "other condition," "everything turns into music and color and rhythm;"[160] as she puts it, "I hear-see a world in which the things stand still and the people move around, just as you've always known it, but in sound that's visible!"[161] Clarisse claims that music connects "with sight [*Gesicht*]." At stake is a listening, that is also a seeing, or rather music-inspired vision. Visions, that play a major role in her delusional system are therefore often connected with musical phenomena. For example, she repeatedly describes Moosbrugger, whom she encounters in the flight of ideas induced by piano music, musically,[162] and in a letter to Ulrich, she also remarks on the sonorous quality of the three syllables in Moosbrugger's name.[163] The acoustic dimension of language furthermore plays a major role in her associations, as noted above ("Snakes, snares, slippery"[164]). When Clarisse speaks, semantics take a secondary position relative to alliteration and assonance (e. g., "Steil steigende Albe leicht"[165]), the "rhythm of arousal, or the "wild refrain" of repeated words.[166] Finally, her letters abound with exclamation marks, italics, and underlining, all meant to convey the sound of spoken language through writing.[167] Ulrich/Anders remarks that the results "[look] like a cryptic musical score."[168]

---

[160] Musil, *The Man Without Qualities*, 719.
[161] Musil, *The Man Without Qualities*, 773–774.
[162] E.g., Musil, *The Man Without Qualities*, 233.
[163] Musil, *The Man Without Qualities*, 773.
[164] Musil, *The Man Without Qualities*, 154.
[165] Musil, *Der Mann ohne Eigenschaften*, 1796.
[166] Musil, *The Man Without Qualities*, 719; 1564.
[167] Cf. her letter, Musil, *Man Without Qualities*, 773–777.
[168] Musil, *The Man Without Qualities*, 1564.

Semantics play hardly any role in the musical rollercoaster of affect that Ulrich/Anders criticizes. Thus, Clarisse's desire to speak with a madman is defined by the "drumming" of the answers he provides, not their content; her own questions then "[make] as little sense as a random sound one might entice from a bugle."[169] Everything that comes out of her is acoustically mediated affect: "she felt tremendously sure of herself as she said it."[170] The letter she writes to Ulrich from the island is confused in its content yet "held together" "by a rhythm of excitement."[171] In an earlier draft, its forceful dynamism is described as follows: "Like an arrow between narrow walls, it shot into the heights, unfolded in a coil; something invisible was rolling, striking, speeding over the tops and crests of the houses."[172] Similarly, Clarisse is credited with a unique understanding for the "dance rhythm" of indigenous peoples.[173]

'Primitive Language'

Oesterreich granted hardly any attention to the language of the mentally ill. Though none are pursued in detail, Bleuler's study does list an array of linguistic symptoms of mania such as logorrhea, the interference of a foreign lexicon (which can go so far as to yield a new language), and messy, disordered writing.[174] Schizophrenia, the author notes, can involve logorrhea, "mutism," abnormal intonation and "speech mannerisms," neologisms, disintegrated syntax, "improper use of words," and "word-salad" or artificial language, as well as writing in an eccentric graphic style with frequent breaks and repetitions, and excessive punctuation.[175] That said, Bleuler does not seek out the laws underlying these peculiarities or assess the relationship of patients to the language they use.

Only Kretschmer, motivated by previous work on 'primitive language,' dedicates attention to those questions and pays close notice to the symbols of schizophrenics. In his estimation, these symbols are to be understood as "preludes to an abstract understanding" whereby the schizophrenic occupies a 'primitive' level of human development. The 'civilized' individual, in contrast, understands such symbols as "translations of fully formed abstractions back again into a sim-

---

169 Musil, *The Man Without Qualities*, 1076.
170 Musil, *The Man Without Qualities*, 1076–1077.
171 Musil, *The Man Without Qualities*, 1627.
172 Musil, *Der Mann ohne Eigenschaften*, 1739.
173 Musil, *Der Mann ohne Eigenschaften*, 1787.
174 Bleuler, *Textbook of Psychiatry*, 468.
175 Bleuler, *Textbook of Psychiatry*, 394–398.

pler kind of picture-language." Kretschmer describes how abstract concepts and syntactical order regress into discrete words, into nonverbal, fantastical agglutinations of images, which, however, for the doctor can be "translat[ed]" back into language.[176] At the same time, he grants that agglutinated images involve combinations of image-words, from which new concepts arise as they gradually shed their visual meaning and sound so that ultimately only an abstract concept remains behind.[177] It remains unclear then whether schizophrenics operate on a prelinguistic level or on that of purely vivid language.

Kretschmer, as I have noted, attends to the laws of agglutination and stylization followed by indigenous peoples and schizophrenics alike, specifically their literal understanding of figurative language.[178] Musil demonstrates the same interest, drawing inspiration from Kretschmer but going much further in outlining Clarisse's character. Three of her features stand out: her handling of figurative language, her reenactment of the phylogenetic development of language, and her identity as an almost-poet.

I have already discussed Clarisse's affinity for the acoustics of language at some length. However, the "snakes, snares, slippery" passage cited in that context offers another item of note. This associative thread derives from a literal interpretation of a metaphor from earlier in the narrative when "the thin gray paths coiled like snakes."[179] Such literalizing exemplifies a process that reveals Clarisse's relationship to language more than any other part of the published novel. A few examples will make as much clear. In Chapter I/54, Ulrich is irritated by the inferred absence of the particle "like" in Clarisse's discourse. When he speaks by "analogy or some 'as if'" governs his propositions, she takes his words literally and in such cases thinks that he really might "be transformed."[180] Chapter I/97 describes the meaning of the abundant quotation marks and underlining of Clarisse's written, spoken, or internal discourse: "the words thus emphasized [tense] up with meaning," that is, they take on a literal meaning. Thus, she takes the expression, *etwas ins Auge fassen* (a figure of speech meaning "to take a [close] look," "consider") literally: as if the eye were able to reach out and grab what it sees as well as be grabbed by the eyes of others. Such (mis)understanding invests the terms with a performative force. When she speaks of "catching someone's eye" with her own, she also acts in this way: the word becomes

---

[176] Kretschmer, *A Text-book of Medical Psychology*, 116.
[177] Kretschmer, *A Text-book of Medical Psychology*, 85.
[178] Kretschmer, *A Text-book of Medical Psychology*, 136.
[179] Musil, *The Man Without Qualities*, 154.
[180] Musil, *The Man Without Qualities*, 233.

"part of the smashing force in her arm," "like a stone to be flung at a target."[181] The same occurs in Chapter I/118. Clarisse's mental emphasis in the phrase, "the situation had come *to a head*," indicates that she actually sees a "head" forming, which yields an actual weapon: "Ulrich [...] was on the other side of the conflict, the side against which this spear*head* would be directed, if there was trouble."[182] Clarisse thus revitalizes the language of metaphors whose tensions with their literal meanings have grown all but invisible over time.[183] At the same time, she activates a magical potential in language, endowing it with a force of action exceeding representation.

A further aspect of her interaction with figurative language concerns the "double word," which plays a key role in Clarisse's system of madness and theory of "double identities."[184] Following this principle, she revitalizes lexicalized metaphors that are compounds of two words – for instance, "birthmark" (*Muttermal*). Clarisse, who has a conspicuous birthmark, reads this "sign" as an indication that she is destined, from birth, to give birth, become a mother (*Mutter*), spe-

---

181 Musil, *The Man Without Qualities*, 475.
182 Musil, *The Man Without Qualities*, 665. Emphases in the original.
183 Scholars have focused on this process almost exclusively in relation to Moosbrugger. See, e. g., Wilhelm Braun, "Moosbrugger Dances," *The Germanic Review* 35 (1960): 220 – 221; Claudio Magris, "Musil und die Nähte der Zeichen," in *Philologie und Kritik. Klagenfurter Vorträge zur Musilforschung*, ed. Wolfgang Freese (Munich: Fink, 1981), 189; and Fred Lönker, "Der Fall Moosbrugger. Zum Verhältnis von Psychopathologie und Anthropologie in Robert Musils *Der Mann ohne Eigenschaften*," *Jahrbuch der Deutschen Schillergesellschaft* 47 (2003): 288 – 289. Lönker incorrectly assumes that "similar phenomena are hardly described in contemporary psychiatric literature" (289), missing their central role for Kretschmer. See also Eberhard Ostermann, "Das wildgewordene Subjekt. Christian Moosbrugger und die Imagination des Wilden in Musils *Mann ohne Eigenschaften*," *Neophilologus* 89 (2005): 608 – 609, and, on the philosophy of language and ethnology, 610 – 611; as well as Robert Krause, *Abstraktion – Krise –Wahnsinn. Die Ordnung der Diskurse in Robert Musils Roman 'Der Mann ohne Eigenschaften'* (Würzburg: Ergon, 2008), 110 – 114. In contrast, Gerd-Theo Tewilt (*Zustand der Dichtung. Interpretationen zur Sprachlichkeit des 'anderen Zustands' in Robert Musils 'Der Mann ohne Eigenschaften'* [Münster: Aschendorff, 1990], 172 – 184) analyzes Clarisse's propensity to take metaphors literally and concludes that she is out to "destroy difference" (173), which makes her own language violent (174); and also Robert Krause, "'Man könnte die Geschichte der Grenzen schreiben.' Moosbruggers wildes Denken und die Kultur des Okzidents," *Musil-Forum 31. Studien zur Literatur der klassischen Moderne: Musil und die Fremdheit der Kultur* (2009 – 2010); this essay situates the killer's "wild thinking" (which is the basis for his idiosyncratic use of language) in the context of Musil's reading of Lévy-Bruhl.
184 Jutta Heinz ("Grenzüberschreitung im Gleichnis," in *Grenzsituationen. Wahrnehmung, Bedeutung und Gestaltung in der neueren Literatur*, ed. Dorothea Lauterbach, Uwe Spörl, and Uli Wunderlich [Göttingen: Vandenhoeck & Ruprecht, 2002]) points out that Clarisse's double words, like Moosbrugger's metaphors, do not serve the purpose of communication (251).

cifically the "mother of God" (*Gottesmutter*).¹⁸⁵ By the same token, when looking for a fitting way to express her relationship to Walter, she takes up the conventional metaphor, *Schirmherr*. The term means "patron" or "protector," but it breaks down into the words *Schirm* ("shield," but also "umbrella") and *Herr* (lord and master), roles filled by Walter as her husband; but this shield becomes a floppy umbrella without the support that Clarisse – the "shaft" – provides.¹⁸⁶

In Chapter II/26, when Clarisse's illness has reached an advanced stage, the meaning of double words for her theory of dual existence is explained. From the existence of double words, she infers dual existences: "The double words were signs [...] to mark a secret path. [...] But a double language means a double life." Hereby, the conventional understanding of the double word is understood as the official meaning, and its literal significance is "secret and personal," which is interpreted further as a meta-sign for the existence of another dimension of a person, of life, and the world. In this light, Clarisse interprets altogether ordinary words on two registers (for instance, "quick" receives a new meaning in becoming a verb: "everything [quicks] in joyful leaps and bounds"¹⁸⁷) and finds meaning in the second half of two-syllable words: "My dar*ling* [...]! Do you know what a *ling* is? I can't work it out. [...] [All the 'lings' were heavily underlined]."¹⁸⁸ This operation combines with the performative process of creating a secret world governed by omnicausality or "delusion of reference,"¹⁸⁹ whereby mere coincidences appear to be matters of fate. Clarisse, for instance, is convinced of a far greater narrative by the sight of a bird eating a caterpillar.

> Fate had placed the two creatures in her path, as a sign that she must act. One could see how the blackbird assumed the caterpillar's sins through its flaming orange-red beak. Wasn't the bird a "black genie"? Just as the dove is the "white spirit"? Weren't these signs linked in a chain? The exhibitionist with the carpenter, with the Master's flight?¹⁹⁰

While the published novel foregrounds Clarisse's use of figurative language, the drafts highlight her invention of her own language. Especially in the "Island" chapter, Clarisse replays the phylogenetic development of language and displays a marked tendency toward gestural communication. An early draft

---

185 Musil, *The Man Without Qualities*, 483.
186 Musil, *The Man Without Qualities*, 716.
187 Musil, *The Man Without Qualities*, 1001.
188 Musil, *The Man Without Qualities*, 773. Square brackets and enclosed language are from the original. Emphasis in the original.
189 Bleuler, *Textbook of Psychiatry*, 94.
190 Musil, *The Man Without Qualities*, 1005.

reads, "A. was accustomed to how hard it was for her to find the right words and how she often tried to seize them with her whole body, so that the meaning for which the words were lacking lay in the movement."[191] Further evidence that words and bodily gestures are directly connected for her can be seen in the passage: "One simply spread out one's arms – and for her that included words [...] – like wings [...]! 'Joyful world aslant' was what she named her wingspread arms and her gaze down the stairwell."[192] The motions she performs and words she speaks are interchangeable; as soon as the gesture is made, its name follows. As early as Chapter I/82, Ulrich remarks that

> [h]er whole slender body was involved; she actually felt everything she wanted to say with her whole body first of all, and was always needing to do something with it. [...] Ulrich [...] now [...] saw Clarisse as a Javanese dancer. Suddenly it occurred to him that he would not be surprised if she fell into a trance.[193]

The chapter "Early-morning Walk" also culminates in Clarisse dancing, expressing her higher insights and high spirits without words. The tendency toward gestural language, the affinity for dance, and above all the references to a trance state call to mind Lévy-Bruhl's writings on the body languages employed by 'primitives.'

The s-drafts of the "Island" chapter, where Clarisse cycles through the phylogenetic development of language, confirm as much.[194] The first stage is based on using objects as symbols.

> Two stones and a feather laid on top, perhaps; that meant: I want to see you, come to me, but you won't find me, as fast as a bird flies. [...] But a piece of coal in the white sand

---

191 Musil, *The Man Without Qualities*, 1541.
192 Musil, *The Man Without Qualities*, 715.
193 Musil, *The Man Without Qualities*, 383–384.
194 Christian Kassung (*Entropie-Geschichten. Robert Musils 'Der Mann ohne Eigenschaften' im Diskurs der modernen Physik* [Munich: Fink, 2001]) likewise examines this process, but he does not see phylogenetic development at work. Instead, from the "standpoint of physics" (429), he identifies the graduated progression from "pressed-together language" (428) to signs not yet amounting to concepts that, through non-grammaticality (429) and analog coding (double words), are supposed to yield an unmediated "onomastic language" (430) and finally a stage of silence punctuated by dashes (431). Claudio Magris ("Musil und die Nähte der Zeichen") also offers a brief discussion of Clarisse's linguistic inventions, focusing on how she forges signs radically detached or independent from context (191) and material language without syntax (192). In the chapter entitled "Word Magic," Genese Grill (*The World as Metaphor* [Rochester, NY: Camden House, 2012]) explores Clarisse's "mystical theory of language" against the background of Musil's readings on the subject (104).

meant: I'm feeling dark today, gloomy and sad. [...] A pepper meant: I'm hot, impatient, and waiting for you.

The thing-symbols that Clarisse constructs function according to the principles of similarity or contiguity. But at the same time, they are strictly speaking neither iconic nor indexical. Instead they are already in such an advanced symbolic stage that Ulrich/Anders must "learn to understand" them.[195] What's more, they are inherently polysemous and relate to multiple referents:

> Something was a stone and meant [Ulrich]; but Cl. knew that it was more than U. and a stone, namely what was rock-hard about [him] and all the heavy matters oppressing her, and all the insight into the world one obtained by recognizing that the stones were like U.[196]

It seems that thing-symbols are the consequence of Clarisse's interest in double words, which possess not double, but rather numerous meanings – as she puts it, "many feelings, which otherwise are separate, crowded around a sign like this"[197] – that tie back to the materiality of their objects. Thus, in the passage above, the symbol "stone" is no longer a word, but an object in its own right that can serve as a weapon. On the basis of their unity, thing-symbols are a communicative medium, but at the same time they are a form of communication with those material things themselves: "Between her and things there existed a continual exchanging of signs and understandings, a conspiracy, [...] heightened correspondence."[198]

The level following thing-symbols is the picture-language Clarisse creates by drawing in the sand: "arrows and circles, a burning heart and a leaping horse, all of them usually hinted at with so few lines that they were comprehensible to the initiated alone." Here the law of stylization Kretschmer develops from 'primitive' picture-language and the law of catathymically motivated condensation are put into striking practice: "a pressed-together language in which the heartbeats are piled on top of one another."[199] Ulrich/Anders credits this language with a magical force that is located in the faculty that stores emotional and spiritual experience. Even though these signs concern others, they are meant above all for Clarisse herself, who finds herself in them.

---

195 Musil, *Der Mann ohne Eigenschaften*, 1741.
196 Musil, *Der Mann ohne Eigenschaften*, 1742.
197 Musil, *Der Mann ohne Eigenschaften*, 1743.
198 Musil, *The Man Without Qualities*, 1559.
199 Musil, *Der Mann ohne Eigenschaften*, 1741.

> When one had forgotten it completely and only through some chance stumbled on it again and suddenly confronted oneself, confronted an instant compressed and full of emotions and thoughts [...]: then [...] the island became populated with many Clarisses [...] it was a lust [...] to run into oneself everywhere.[200]

The magic of this picture-language lies in its affinity to the narcissistic multiplication of the ego, or manifold personality splits. Accordingly, the "continuous exchange of signs" is glossed elsewere as "Clarisse [thinking that] she was being torn out of her slender body," which allows her to fly over the island as a witch while still lying calmly in the sand.[201]

Finally, at the third stage of linguistic phylogeny, Clarisse moves from picture-language to a development of words and even begins "to express her life in poems."[202] In so doing, she seeks to free herself from conventional language. For instance, she breaks up concepts by inventing new compound words and ruptures syntactic coherence, thus relieving words from their wonted connections. Means to this end include exclamation marks or repetitions that make the weight of a word so heavy that it outbalances the force of the old syntactic order.[203] Asyntactic sequences of coinages ("Ichrot"– formed from *ich* [I] and *rot* [red], for example) are the result. The resemblance of these linguistic products to the anomalous language use of manic or schizophrenic patients described by Bleuler and Kretschmer is obvious. In contrast to them, however, Clarisse reflects on and explains her rupture of syntactical restraints to the narrator, who has his own insights into her motivations.

Kretschmer mentions an unusual case of schizophrenia, in which the patient likewise has insight into the process of creating symbols and their meaning. This case, for him, calls to mind Expressionism:

> A single example of this kind suffices to provide a clear explanation of the modern tendency in art known as "*expressionism.*" If we think of our patient's inner 'picture show' as a painting with a title, "The Infinity of Space," underneath, we can exactly understand the principles underlying expressionistic pictures in which the artist seeks to set down his inner feelings and ideas.[204]

---

200 Musil, *The Man Without Qualities*, 1554–1555.
201 Musil, *The Man Without Qualities*, 1559.
202 Musil, *The Man Without Qualities*, 1564. Roger Willemsen ("Dionysisches Sprechen," 104–135) sees here the "program of expanding the revealed name" (128), which amounts to an "erotic or, more still, orgiastic relationship to language" (129) that has returned to its origins in "image and stimulation" (130).
203 Musil, *The Man Without Qualities*, 1564.
204 Kretschmer, *A Text-book of Medical Psychology*, 137. Emphasis in the original.

In this light, and given Clarisse's insight into the process by which she produces certain linguistic effects, she is akin to Expressionist artists.[205]

On this score, a further point of contact with primitivist discourse emerges. Not just Kretschmer, but other contemporary psychologists viewed schizophrenia, 'primitive thinking,' and Expressionism in a single phenomenological and genealogical context – one that had been topical at least since *Der Blaue Reiter* exhibition. Thus, Kretschmer affirms that creative states are animated by 'primitive thinking': in "men and women of creative genius," a condition of "lessened consciousness" is at work, as well as "primitive phylogenetic tendencies toward rhythm and stylization with elemental violence" characterized by spoken words "surrounded by a nebulous constellation of imaginal agglutations and strong affective currents."[206]

However, Musil's profile of Clarisse as an Expressionist holds only for the earliest drafts of the novel. Later notes reveal that the author refused to identify poets with the mentally ill. At this point, Musil instead portrays this equation itself as the product of pathological thinking, the wishful thinking of relatives (Walter), and as a cliché widespread in psychology and art theory during the 1920s (as I demonstrated in Chapters 3 and 4). In keeping with Musil's shift in thought, Ulrich grows increasingly removed from Clarisse as Musil's revisions proceed. Whereas in the early drafts of the "Island" chapter he still allows himself to be absorbed by her delusions entirely, Ulrich later takes distance from Clarisse's delusions and shows greater inclination to reflect on the differences he observes of their two conditions.

My discussion of mania, schizophrenia, music, and language reveals the primitivist contour of Clarisse's character and therefore the central role that primitivist motifs play in *The Man Without Qualities*. But the features I have noted can also be read as signs of a primitivist aesthetic, at the core of which stand a participative (in the ethnological sense of the word) approach to the world as well as a symbolic language concentrated more on vividness than ab-

---

[205] Michael Jakob ("Von der 'Frau ohne Eigenschaften' zum 'Mann ohne Eigenschaften,'" in *Robert Musils 'Kakanien' – Subjekt und Geschichte*, ed. Josef Strutz [Munich: Fink, 1987]) reads Clarisse in the early drafts as the portrait of a "classical creative genius" (130) when she draws and writes in the throes of the "other condition" (124); in the published novel, however, such activity amounts only to pseudo-genius (121). On Clarisse's "identificatory reading of Nietzsche" between Walter's cultural conservatism and Ulrich's "personification" of the "ambivalence" that results from the dissolution of traditional values, cf. Alexander Honold, *Die Stadt und der Krieg. Raum- und Zeitkonstruktion in Robert Musils Roman 'Der Mann ohne Eigenschaften'* (Munich: Fink, 1995), 394–395.

[206] Kretschmer, *A Text-book of Medical Psychology*, 125–126.

straction. At the same time, it becomes clear that this aesthetics is not informed by any foreign practices or alien forms of thought so much as by tropological language and musical procedures native to Europe. Clarisse gets caught up in the logics of such procedures and thus risks going insane. She is an explorer so influenced by her objects of study that her expedition morphs into self-experiment. Under her influence, Ulrich (Musil's alter ego) also ventures into a primitivist realm that presents itself as an alienated version of the familiar. And with that, I turn to the second part of my thesis: the movement from expedition to self-experiment in *The Man Without Qualities*.

Regression

Clarisse shares an historical and biographical place, time, and social background with Ulrich as he embarks on his expedition. She is also his friend, potential lover, and, originally, sibling (indicated by his addressing her as his "little sister"[207] in the letter to Alice, in whom Agathe and Clarisse are fused at an earlier draft stage). Furthermore, Clarisse remains fixated on Ulrich and Moosbrugger, to whom Ulrich (and his draft predecessors, Achilles and Anders) is bound by an enigmatic sympathy.[208] Many passages featuring Clarisse point to parallel interests, views, and experiences with Ulrich; often, these are connected to the "other condition." Thus, in the chapter entitled "The Turning Point," Ulrich notes, "during her attack [...] she had said things that were too close for comfort to much that he had occasionally said himself."[209] From this perspective, Clarisse can be understood as an outsourced projection of Ulrich's own self-experiment. Evidence of this is also found in the C- and s-drafts, where Ulrich (or figures who would become Ulrich), like the protagonists of Musil's early expedition narratives, is infected by Clarisse and enticed into self-experiment. A sequence in an early draft of the "Island" chapter is striking in this regard:

> *Délire à deux:* It's a question of two people, one of whom is insane and the other predisposed to insanity. [...] Through constant contact, by being constantly bombarded with confused and inchoate ideas, the predisposed person ends up acting like his companion, and gradually the same madness shows up in him.[210]

---

207 Musil, *KA, Transkriptionen*, Mappe IV/3/450.
208 Regarding the genealogical relationship and structural similarity between Ulrich and Moosbrugger, see Ostermann, "Das wildgewordene Subjekt," 605–606.
209 Musil, *The Man Without Qualities*, 722.
210 Musil, *The Man Without Qualities*, 1585–1586.

However, the more Musil developed his material, the more Clarisse lost such direct influence on Ulrich. Instead, a distanced, critical view came into focus – evident, for instance, in the distinction between delusion and poetry absent from early drafts. In a 1930 note to himself, the author declares, "One who is mentally ill isn't a poet, after all!"[211] A brief sketch of this progressive uncoupling follows below.

In a 1919–1920 C-draft of the "Island" chapter, "the same madness" affecting Clarisse strikes Achilles (another forerunner of Ulrich): "He lives through the essence of Expressionism. He, who is so precise, writes such poems. At that time poetry had not got to that point."[212] Here, the reflections on language later attributed to Clarisse occur to Ulrich's precursor. Similarly, the signs Clarisse later scatters over the island – which Anders later gives up on deciphering – are put there by her *and* Achilles.[213]

When the novel was still called *Zwillingsschwester* (Siamese Twins, 1923–1926), Anders stops personally identifying with Clarisse in the Island episode. However, he learns to understand and participate in her languages (they "arrange signs in the sand"[214] together), is able to follow the course of her thoughts, and even justifies her delusions ("For a while, Cl. saw things that one otherwise doesn't see. A. could explain it readily"[215]). He invokes the relativity of all human perception determined by various interests, affects, and mediums and even cites his own experience of hallucination: "So unreliable and extensive is the boundary between insanity and health."[216] A few drafts later, Clarisse's delusions are appreciated as superior "insights."[217] In other words, Clarisse is credited with the discovery that Musil had formulated years earlier in his apperceptor theory and on which Ulrich reflects so extensively in later chapters on emotional psychology: "Cl. [...] recognized that feelings change the world."[218] Living out this insight provides the basis for her definition of genius and affirms her own identity as one.[219] The narrator grants that an "obscure [...] charm emanates" from her poems, "something with the glowing fire of a volcano, as if one were looking into the bowels of the earth." Thereby, she is said to anticipate what

---

211 Musil, *Der Mann ohne Eigenschaften*, 1377.
212 Musil, *The Man Without Qualities*, 1586.
213 Musil, *Der Mann ohne Eigenschaften*, 1797.
214 Musil, *Der Mann ohne Eigenschaften*, 1741.
215 Musil, *Der Mann ohne Eigenschaften*, 1743.
216 Musil, *The Man Without Qualities*, 1556.
217 Musil, *The Man Without Qualities*, 1564.
218 Musil, *Der Mann ohne Eigenschaften*, 1749.
219 Musil, *Der Mann ohne Eigenschaften*, 1750.

will soon become "fashion among the healthy": "Flakes of fire would be stolen by poets from the volcano of madness."[220] Clarisse appears as a madwoman, but she is at the same time a fiery muse for later poetry – as was quite topical in the discourse of artistic and psychological theory of the 1920s (see Chapters 4 and 5).

In the published novel (1930, 1932), Clarisse remains convinced that insanity expresses intellectual superiority: "'Crazy' to her meant being something like [...] enjoying so extraordinary a degree of health that it frightened people; it was a quality her marriage had brought out in her, step by step, as her feelings of superiority and control grew."[221] However, now Walter, not Ulrich, justifies her mental state as an artistic one. When Clarisse's doctor declares her visions and sense of omnicausality to be symptoms of mental illness, her husband justifies them as typical for an artistic gaze that only sees those parts of the outside world that fit its own picture and artistic palette. He disparages the physician for his blindness to creativity:

> You're not a creative man, after all; you've never learned what it means to "express oneself," which means first of all, for an artist, to *understand* something. [...] Of course you'll say it's paradoxical, a confusion of cause and effect; you and your medical causality![222]

Along the same lines, Walter (and no longer Ulrich) acknowledges Clarisse's explanation of her own madness in the chapter "Armistice"; he elevates her delusion to a "prophecy," deems it a creative gift that he had also once possessed ("he, too, had been this full of images once, he persuaded himself"), and ultimately rejects any need for elucidation in favor of a purely affective response: "He found this image magnificent. Of course it did not explain anything, but what good is explanation?"[223]

By introducing distance in this way, Musil follows guidelines he outlined in 1930 for the continuation of the novel, where he mentions the proximity of Clarisse's delusions to the artistic productions of the day in order to critique them:

> Green states even have their composers, who set them to music; these days sounds are painted, poems form sensory spaces [...]: this is a vague kind of associating that has become popular because thinking has lost its authority; it's about one eighth sensible and seven eighths nonsensical.[224]

---

[220] Musil, *The Man Without Qualities*, 1565.
[221] Musil, *The Man Without Qualities*, 715–716.
[222] Musil, *The Man Without Qualities*, 1007.
[223] Musil, *The Man Without Qualities*, 1381.
[224] Musil, *The Man Without Qualities*, 1624.

Accordingly, the narrator makes it plain that Clarisse is indeed suffering from delusions but cannot recognize as much, thus treating her character with an ironic tone not found in earlier drafts. In other notes made at the time, Musil stresses the difference between delusion and health, and therefore between delusion and poetry, stating that madness involves the "absence of the possibility of [drawing] logical consequences" from ideas.[225]

In sketches for the further course of the novel made in 1936, Musil held on to Clarisse's institutionalization, travels, and the "Island" chapter with Ulrich. The latter, however, was to undergo major changes. According to note II/7/102, Ulrich would only interrupt his vacation with Agathe for half a day. Clarisse would try to seduce him, but it "probably won't lead to coit [sic]."[226] The centerpiece of the episode would be Walter's reckoning with Ulrich when the latter comes to pick up Clarisse. There are no indications of what role her symbolic use of objects, picture-language, and lyrical flights – and therefore her affinity with Expressionism – would play.

In parallel to these changes, Ulrich's self-experiment shifts from Clarisse to Agathe, that is, to the "other sister" who progressively splits off from the former in the successive drafts.[227] This shift suggests that the novel's construction of the 'primitive' does not represent the boundary separating Clarisse's delusion from

---

225 Musil, *The Man Without Qualities*, 1711.

226 Musil, *KA, Transkriptionen*, Mappe II/7/102.

227 On "Clarisse's world" as the "antisocial parallel of the asocial world of the siblings" and the "difference" between "delusional" and "utopian" totality it implies, see Willemsen, *Das Existenzrecht der Dichtung*, 315. Philip H. Beard ("Clarisse und Moosbrugger vs. Ulrich/Agathe: Der 'andere Zustand' aus neuer Sicht," *Modern Austrian Literature* 9, no. 3 [1976]) stresses the sober testing that Ulrich and Agathe practice with regard to the "other condition," as well as this state's relationship to reality and alterity, which represents a "quasi-scientific" mode of monitoring and control (120–121). Likewise, Beard stresses that the "other condition" in Moosbrugger and Clarisse follows from isolation and compulsion, whereas for Agathe and Ulrich it leads to a broader understanding of life and greater self-control (125–127). Emphasizing the latter point differently, Richard E. Hartzell ("The Three Approaches to the 'Other' State in Musil's *Mann ohne Eigenschaften*," 217) and Maximilian Aue ("'Pandämonium verschiedener Formen des Wahns?' Vom Wahnsinn und seinen Grenzen in Musils *Mann ohne Eigenschaften*," in *Literatur und Kultur im Österreich der zwanziger Jahre. Vorschläge zu einem transdisziplinären Epochenprofil*, ed. Primus-Heinz Kucher [Bielefeld: Aisthesis, 2007], 139–141) draw attention to the violence with which Moosbrugger and Clarisse experience the "other condition." Ostermann ("Das wildgewordene Subjekt," 619–621) also stresses that for Agathe and Ulrich the "other condition" involves mutual recognition, not the isolation or violence Moosbrugger and Clarisse experience. Hartzell places Ulrich's sobriety and the siblings' combination of an objective and subjective approach in opposition to Clarisse's intuitive and emotional encounter with the "other condition" ("The Three Approaches to the 'Other' State in Musil's *Mann ohne Eigenschaften*," 206–213).

incestuous experiments. Instead, the 'primitive' emerges in two variations: one that is associated with the "other condition" of supposed prehistoric cultures, and the other with a regressive move leading back either to childhood (for Ulrich and Agathe) or to madness (for Clarisse).

Apropos of the first, the narrator "assume[s] the existence of a certain alternative and uncommon condition [...] which has deeper origins than religions,"[228] having existed "thousands of years" ago in "primordial" times.[229] Elsewhere, there is also talk of a "society of savages" now deprived of the "other condition," which has been replaced by "properly regulated and intelligible morality."[230] However, the reference to prehistory does not simply equate the former with the "other condition"; instead, it identifies the latter as a capacity distinguishing humans as such, that is, as something innate that exists independent of varying historical and cultural standpoints. Ulrich affirms this anthropological understanding not only by invoking prehistory but also by citing mystical testimonies from very different ages, all of which are meant to prove that the condition is real. And he even describes the culture of his own day not only as the "asylum," but also as the "temple" of the "other condition" – which, however, is neglected and rejected as delusion by that same culture.[231]

Musil even looks for the "other condition" in the future. He puts the words of the seventeenth-century mystic Jean-Joseph Surin (quoted by Oesterreich) in Clarisse's voice when she tries to describe her condition to Walter:

> I can't describe what happens to me then, and how this spirit unites itself with mine – without depriving me of awareness or the freedom of my own mind, but nevertheless working like another person [ein anderes Ich], as if I had two souls.[232]

Clarisse ties these symptoms with the idea that she has achieved "the thought of God or the next level of human evolution."[233] In his notes on Oesterreich, Musil writes,

> Cl.: A conception of self [ein Ich] floats before her eyes, that she is capable of devoting herself to two or more sequences of actions or successions of thought with equal levels of at-

---

228 Musil, *The Man Without Qualities*, 832.
229 Musil, *The Man Without Qualities*, 833.
230 Musil, *The Man Without Qualities*, 835.
231 Musil, *The Man Without Qualities*, 834.
232 Oesterreich, *Die Phänomenologie des Ich*, 435; Musil, *KA, Transkriptionen*, Mappe, II/1/80.
233 Musil, *KA, Transkriptionen*, Mappe I/5/154.

tention and involvement. Two or several modes of functioning [*Funktionsströme*] flow from it. It can also have a male and female personality simultaneously.²³⁴

This is not the author's own original thought, but Oesterreich's, who writes,

> [i]n such multiplicity, but taken *ad infinitum*, must the thought of God be conceived, if one accepts that it exists. – Perhaps it might also be the next-highest level of evolution beyond human thought, when manifold thinking of this kind occurs.

The fascination that this idea held for Musil probably stems from the fact that Oesterreich actually considers such a development to be really possible: "the concept of subjectivity contains nothing that would exclude phenomena of this kind."²³⁵ From ethnology to psychopathology, Musil's engagement with Oesterreich thus led him to science-fiction anthropology – terrain he had already approached in *Land über dem Südpol* – which follows the model of an avant-garde regression into the future (reminiscent of Brandlberger's project in *Tropen* and in contrast to the anti-modern regression characteristic of Walter).

At the same time, Musil's notes also establish his distance from Oesterreich. The author calls the notion "Oesterreich's invention" and only has Clarisse, who is mad, express it. Musil goes on to give an even more ironic treatment of the thesis Oesterreich subsequently develops in *Die religiöse Erfahrung* (1915) – that the immanence of the divine is still to be observed as an immediate experience in the modern world – which Musil criticizes as both naïve and academic.²³⁶ Like Oesterreich, Musil favors an anthropological perspective on schizophrenic and ecstatic splits of personality. At the same time, however, he does not aim at a philosophical or even religious postulation of a different way of thinking and relating to the world. Instead, he aspires to study it using general psychological research, raw data from ethnology, history of religion, psychopathology, and the methods of literature.

In his conversations with Agathe, Ulrich claims that the "other condition," subjected to centuries of neglect and repression, has been kept from developing and, accordingly, "never got beyond […] primordial disorder and incompleteness"²³⁷; alternatively, it has been reduced to a purely irrational faculty or minimized as affected sentimentalism. Either way, no fitting contemporary expression of the "other condition" exists for the characters, as it did at other places

---

234 Musil, *KA, Transkriptionen*, Mappe II/1/81.
235 Oesterreich, *Die Phänomenologie des Ich*, 445.
236 Musil, *KA, Transkriptionen*, Mappe II/1/76.
237 Musil, *The Man Without Qualities*, 833.

and times. It may be because of this non-cultivation or even repression of the "other condition" that its second form in *The Man without Qualities*, the form of regression, appears hand in hand with madness, blind activism, and the propensity to violence.[238] As such, like the regressed post-civilized characters in the early expedition stories of Musil and his contemporaries, it harbors a potential for destruction and inhumanity, expressed in *The Man Without Qualities* in the affinity of Clarisse's delusion to the delusional mass enthusiasm for World War I, as well as in hints of the mass delusion of Nazi ideology.[239]

In contrast to psychological research of the time, Musil's characterization of Clarisse also points to the importance of preformed collective representations for the delusional systems held by manic schizophrenics. In this case, "collective" refers to the highly educated bourgeois society from which she comes. Hence, alongside myths we can include any number of canonical and fashionable writings, in particular those by Nietzsche, which exercise a marked influence on Clarisse's delusional system, and from which she draws mythological references (e. g., the connection between goat and god in the satyr). She ignores overall context, however, and picks and chooses only those passages that offer intellectual support for her moods and fancies. Accordingly, her thoughts are a patchwork of distorted textual fragments.

> Sometimes she's so hopelessly conventional, [Ulrich] thought; it's like coming upon a page from another book bound in what one is reading. [...] [It gave] the uncanny impression that she herself consisted of many such misplaced texts.[240]

Clarisse's delusional thought is not from lack of culture, then, as one might claim of the homicidal Moosbrugger. On the contrary, it comes from a naïve subscription to a mode of education that does not critically reflect on its content and

---

238 Eberhard Ostermann observes, apropos of the underlying logic: "It becomes clear that excluding the power of transgression represented by Moosbrugger – after all, he crosses the border between inside and outside, sense and madness, norm and crime – by the systems of law, morality, and science produces the opposite effect. The repressed returns with all the more vehemence from inside these systems and topples them" ("Das wildgewordene Subjekt," 616).
239 See especially Kappeler's discussion of Musil's critique of irrationalism in the anti-modern movements of his day (*Situiertes Geschlecht*, 269–295), e. g., the "group around Hans Sepp," which was part of "the German nationalist and *völkisch* opposition" (269), and the persistence of "pre-logical" thinking in the sphere of science and technology (289). Along these lines, Wolf observes that "Musil is not at all subject to the 'secret yearning for the return of a mythical time' (Götz Müller). [...] On the contrary, he is interested in understanding as precisely as possible different ways of *modern* thought and perception," and therefore figures "practicing *wild thinking* to various degrees" ("Das wilde Denken und die Kunst," 389).
240 Musil, *The Man Without Qualities*, 716.

judgments, but is simply accepted from childhood on as self-explanatory and as natural as the language one speaks. Thus, Clarisse takes Nietzsche's writings literally. It never occurs to her to take distance from them and ask about their figurative meaning. She reads the same way she listens to music, aligning her own ideas with the stipulations of the text. Clarisse thus practices a modified form of identificatory reading whereby she does not lose herself in the represented material, but the reverse, that material is drawn into the ego, where the ego is formed by it. And since Clarisse's delusion is founded on collective representations, it appears to be much more valid and is therefore more dangerous than Moosbrugger's madness:[241] Post-civilized society and its members, caught up in its regressive primitivism, are heading for catastrophe. The parallels to Nazi ideology, which likewise made massive use of Nietzsche's mythical motifs and slogans in order to promote and justify itself, are abundantly clear.[242]

Regression never really leads back to the "other condition," but only to its distorted image. The same holds for the ontogenetic return to the 'primitive' that takes place in Musil's novel: the regression to childhood. Ulrich and Agathe's efforts to momentarily access the "other condition" through sibling-love are shaped by two retrograde movements. The first involves a return to mysticism, which has been discussed often enough to not warrant treatment here,[243]

---

241 That said, Ostermann rightly observes that Moosbrugger has "*grown* wild" ("Das wildgewordene Subjekt," 615); his murders represent "falling back to archaic, quasi-precivilized behavior." In this sense, Ostermann also sees a foreshadowing ot the "collective eruption of violence" in the First World War (608; cf. 616).

242 The logic, developed by observing 'modern primitives,' that regression never leads back to the "other condition" as it once existed, but instead – on the basis of repression or simply the process of regression itself – to a distorted version of what it was, is clear in the example of mysticism. As I have noted, Clarisse stresses that Saint Francis of Assisi was able to lead a normal life, whereas today his counterparts exhibit only "religious mania" (Musil, *The Man Without Qualities*, 833).

243 Cf. Ursula Reinhardt, *Religion und moderne Kunst in geistiger Verwandtschaft. Robert Musils Roman "Der Mann ohne Eigenschaften" im Spiegel christlicher Mystik* (Marburg: Elwert, 2003); Genese Grill, "The 'Other' Musil: Robert Musil and Mysticism," in *A Companion to the Works of Robert Musil*, ed. Philipp Payne, Graham Bartram, and Galin Tihanov (Rochester: Camden House, 2007); Norbert Christian Wolf, "Salto rückwärts in den Mythos? Ein Plädoyer für das 'Taghelle' in Musils profaner Mystik," in *Profane Mystik? Andacht und Ekstase in Literatur und Philosophie des 20. Jahrhunderts*, ed. Wiebke Amthor, Hans R. Brittnacher, and Anja Hallacker (Berlin: Weidler, 2002); Niklaus Largier, "Mystik als Medium. Robert Musils 'Möglichkeitssinn' im Kontext," in *Intermedien. Zur kulturellen und artistischen Übertragung*, ed. Alexandra Kleihues, Barbara Naumann, and Edgar Pankow (Zurich: Chronos, 2010); Robert Leucht and Susanne Reichlin, "'Ein Gleichgewicht ohne festen Widerhalt, für das wir noch keine rechte Beschreibung

and the second a return to childhood. Already with his lover Diotima, Ulrich yearns for the "tender emotions of that age [which] can in a single moment of yielding cause the whole, still-tiny world to burst into flames, since they have neither an aim nor the ability to make anything happen."[244] But only the journey to his parents' house in fact opens this possibility to him. Once more the narrative draws on a topos of expedition literature when a "shipwreck strands [the siblings] back on the lonely island of their childhood,"[245] a land of the forgotten self. In the "Siamese Twins" chapter, Ulrich describes his wish to regain the participative perception of his childhood:

> When I remember as far back as I can, I'd say that there was hardly any separation between inside and outside. [...] When something important happened, the excitement was not just in us, but the things themselves came to a boil.

He reflects that no child can have awareness of this state, so that it seems like a lost paradise only in hindsight to the adult, who must "see himself from the outside like a thing." All the same, when gathered together with Agathe, he wishes to reverse this development.

> Everything you touch, including your inmost self, is more or less congealed from the moment you have achieved your "personality," and what's left is a ghostly hanging thread of self-awareness and murky self-regard, wrapped up in a wholly external existence. What's gone wrong? There's a feeling that something might still be salvaged. Surely you can't claim that a child's experience is all that different from a man's?[246]

But Agathe, not Ulrich, exhibits childish traits in this context:[247] "his young sister's questions sometimes seemed to [him] [...] like the questions of a child,

---

gefunden haben.' Robert Musils 'anderer Zustand' als Ort der Wissensübertragung," in *Medien, Technik, Wissenschaft. Wissensübertragung bei Robert Musil und in seiner Zeit*, ed. Ulrich Johannes Beil, Michael Gamper, and Karl Wagner (Zurich: Chronos, 2011); Ritchie Robertson, "Everyday Transcendence? Robert Musil, William James, and Mysticism," *History of European Ideas* 43, no. 3 (2017).

244 Musil, *The Man Without Qualities*, 310.
245 Musil, *The Man Without Qualities*, 782.
246 Musil, *The Man Without Qualities*, 979.
247 Clarisse also embodies childishness insofar as she is described as a boyish, androgynous being whose behavior seems game like (e.g., "behind the hills, Cl. gamboled in the thistles, playing like a child" [Musil, *Der Mann ohne Eigenschaften*, 1745]) – but for the fact that it is consistently supported by a complex system of delusion. The child in Clarisse displays animalistic behavior that Ulrich describes as "reach[ing] out briskly for everything and set[ting] about everything ... rush[ing] over obstacles like a torrent, or foaming into a new course; [its] passions are strong and constantly changing" (*The Man Without Qualities*, 1334).

which are as warm as the little hands of these helpless beings." Agathe asks both naïve and wise questions, and Ulrich gives answers;[248] she thinks in eidetic images, while he does so with concepts; her inclinations are effusive, and his sober;[249] Agathe is prepared to engage thoughtlessly with "the other condition," whereas Ulrich tends to hesitate, doubt, and theorize.[250] Like Clarisse, Agathe suffers from a psychic disorder, depression. In depression, however, no 'primitive' state of mind is realized; rather, it goes hand in hand with the sentimental longing for an "other condition" modeled after childhood and the insight into the impossibility of its realization. Musil's notes read,

> Agathe: describe a deep depression. It's as if a secret drawer inside had been turned upside down, revealing contents never seen before. – All is darkened. Little reflection, actually an incapacity to reflect. The idea: I must kill myself, is only there in the form of this sentence, unvoiced, uncannily manifest [bewußt] in her presence, fills out the dark emptiness more and more. [...] "Let's kill ourselves," said Agathe. "We're the unfortunate, who bear the law of another world within without being able to carry it out! We love what is forbidden and will not defend ourselves."[251]

Regression to childhood then leads at best back to a sentimental, kitschy, and highly disappointing caricature.[252] In the late galley-proof chapter, "Love Blinds" (1937–1938), this caricature is symbolized in the sticky grains of sugar the siblings discover inside a confectioner's horse, whose belly had always seemed to held a great secret to the children:

> the confectioner's horse constituted part of the large family of children's fancies which are always chasing their desires with the zigzag flight of a butterfly, until at last they reach their goal only to find a lifeless object.[253]

---

[248] Musil, *The Man Without Qualities*, 810.
[249] Musil, *The Man Without Qualities*, 819.
[250] Musil, *The Man Without Qualities*, 834.
[251] Musil, *KA, Transkriptionen*, Mappe I/8/59.
[252] In light of this critical assessment, it is mistaken to describe Musil's novel as a "leap backwards into regression," as Wolfgang Riedel does ("Robert Musil: Der Mann ohne Eigenschaften," in *Lektüren für das 21. Jahrhundert. Schlüsseltexte der deutschen Literatur von 1200 bis 1990*, ed. Dorothea Klein and Sabine M. Schneider [Würzburg: Königshausen + Neumann, 2000], 278).
[253] Musil, *The Man Without Qualities*, 1207.

Psychology as the Springboard for Literature

The progressive dissociation of Ulrich from Clarisse and the differences between him and Agathe correspond to his insight that the "other condition" cannot be regained through regression, which instead leads only to pathological results. Musil, like the protagonist of his novel, thus chose a different path. As early as 1913, the author expresses the view that literature should apply scientific rigor, "with its claims to profundity, boldness, and originality," to the realm of feeling,[254] that is, to what he would come in 1918 to call the "nonratioid."[255] At the same time, he underscores that the writer should not be induced to abandon thought so much as to embrace other, additional qualities. Critiquing Walter Rathenau in 1914, Musil declares that those who are prone to investigate "experience" (*Erlebnis*) and "mysticism of feeling" lack "the virtues of method and precision," whereas those who possess these virtues "have no idea" what there is to gain from a scientific investigation of these non-rational phenomena.[256] In 1918, he stresses that a writer is "neither 'madman,' nor 'visionary,' neither 'the child' nor any other deformation of reason." Indeed, he does not "apply any different kind of reason than the rational person," but with one exception: he "commands the greatest factual knowledge *and* the greatest degree of rationality in connecting the facts."[257] Three years later, Musil insists that "where everything is flowing, [reason] must grasp and discriminate that much more sharply" in order "to investigate the logic of the analogical and the irrational."[258] Taking aim at the anti-intellectual and irrationalist positions of his contemporaries, he observes in 1922, "we do not have too much intellect and too little soul, but too little intellect in matters of the soul."[259] In other words, Musil had by

---

254 Musil, "The Mathematical Man," in *Precision and Soul*, 43.
255 Norbert Christian Wolf explores this critical assessment of regression in an essay that explicitly takes up Riedel's formulation above. In particular, he points to the failure to conjure up the paradise depicted in the novel, Musil's distance from (neo-)Romanticism, his use of concepts borrowed from Lévy-Bruhl to analyze the modern world, and his call for precision relative to the "same ontological value" attached to "normal" and "other" conditions where "anthropological insight" stands at issue ("Salto rückwärts in den Mythos?" 264). See also Eckhart Goebel, *Konstellation und Existenz. Kritik der Geschichte um 1930. Studien zu Heidegger, Benjamin, Jahnn und Musil* (Tübingen: Stauffenburg, 1996), 197–219.
256 Musil, "Commentary on a Metaphysics," in *Precision and Soul*, 57.
257 Musil, "Sketch of What the Writer Knows," in *Precision and Soul*, 64.
258 Musil, "Mind and Experience," in *Precision and Soul*, 142.
259 Musil, "Helpless Europe," in *Precision and Soul*, 131. The same is also quoted by Kappeler (*Situiertes Geschlecht*, 266), who concludes, "Although Musil is an adherent of progress through

then already arrived at the critical perspective presented in "Toward a New Aesthetic" (1925): the artistic quest for the "other condition" should not concern liberation from thought, but rather the break up of normative forms of experience, fleeting moments of interruption that do not produce any kind of totality of the "other dimension."[260]

The same stance is taken in *The Man Without Qualities* by Ulrich, who eschews regression and seeks to put an end to centuries of neglecting the "other condition" by cultivating it in a manner suited to the times. In exchanges with Agathe, Ulrich repeatedly claims to regard the "other condition" as soberly as possible and to conduct an exact investigation[261] in order to achieve a real understanding of it based on demonstrable and verifiable facts. Accordingly, his considerations of the matter are a tireless struggle for concepts and explanations, striving for the accuracy of the scientific procedure:

> He himself really ought not to have thought it either: the scientific procedure – which he had just finished explaining as legitimate – consists, aside from logic, in immersing the concepts it has gained from the surface, from "experience," into the depths of phenomena and explaining the phenomena by the concepts, the depths by the surface; everything on earth is laid waste and leveled in order to gain mastery over it, and the objection came to mind that one ought not extend this to the metaphysical. But Ulrich now contested this objection.[262]

Both with and without his sister's help, Ulrich encounters one paradox after the next and abandons one scheme of explanation for another, leading to the galley-proof chapters on emotional psychology (chapters II/52 to II/58), where he provides an account of the "other condition" by cycling through old and new theories on the relationship between sensation and reality. In contrast to Clarisse, Moosbrugger and Agathe, Ulrich does not simply yield to this state but also reflects on what it involves. In his estimation, the "other condition" is not a capacity located beyond thought so much as a matter of "alter[ing] [...] consciousness."[263] He turns against the age-old "irrationalization" of this frame of mind not only to have an object of scientific knowledge but also to glimpse the subjectivity of another, but still rational mode of knowing.

---

knowledge and rationality, he takes the critique of one-sided rationalism seriously and calls for the 'embedding of thought in the emotional sphere'" (132).
260 Musil, "Toward a New Aesthetic," 206.
261 Musil, *The Man Without Qualities*, 831–832.
262 Musil, *The Man Without Qualities*, 1188.
263 Musil, *The Man Without Qualities*, 833.

In the chapters on emotional psychology, Ulrich declares that a "certain state" – sobriety – is the precondition for scientific insight. The "other condition" arises when this sober state of mind is replaced by another emotional state, which would make visible not only another image of the world but another world altogether.[264] This other emotional state is characterized by indeterminate sensations that do not entail any particular behavior but simply bathe the world in a different light: "the nonspecific emotion changes the world in the same way the sky changes its colors."[265] In other places Ulrich compares the sensation with a certain form of love:

> An emotion that is not an emotion *for* something; an emotion without desire, without preferment, without movement, without knowledge, without limits; an emotion to which no distinct behavior and action belongs [...]. Love is already too particular a name for this, even if it is most intimately related to a love for which tenderness or inclination are expressions that are too obvious.[266]

Finally, in the "Breaths of a Summer Day" chapter, he calls this feeling "vegetative" and finds something "feminine" about it, opposed to the "appetitive" and "animal" feeling, which is coded as masculine and in his eyes underlies the drive for science:

> The world has the appetitive part of the emotions to thank for all its [...] progress. That man should thank for his progress precisely what really belongs at the level of the animal is, at the very least, unexpected. [...] Doubtless there are at its core the same few instincts as the animal has.[267]

In sum, Ulrich's cultivation of the "other condition" moves on two parallel tracks. On one path, it brings together the worldview tinted by vegetative feeling with the investigation of it. On the other, this investigation is based on the (completed) cultivation of animalistic sentiment, which yields the ethos of the scientific explorer. Ulrich's metaphors reveal this double movement to be another form of hermaphroditism. Instead of involving a transcendence of bestial sexuality into genius as Clarisse envisions, it concerns the rational comprehension of a 'feminine' emotional state and its corresponding world view.[268]

---

[264] Musil, *The Man Without Qualities*, 1444–1446.
[265] Musil, *The Man Without Qualities*, 1305–1306..
[266] Musil, *The Man Without Qualities*, 1279.
[267] Musil, *The Man Without Qualities*, 1332.
[268] In his 1919 essay, "Abbau der Sozialwelt" (Breakdown of the Social World), Müller likewise speaks of a "vegetative" existence at the beginnings of human history, claiming that the "vege-

It would be mistaken, however, to conclude that Ulrich, who thinks at length about the psychology of emotion and values scientific ethos, is an advocate of psychology. Instead, the novel casts the discipline in a very critical light. This critique, presented in the voices of the characters as well as the narrator, bears on the pathologization of norm-breaking thought as well as on the language of science and the discursive power it exerts.

The person of Moosbrugger plays a central role in this critique.[269] Having "discovered that it was the possession of this scientific language that gave those in power the right to decide his fate with their 'findings,'"[270] the criminal seeks to include foreign words and technical terms in what he says. These efforts at self-assertion are directed above all at "medical diagnoses," the "psychiatrists" who make them, and the "science" upon which they rely. Using their own weapons against them, Moosbrugger wishes to counter those who "dismiss his whole complex personality with a few foreign words."[271] The narrator also finds fault with forensic psychology's blindness to the particular case and the individual behind the case: "the cruelty of a mind that shuffles concepts around without bothering about the burden of suffering and life that weighs down every decision."[272] In contrast to the "angel of medicine," this type of psychology no longer aspires to the (impossible goal of) healing each and every patient, but instead seeks only to categorize their cases.[273] The narrator's insight that there is no knowledge and no truth beyond the system of categorization put in place and maintained by the power of discourse (figuratively expressed

---

tative man of the senses" is followed by "civilized man [*Kulturmensch*]"; ultimately, the process will conclude in the "redemption of the "vegetative man of spirit [*Geistmensch*]" (358). In spite of superficial similarities, there are major differences: Müller relies on an evolutionary model, whereas Musil posits two sensibilities that are (supposed to be) cultivated; Müller's aims remain esoteric, while Musil seeks rational understanding.

269 Cf. Ostermann, "Das wildgewordene Subjekt," 611–614; and, on Musil's intensive engagement with psychology, Sandra Janssen, *Phantasmen. Imagination in Psychologie und Literatur 1840–1939. Flaubert – Cechov – Musil* (Göttingen: Wallstein, 2013), 413–422 (with particular bearing on the writer's early works); Norbert Christian Wolf, "Wahnsinn als Medium poet(olog)ischer Reflexion. Musil mit/gegen Foucault," *Deutsche Vierteljahrsschrift für Literaturwissenschaft und Geistesgeschichte* 88 (2014); Maximilian Bergengruen, "Moosbrugger oder die Möglichkeiten der Paranoia. Psychiatrie und Mystik in Musils 'Der Mann ohne Eigenschaften,'" *Zeitschrift für deutsche Philologie* 135, no. 4 (2006). See also Inka Mülder-Bach, "Der Fall Moosbrugger," in *Was der Fall ist. Casus und Lapsus*, ed. Inka Mülder-Bach and Michael Ott (Paderborn: Fink, 2014).
270 Musil, *The Man Without Qualities*, 71.
271 Musil, *The Man Without Qualities*, 72.
272 Musil, *The Man Without Qualities*, 580.
273 Musil, *The Man Without Qualities*, 262.

as a "sea of scientific papers") positions him as a practioner of discourse analysis *avant la lettre*.[274]

Unlike the forensic psychiatrists and lawyers who only look for somewhere to file Moosbrugger, Ulrich and the narrator display an understanding of him. When the judge casts a "net woven from incomprehension" for Moosbrugger to get tangled up in – because he has no language to unravel it – the narrator adds what it looks like inside Moosbrugger's mind, lending him to a certain extent the language he lacks. Likewise, Ulrich attributes Moosbrugger's rage to his lack of education, [275] which the judge sees merely as an excuse.

The "Visit" chapter presents three representatives of forensic psychiatry. From the start, the narrator treats it with irony, calling it a pseudo-science similar to art and theology:

> In science the slighter the success in precision, the greater, generally speaking, is the artistic component, and up until a few years ago psychiatry was by far the most artistic of all modern sciences, with a literature as ingenious as that of theology and a success rate that could not be discerned in the earthly realm here below.[276]

One representative of the field, Dr. Friedenthal, is portrayed as a sorcerer obsessed with his effect on others and who enjoys dealing with demonic auras. Another, Dr. Pfeifer, is a collector of victims who takes "dangerous" pleasure in "judicial murder[s]" that he supports with expert testimony.[277] Only the unnamed "young physician," who wants to secure a diagnosis of insanity for Moosbrugger, is charactized as sympathetic and having integrity. Nonetheless, he finds himself in a macabre competition with Pfeifer over Moosbrugger's head (and for the professional recognition each doctor wishes to win as a result) by means of a card game between the inmate, the psychiatrists, and a clergyman. The game is in fact a secret observation meant to collect symptoms and arrive at a diagnosis. Friedental, who represents the status quo of the field, justifies the practice along pragmatic and bureaucratic lines.

In passages like these, Musil distances himself from the approach and language of psychologists, especially those practicing forensic medicine, which is corrupted by jurisprudence. As the novel develops, its critique of the pathologization of norm-breaking behavior and thinking shifts to targetting the classify-

---

[274] I owe this suggestion to Norbert Christian Wolf ("Wahnsinn als Medium poet[olog]ischer Reflexion").
[275] Musil, *The Man Without Qualities*, 75.
[276] Musil, *The Man Without Qualities*, 1542.
[277] Musil, *KA, Lesetexte*, vol. 3, *Kapitelkomplex Clarisse. Besuch*, 27.

ing approach and scientific language that mispresent individuals. What is now sought is not an alternative to psychology, but an alternative way of dealing with psychological phenomena. Thus, Musil's relationship to psychology cannot be described as a transfer (from the realm of science to the realm of literature), but as critical reflection first, and only then as a transformation.

In a fragment from around 1920, Musil declares that "there never has been a literary psychology,"[278] for psychology differs from literature in that it strives for insight by seeking out "the relatively general in the individual case."[279] It follows that literature can never do justice to the claims of psychology: the conclusions it offers remain "stuck" and prove to be mere "pseudoexplanation[s]." Yet Musil still locates a literary dimension in psychological case histories, which "are depictions of pathological processes of the soul that are marvelously penetrating, and so strongly metaphorical (for the 'normal' reader) that the addition of interpretation that would make them into great literature is hardly missed."[280]

Musil's notes, especially on Oesterreich, assign a major part to these same features. The task is not to compete with the explanations of psychology so much as to approach its poetic quality and take up its case histories as a springboard for creating literary art. Musil did not set out to write "a pseudoscientific novel," but "really to go all the way to the end of the trampoline of science and only then to jump."[281] For him, "great literature" begins precisely where science ends; it is a land that science abuts but can never reach. Inspired by the poetic aspects of psychology, the writer takes the final step (or jump) from the one territory to the next.[282] Unlike the psychologist, he has no interest in arriving at a final diagnosis. Instead, he sets out "to discover ever new solutions, connections, constellations, variables, to set up prototypes of an order of events, applying models of how one can be human, to *invent* the inner person."[283] In contrast to scientific experimentation, literary experiment does not focus on the rule but the exception, not what repeatedly happens but what is new. Also, the writer's work resounds with the hermeneutic insight that interpretion is never final and requires a subject to carry it out.

---

278 Robert Musil, "Psychology and Literature," in *Precision and Soul*, 66.
279 Musil, "Psychology and Literature," 66.
280 Musil, "Psychology and Literature," 67.
281 Musil, "Psychology and Literature," 67.
282 "*Psychologia fantastica:* summarize Klages, part. Freud, Jung … like this. My instinctive hostility: because they're pseudo-poets and deprive literature of the support of psychology!" (Musil, *KA, Transkriptionen*, Heft 30/110).
283 Musil, "Sketch of What the Writer Knows," in *Precision and Soul*, 64.

In Chapter 62 of *The Man Without Qualities*, Ulrich assigns a genre to this kind of literature. He reflects, "A man who wants the truth becomes a scholar; a man who wants to give free play to his subjectivity may become a writer; but what should a man do who wants something in between?"[284] Literary craft starting from psychology and invention/interpretation, as its fraternal twin, springs from this "in between." At stake is the establishment of an inter-genre and an inter-language that operate between psychological research and subjective literary art. Ulrich finds this "intermediate space" in the essay genre, for its "domain lies [...] between example and doctrine, between *amor intellectualis* and poetry."[285] In literary terms, the essay is the "unique [...] form" that the "inner life of a human being" takes at a certain point in time; in scientific terms, it takes the form of a "thought" that is subject to "laws that are [...] strict."[286] The essay's strength is that it expresses both subjectivity and truth by representing the emotionally colored realities of each individual. Its cases are more exemplary than purely subjective fictions and at the same time more individual than psychological examples.

By demanding a scientific approach, Ulrich removes himself from any form of regressive movement. At the same time, and like the author of *The Man Without Qualities* himself, he holds psychology at arm's length while embracing a literary method of investigation inspired by yet distinct from psychology – a procedure that is just as rational as the scientific method but more inventive, flexible, and variable. What do these reflections about an inter-genre imply about the narration of the *Man Without Qualities?* In what way does it differ from the writing of psychologists on the one hand and from regressions to the 'primitive' on the other?

Primitivistic Narration

Following a chapter from which Musil made extensive excerpts, Oesterreich advances the thesis that not only schizophrenics but also "novelists" are predestined for the "development of an inner double consciousness."[287] Although Musil took no notes on this passage, it resonates with the way he brings together

---

[284] Musil, *The Man Without Qualities*, 274.
[285] Musil, *The Man Without Qualities*, 273.
[286] Musil, *The Man Without Qualities*, 253.
[287] Oesterreich, *Die Phänomenologie des Ich*, 449.

madness and art in early sketches of Clarisse. Can Clarisse be understood as a figure for the novelist?[288] There is no doubt that she, as Walter Fanta has demonstrated, pictures "illusionary realities" and devises "fantastic sign-systems" that may be viewed as yielding a "metafictional structure." What's more, she finds eager readers, or rather listeners, in the person of Ulrich and his predecessors of the early drafts, as well as Walter and the General in later ones.[289] Indeed, the General lauds her capacity for storytelling ("You speak so vividly [*erzählen so plastisch*] that one understands everything"[290]). Does Musil differentiate his stylistic treatment of Clarisse from the style of psychological texts because he models his own authorship after her primitivisms? Can the in-between-space between science and literature also be understood as an interspace between the psychologist and Clarisse?

What Musil writes about Clarisse differs in narrative perspective from the psychological studies he consulted, as he often adopts her point of view. In this respect, the author follows his character's dictum that one must "participate" in the other in order to understand her.[291] Many chapters shift from a third-person omniscient narrative perspective to a third-person limited one, focalized through a single character, with the result that no external view of Clarisse is available to the reader. A case in point is the chapter, "Clarisse and Her Demons." Just after the narrator's distanced description of the abrupt shifts in topic in Clarisse's mental process, a direct demonstration of just that type of thinking follows for three full pages, until the perspective shifts yet again, this time to Walter. This move from an external perspective onto Clarisse to an internal one is particularly evident in passages where quotation marks first enclose one of her thoughts, but then further thoughts of hers proceed without such marks. The same holds, for instance, for the following, largely incomprehensible sentence shaped by Clarisse's delusional logic:

> But it was also like a metaphor [*Gleichnis*], where the things compared are the same yet on the other hand quite different, from the dissimilarity of the similar as from the similarity of the dissimilar two columns of smoke drift upward with the magical scent of baked apples and pine twigs strewn on the fire.[292]

---

[288] This thesis has already been proposed, albeit without reference to Oesterreich. See Walter Fanta, "Die Spur der Clarisse in Musils Nachlass," *Musil-Forum* 27 (2001/2002): 283–285.
[289] Fanta, "Die Spur der Clarisse," 284.
[290] Musil, *KA, Transkriptionen*, Mappe V/4/212.
[291] As Philip Payne observes, this also holds for the narrational treatment of Moosbrugger ("Musil erforscht den Geist eines anderen Menschen – zum Porträt Moosbruggers im *Mann ohne Eigenschaften*," *Literatur und Kritik* 11, no. 106–107 [1976]: 392–396).
[292] Musil, *The Man Without Qualities*, 152–153.

Even in context, it is unclear to what "it" refers. Might it be the "orderliness [that] long[s] to be torn apart," which is "provoked by Moosbrugger"? Or the "thunder of the music [...] inwardly eating away at the timbers"? Most of all, however, the metaphor for the metaphor remains incomprehensible because its details establish neither similarity nor how things might fit together. In spite of the delusional logic of this passage, it is not set off in quotation marks – unlike the thoughts that precede and follow it. This indicates that at this point the narrative is focalized only through Clarisse's perspective. The early drafts sometimes even feature first-person narration (e.g., "The most majestic sight for me was Nietzsche's psyche – in the form of the head doctor at the institute"[293]).

In passages where the narrative is focalized through Clarisse's limited perspective, the narrative takes on a paratactic and elliptical style, the latter in relation to the correlation among her thoughts. According to Kretschmer, both of these qualities typify manic thinking.[294] An impressive example occurs at the beginning of Chapter 97, entitled "Clarisse's Mysterious Powers and Missions." If one compares the first one and a half pages with the relatively conventional format of the final part of the preceding chapter, it is striking how many paragraphs visually break up the left margin of the text, pointing to the disjointedness of the thoughts recorded there. If one looks at the first, comparatively long paragraph, one also notices the predominance of parataxis. Eight out of ten sentences or sub-sentences are short and paratactic constructions – a striking contrast to the last page of the previous chapter.

Furthermore, metaphorical language saturates narrative sequences regarding Clarisse. In extreme instances, it yields unintelligible comparisons like the one cited at length above. Alternatively, the literal meaning of metaphors turns into the associative principle ordering the text. The birthmark Clarisse calls the "Devil's Eye" is a case in point: as mentioned above, the phrase leads to the expression *etwas ins Auge fassen*, which figuratively means "to take a close look." Yet here it is taken literally and leads to the image of physically grabbing and throwing a rock.[295] These passages may still be understood as a kind of inner monologue of Clarisse's, and the pronounced use of metaphorical language in them may therefore be seen as a symptom of her insanity. Nonethless, the text adopts the same procedure in passages that do not appear to be written from her internal perspective.

---

[293] Musil, *Der Mann ohne Eigenschaften*, 1780.
[294] Kretschmer, *A Text-book of Medical Psychology*, 144, 148.
[295] Musil, *The Man Without Qualities*, 475.

This applies, for example, to the passage leading up to the change in narrative perspective in Chapter 38. First, Clarisse's thoughts are described in terms of "images [springing] up, [...] fused" in "fluttering mists" and "lurking [...] demons behind the stage."[296] Then, when the narrative is focalized through Clarisse's limited perspective, these "demons" are taken literally and materialize in the form of Moosbrugger, Ulrich, and Walter. The same tendency is also evident where the narrative has no direct relation to Clarisse's subjective experience. Dietrich Hochstätter draws attention to a scene in Chapter 84:[297] "Clarisse, looking like a little angel in the long nightgown that covered her feet, had stood on her bed declaiming Nietzschean sentiments, with her teeth flashing [...]. In the twilight of the bedroom this had made a rather gruesome spectacle."[298] In German, the first sentence reads, "Clarisse, im langen, die Füße bedeckenden Nachthemd wie ein kleiner Engel anzusehen, stand aufgesprungen im Bett und deklamierte mit blitzenden Zähnen frei nach Nietzsche";[299] *aufgesprungen*, which is lost in translation but means something like "bolt upright" (literally, "jumped-up" as well as "sprung-open"), is deployed to two ends. It not only refers to standing up quickly but also to the sudden rift in her personality that brings forth an angel *and* a beast (her "teeth flashing") and provokes a "gruesome" feeling in the onlooker. The third-person limited narrative thus exhibits some characteristics of schizophrenic or manic 'primitive' language, but so does the language used by the omniscient narrator when he talks about Clarisse. In a way, he assimilates his style to those about whom he speaks.

This quality colors other parts of the novel as well, where Musil employs similar procedures. Clarisse's "double beings" and "double words" find their counterpart in the interest Ulrich takes in Siamese twins and use of metaphorical operations.[300] His – and the narrator's – efforts to describe the "other condition" he experiences with Agathe rest on metaphors. In the galley-proof chapter, "Beginning of a Series of Wondrous Experiences," the siblings experience an "instant in the midst of that shared condition," which, as elsewhere, is likened to the existence of Siamese twins: "The fraternal stature of their bodies communicated it-

---

[296] Musil, *The Man Without Qualities*, 152.
[297] Dietrich Hochstätter, *Sprache des Möglichen. Stilistischer Perfektionismus in Robert Musils 'Mann ohne Eigenschaften'* (Frankfurt am Main: Athenäum, 1972), 131.
[298] Musil, *The Man Without Qualities*, 400.
[299] Musil, *Der Mann ohne Eigenschaften*, 368.
[300] Musil employs the term *Gleichnis* as a synonym for other tropes. Cf. Inka Mülder-Bach, *Robert Musil. Der Mann ohne Eigenschaften. Ein Versuch über den Roman* (Munich: Hanser, 2013), 333–346, and "Allegorie und Gleichnis im 'Mann ohne Eigenschaften,'" in *Allegorie. DFG-Symposion 2014*, ed. Ulla Haselstein (Berlin: De Gruyter, 2016).

self to them as if they were rising up from a single root."³⁰¹ It is described as a "shadowy union, of which they had already had a foretaste as in an ecstatic metaphor."³⁰² The Siamese twins function as a metaphor for the metaphor of a union that defies conceptual understanding.³⁰³ The metaphorical quality of experience is reinforced by the fact that Ulrich can only express the state in a "senseless" declaration: "You are the moon – [...] You have flown to the moon and it has given you back to me again."³⁰⁴ He qualifies his remark as a "metaphor" and an "impossible" one; even so, "the exaggeration was quite small and the reality was becoming quite large."³⁰⁵

The linguistic advance toward the "other condition" thus takes place by means of metaphorical operations, which are credited with the ability to register a certain dimension of reality.³⁰⁶ At play here is the reality of emotional states or, more precisely, states determined by feeling and sensation (as Ulrich explains with the example of the moonlit night). Because they are subjective and objective at once (inasmuch as they color one's view of the world), truth and untruth are "inextricably bound up with each other."³⁰⁷ In this context, Ulrich also mentions dreams, art, and religion as spaces of such experiences.³⁰⁸ Elsewhere, however, a differentiation between these states and metaphor is made. While Chapter I/116 declares that "metaphor is like the image that fuses several meanings in a dream,"³⁰⁹ the preceding chapter specifies that metaphor is the "*relationship* between a dream and what it expresses."³¹⁰ Dreams and art are likened to a use of metaphor that dispenses with interpretation and proceeds literally. In this sense,

---

301 Musil, *The Man Without Qualities*, 1177.
302 Musil, *The Man Without Qualities*, 1178. Cf. Musil, *KA*, *Transkriptionen*, Weitere Mappen, Gelbe Mappe/4545: "We have examined whether it might not be possible, all the same, to be wholly oneself [*eins zu sein*] and to live as two with one soul. We hinted at all sorts of answers, but I forgot the simplest one: that the two people could be well-disposed toward each other and able to accept everything they experience as a mere likeness [*Gleichnis*]! Consider that every comparison is ambiguous for the mind, but unambiguous for sentiment."
303 Riedel draws attention to how the twin relationship transfers to thinking and speech through Ulrich ("Robert Musil," 279).
304 Musil, *The Man Without Qualities*, 1178.
305 Musil, *The Man Without Qualities*, 1179.
306 "Similes refer to what really is similar [*Gleichnisse bezeichnen wirklich Gleiches*]" (Musil, *KA*, Transkriptionen, Mappe II/8/58); cf. Kerstin Schulz, "'Als wäre mein Mund so fern von mir wie der Mond,'" in *Denkbilder. Wandlungen literarischen und ästhetischen Sprechens in der Moderne*, ed. Ralph Köhnen (Frankfurt am Main: Lang, 1996), 123–127.
307 Musil, *The Man Without Qualities*, 634.
308 Musil, *The Man Without Qualities*, 634–635.
309 Musil, *The Man Without Qualities*, 647.
310 Musil, *The Man Without Qualities*, 634. Emphasis added.

they act analogously to delusion, which, as Ulrich states elsewhere, involves the same relationship to metaphor.[311] This stance contrasts with a scientific view that takes interest only in the "truth" and therefore deprives metaphors of their true potential. Ulrich is looking for a third possibility – for which, however, so far only the metaphor itself, representing a state of limbo between science and art, between a figurative and a literal meaning, is available.

As the preceding pages have shown, the *twin* serves as a key metaphor for grasping the "other condition." But, in contrast to other tropes, the metaphor, i.e., the type of trope itself, is also a central metaphor for the experience of the "other condition";[312] "Every analogy contains a remnant of that magic of being identical and not identical," Ulrich tells Agathe.[313] Figurative language does not just *represent* the "other condition" then. Instead, to receive a metaphor means to experience a version of the "other condition." A brief but telling note by Musil reads: "metaphor [*Gleichnis*] as second state."[314] Metaphor does not portray the other condition so much as it *induces* it in the first place, forging a momentary union between what is otherwise separated forever, between dream and truth.[315] Further evidence for this is provided in "Toward a New Aesthetic." The reception of a metaphor opens onto the experience of "another condition" be-

---

[311] Musil, *The Man Without Qualities*, 957.
[312] On the fusion of metaphor and the image of twins, see Jörg Kühne (*Das Gleichnis. Studien zur inneren Form von Robert Musils Roman 'Der Mann ohne Eigenschaften'* [Tübingen: Niemeyer, 1968], 155–166), who discusses the comparison of the two trees with the twins; as does Tewilt (*Zustand der Dichtung*, 132–171). Riedel ("Robert Musil," 265–285) points to Ulrich's transfer of the twin relationship to thought and language, drawing a connection to theories of "mythical" or "archaic" thinking proposed by Vischer and Lévy-Bruhl, as well as to Jung and Freud's theories of dreams (279). The interconnection between figurative language and the "other condition" is at most hinted at by scholars; cf. Willemsen, "Dionysisches Sprechen: Zur Theorie einer Sprache der Erregung bei Musil und Nietzsche," *Deutsche Vierteljahrsschrift für Literaturwissenschaft und Geistesgeschichte* 60 (1986): 117; and Jutta Heinz, "Grenzüberschreitung im Gleichnis. Liebe, Wahnsinn und 'andere Zustände' in Robert Musils Mann ohne Eigenschaften," *Grenzsituationen. Wahrnehmung, Bedeutung und Gestaltung in der neueren Literatur*, ed. Dorothea Lauterbach, Uwe Spörl, and Uli Wunderlich (Göttingen: Vandenhoeck & Ruprecht, 2002), 254–255.
[313] Musil, *The Man Without Qualities*, 983. Cf. the author's notes: "Schwester? Ein Gleichnis … ein schönes Gleichnis…" (Musil, *KA, Transkriptionen*, Mappe II/1/227); "Schwester ein Gleichnis des Bruders" (Mappe II/4/114).
[314] Musil, *KA, Transkriptionen*, Mappe II/3/54.
[315] Roger Willemsen also indicates as much in "Dionysisches Sprechen," 117. Likewise, Heinz also speaks of how "in the reception of metaphor […] the limit that had been initially postulated between the ratioid and non-ratioid realms is selectively overcome" (Jutta Heinz, "Grenzüberschreitung im Gleichnis," 255).

cause, as Musil writes, it explodes the formulaic schematism of standing concepts:

> In [the process of expansion and contraction] art has the task of ceaselessly reforming and renewing the image of the world and of our behavior in it, in that through art's unique experiences [*Erlebnisse*] it breaks out of the rigid formulas of ordinary experience [*Erfahrung*]; [...] literature [does so] most aggressively and directly because it works without mediation with the material of formulation itself.[316]

In other words, metaphors make possible a linguistic experience of the "dissimilarity of the similar" and the "similarity of the dissimilar" (*Ungleichnis des Gleichen* and *Gleichnis des Ungleichen*) (which Clarisse can only put as another metaphor: "it was [...] like a metaphor, where [...] from the dissimilarity of the similar as from the similarity of the dissimilar columns of smoke [drifting] upward with the magical scent of baked apples and pine twigs strewn on the fire"[317]).

Thus, Musil's writing moves from the outward expedition to self-experimentation. The author follows Clarisse's lead insofar as his style of writing takes on traits that the novel and his discursive contexts would code as 'primitive,' i.e., proceeding paratactically, piling up ellipses, and being shaped by metaphors and figurative language. The novel even appropriates the performative magic of turning representation into production that Musil, with Hornbostel, ascribes to 'primitives' and designates as a potential model for modern literature.[318]

Another Primitivist Aesthetics?

Still, Musil's method of narration cannot be defined as primitivist in precisely the *same* sense as Clarisse's. From the outset, the Clarisse passages are commented upon by the narrator. The form of these comments changes between the early drafts and the published text. The omniscient narrator of the early drafts remarks directly on Clarisse's thoughts and actions. Later in the writing process, the content of this same commentary is relocated to dialogues, voiced in Ulrich's reflections, or presented as indirect, general essayistic reflection. Paradoxically, these changes increase the narrator's distance from Clarisse. This is exemplified in the earlier mentioned turn of phrase, "vom Hundertsten ins Tausendste" ("from the hundreth to the thousandth"), words borrowed from Bleuler,

---

[316] Musil, "Toward a New Aesthetic," 206.
[317] Musil, *The Man Without Qualities*, 152–153.
[318] See Musil, *KA, Lesetexte*, vol. 12, "Literat und Literatur"; *KA, Transkriptionen*, Mappe VI/3/6.

recorded both in early versions of the chapter[319] and in the notes Musil made while writing it.[320] The drafts record a steady removal of reflections voiced by the omniscient narrator on Clarisse's reasoning, which are then transferred to an exchange between her and the General:

> "Has your mind ever raced with a thousand and one things [*Sind Sie schon einmal vom Hundertsten ins Tausendste gekommen*], General?" she asked, and he had to reply in the affirmative. "And has it ever gone the other way around [...] from a thousand and one things back down [*vom Tausendsten ins Hundertste*]?" she asked further and [the General] was even less willing to answer in the negative, because a man takes pride in thinking things through, even down to that one thing called "truth." [...] But Clarisse concluded: "You see, that's nothing but cowardice – always thinking in an orderly and deliberate way!"[321]

In contrast, the earlier draft features the same expression to describe what Clarisse herself thinks and how.

> When she gave herself over to reflection, a thousand and one things occurred to her. For example, just as she could see herself addressed as a man [...] and feel herself a man and a woman at once, and therefore really [...] a double being, [...] she could feel herself also to be related to the [mentally] ill, for they are double beings, too [...]. But relations also extended in many other directions. [...] The reciprocal relations yielded a whole, and new points of departure emerged in almost unlimited number.[322]

Against the backdrop of Musil's reading of Bleuler, the early draft introduces a diagnostic distance between the narrator's voice and the figure of Clarisse. At the same time, however, the narrator takes up a position similar to Clarisse's in the later draft: just as Clarisse criticizes men's overly focused way of thinking in the later draft, the narrator in the earlier draft criticizes focused thinking as "all too orderly" and thus normalizes Clarisse's way of thinking.

> Incidentally, no one thinks any differently than Clarisse, as soon as one's thoughts go from one thing to a thousand-and-one of them [...]; it's just that [...] another, less personal form of thinking is applied in order to get there, [...] or back to what is called "truth." But Clarisse had started to hold such overly orderly and deliberate thinking in contempt.[323]

---

[319] E.g., Musil, *KA, Transkriptionen*, Mappe II/1/12 and Mappe V/4/214.
[320] Musil, *KA, Transkriptionen*, Mappe I/5/108.
[321] Musil, *KA, Transkriptionen*, Mappe V/4/214.
[322] Musil, *KA, Transkriptionen*, Mappe II/1/12.
[323] Musil, *KA, Transkriptionen*, Mappe II/1/13.

Whereas the later draft can be read to present a typical symptom of pathological speech, not just because of the copious figurative language but also because it is reversed ("vom Tausendsten ins Hundertste") and taken literally, in the earlier draft the narrator practices this approach to language himself and in this way resembles Clarisse.

In the published text and the later drafts, the scenes featuring Clarisse thus receive stronger commentary through dialogue, Ulrich's thoughts about their exchanges, and generally contemplative passages. A good example of this tendency is the transition from Chapter 38, which is devoted almost entirely to Clarisse's disordered thoughts narrated from the subjective third-person position, to Chapter 39, where the narrator and Ulrich reflect on a "world of qualities without a man" in an essayistic style. The latter can be read as an evaluation of what Clarisse thinks and says in the former, namely that human beings are pushed and pulled around by experiences and feelings for which they bear no responsibility.[324] In a similar way, the chapter, "Moosbrugger Thinks," told largely through the inner perspective of Moosbrugger, is related to the following essayistic chapter, which reflects on Moosbrugger and his treatment by jurisprudence and forensic medicine.

This interweaving of narration, metaphorical language, and essayistic reflection applies all the more to the passages involving sibling-love and the "other condition." As I noted above, Ulrich calls for "real understanding" of the "other condition" instead of irrational effusions; he objects to the "other condition" being positioned in opposition to thought and wishes that it be seen as a "peculiar change in thinking" instead.[325] Accordingly, in its investigation he calls for a combination of "visions" and "exact research."[326]

While the largely narrative Clarisse passages are supplemented in this way by reflective chapters, the reverse applies to the contemplative passages: they bleed into narrative sections and abound with metaphorical language.[327] Thus, the last paragraph of the essay chapter slides into a narrative sequence describing Ulrich's nighttime walk in the garden. The passage also takes up a metaphor

---

324 Musil, *The Man Without Qualities*, 158–159; cf. Simon Jander, "Die Ästhetik des essayistischen Romans. Zum Verhältnis von Reflexion und Narration in Musils *Der Mann ohne Eigenschaften*, und Brochs *Hugenau oder die Sachlichkeit*," *Zeitschrift für deutsche Philologie* 123, no. 4 (2004): 533.
325 Musil, *The Man Without Qualities*, 831.
326 Musil, *The Man Without Qualities*, 820, 831.
327 Convincingly demonstrated, most recently, by Jander, "Die Ästhetik des essayistischen Romans."

("milky foam of the mist outside") that plays a part in the reflection on essayistic form two pages earlier ("soothing mother's milk").[328]

It follows that the linguistic style of passages involving Clarisse and the metaphorical language of those passages reflecting on the "other condition" represent only a part of what can be mustered to express the "nonratioid." They are flanked by reflections and essayistic passages also containing narrative elements and figurative language. Simon Jander argues that this combination of reflection, narration, and metaphor itself displays a metaphorical quality inasmuch as the vividness of the particular and the more general conclusions afforded by interpretation are fused or held in suspense.[329] What Musil has to say in an earlier text supports such a view:

> Poets are analytical. Because every comparison [Gleichnis] is an unintentional analysis. And one understands one phenomenon by recognizing how it arises or is composed, related, connectable with others. Of course, one can just as well say, every comparison is a synthesis, all understanding is one. Of course; they are two halves of the same activity.[330]

Only in this abstract sense – that the merging of reflective analysis, narrative, and figurative synthesis has a metaphorical quality – one could assert that metaphor is the appropriate linguistic form for the "nonratioid."

The form of primitivist narration oriented on Clarisse is thus counteracted by the reflective passages of the novel – in contrast to Müller's novel, where the commentary is reserved solely for the preface and remains ambivalent. This distinction introduces a different understanding of how Musil may have viewed narrational acts modeled after the 'primitive.'[331] It does not involve a mere assimila-

---

[328] Musil, *The Man Without Qualities*, 277, 279. In this sense, the essay already combines narration, figurative language, and reflection. See Birgit Nübel, *Robert Musil – Essayismus als Selbstreflexion der Moderne* (Berlin: de Gruyter, 2006), who describes it as "another reason" and "always simultaneously occupying the space of experience and reflection" in her chapter headings. See also Wolf, *Kakanien als Gesellschaftskonstruktion*, 211–257.

[329] Before Jander, this feature was suggested by Gilbert Reis ("Eine Brücke ins Imaginäre. Gleichnis und Reflexion in Musils *Der Mann ohne Eigenschaften*," *Euphorion. Zeitschrift für Literaturgeschichte* 78 [1984]: 154) and Jutta Heinz ("Grenzüberschreitung im Gleichnis," 256). See also Vatan ("'Und auch die Kunst sucht Wissen,'" in *Aisthesis und Noesis. Zwei Erkenntnisformen vom 18. Jahrhundert bis zur Gegenwart*, ed. Hans Adler and Lynn L. Wolff [Munich: Fink, 2013]) for discussion of the significance of metaphors and figurative language in efforts to convey the nonratioid (123–126).

[330] Musil, *KA*, *Lesetexte*, vol. 12, Analyse und Synthese, n.p.

[331] Elsewhere, Musil defines the "elementary, narrative mode of thought" (*das primitiv Epische*) (*The Man Without Qualities*, 709) as a linear sequence of related events, remarking that narration of this particular kind has gone missing in modernity because of the complexity

tion to the 'primitive' (as occurs in the primitivistic narration focalized through Clarisse).[332] Were this the case, it would be a matter of primitivistic aesthetics in the mimetic sense described above and criticized by Ulrich. But Musil has something else in mind. As I have already noted, in the context of his essay, "Literati and Literature," Musil's detailed notes on Hornbostel's lecture on ethnology focused on two particular aspects: First, Musil notes, the poetry of indigenous peoples is ritual song, which does not represent events so much as bring them about.[333] Second, he reflects on the potential that this performative quality – and its intermingling of form and content, means and ends – might hold for contemporary literature.[334] In this way, Musil deploys what he considers the 'primitive' principle of production (instead of representation) in *The Man Without Qualities:* To begin with, Clarisse's madness is not simply depicted; rather, the narrative *is* the madness it describes insofar as it bears the features of insane discourse – including the linearity of what Musil calls "the primitively epic" (*das primitiv Epische*), which characterizes Clarisse's delusional outlook because of its lack of complexity and purely additive stringing together of thoughts.[335] Second, and as noted above, the "other condition" is not simply represented by met-

---

of the world in which each individual life is caught up. Unlike Ulrich, whose lack of qualities represents a symptom of this diagnosis, Clarisse inhabits a delusional world without much complexity insofar as only her own perspective prevails and contradictions or conflicts are barely noticed; as such, it could be described as elementary. However, Clarisse's subjective narratives are not linear at all, in the sense of providing a comprehensible series of events. Without logical structure, they are erratic and revolve around fixed points instead of developing. At most they are sequential in a chronological way: paratactical with purely additive conjunctions (such as "and"). In a sense, this structure holds for the novel as a whole. In a letter to Guillemin, Musil states that he is not aiming for a causal sequence of events, especially in the first book (26 January, 1931, in Musil, *KA, Lesetexte,* vol. 19, 1931, n.p.). Instead, and in the manner of Clarisse's catathymic thoughts, he aligns himself with the logic of feeling. Readers can grasp why one event follows another only if they can grasp the underlying sentiments and their connection – that is, by engaging with characters' specific, emotionally conditioned worldviews. Otherwise, events in the novel seem random. Gilbert Reis has suggested that "epic naïveté" is first achieved in the second book ("Eine Brücke ins Imaginäre," 150), but by this definition, it is already evident in the first. On this problematic, see also Kappeler, *Situiertes Geschlecht,* 313–314.

332 Wolfgang Riedel ("Robert Musil") explains the "non-narration" Musil sought to practice in terms of "a) insufficient 'linearity,' b) the author's critique of 'mimetic' narration, and c) abandonment of plot" (266).

333 On Musil's reading of Hornbostel in the context of "Literati and Literature," see Bonacchi, *Die Gestalt der Dichtung,* 292–300.

334 Musil, *KA, Transkriptionen,* Mappe VI/3/41.

335 Musil, *The Man Without Qualities,* 709. Also, the narrative *is* the emotion-tinted reality of other characters, which comes out in the affective structural logic of the novel. On both, see Musil's letter to Bernard Guillemin, 26 January 1931, in *KA, Lesetexte,* vol. 19, 1931, n.p..

aphors; it is actualized and simultaneously, for a brief moment, made manifest in the act of reading, i.e., in making sense of tropological language. Finally, the novel not only thematizes the effort to formalize the nonratioid (the goal of modern literature, for Musil); it already *is* this formal language, thanks to the reflective procedures I have described.

When Musil demands that the literature of his day should produce a "certain kind of spirit," his main point is the observation that writers lack the formal means for nonratioid expression and that such a language needs to be developed. *This* is the "spirit" he seeks, and it cannot be gained by returning to some sort of 'primitive' stage. For this kind of spirit, language cannot simply be identical with the non-ratioid itself but must at the same time perform the latter's sensitive reflection. *The Man Without Qualities* realizes this goal by interweaving primitivistic Clarisse-oriented narration and essayistic reflection. When Musil speaks in "Literati and Literature" of the "magic" of contemporary literature, he thus envisions literature creating itself as a form in which it is possible to speak and think about nonratioid matters. Accordingly, Musil's novel not only thematizes the search for the formal language of the nonratioid but already *realizes* this language. *This* can be grasped as primitivistic narration in the modern sense, meaning in other words that the motion from expedition to self-experiment breaks away from the problem of regressive mimesis and instead leads to self-experiment as a critical exploration of and sensitive confrontation with the alienated self and its social conditions.

Chapter 9
The Dialectical Turn of 'Primitive Thinking': The Child and Gesture in Walter Benjamin

In "Walter Benjamin und sein Engel" (1967; "Walter Benjamin and his Angel," 1991), Gershom Scholem recalls that "it is one of Benjamin's most important characteristics that throughout his life he was attracted with almost magic force by the child's world and ways."[1] Benjamin's writings, especially those from the mid-1920s onwards,[2] reveal a marked interest in children's activities and objects, their games, toys, and books, which, beginning in 1924, he addresses in a wide range of reviews and then from 1931 in reflections on his own childhood memories. This interest is also manifest in his writings on contemporary literature and on the philosophy of history formulated in the context of his *Passagenwerk* (1982 [1927–1940], *The Arcades Project*, 2002).

Benjamin began collecting children's books as early as 1918 – likely sparked by his son's birth. In drafts of the *Arcades Project*, Benjamin notes that the occasion for immersion in and awakening from the dreamworld of childhood is one's own children.[3] Similarly, Scholem posits that Benjamin's "profound interest and absorption in the world of the child" was related to his own son's childhood.[4] However, Benjamin's interest in childhood had already been evident in his writings on fantasy and color from the mid-1910s, which repeatedly reference "the

---

1 Gershom Scholem, "Walter Benjamin," in *On Jews and Judaism in Crisis: Selected Essays*, ed. Werner J. Dannhauser (Philadelphia: Paul Dry, 2012), 175.
2 Parts of this chapter have been published as Nicola Gess, "Magisches Denken im Kinderspiel. Literatur und Entwicklungspsychologie im frühen 20. Jahrhundert," in *Literatur als Spiel. Evolutionsbiologische, ästhetische und pädagogische Aspekte. Beiträge zum Deutschen Germanistentag 2007*, ed. Thomas Anz and Heinrich Kaulen (Berlin, New York: De Gruyter, 2009); "Walter Benjamin und 'die Primitiven.' Reflexionen im Umkreis der Berliner Kindheit," *Text+Kritik. Zeitschrift für Literatur. Walter Benjamin* nos. 31–32 (2009); "Gaining Sovereignty: The Figure of the Child in Benjamin's Writing," trans. Joel Golb, *Modern Language Notes* 125, no. 3 (2010); and "'Schöpferische Innervation der Hand.' Zur Gestensprache in Benjamins *Probleme der Sprachsoziologie*," in *Benjamin und die Anthropologie*, ed. Carolin Duttlinger, Ben Morgan, and Anthony Phelan (Freiburg: Rombach, 2012). The translation of this chapter is indebted to the article above translated by Joel Golb and is a product of both Golb's and Erik Butler's and Susan L. Solomon's translation efforts.
3 Walter Benjamin, *The Arcades Project*, trans. Howard Eiland and Kevin McLaughlin (Cambridge, MA: Harvard University Press, 1999), 390.
4 Gershom Scholem, *Walter Benjamin: The Story of a Friendship*, trans. Harry Zohn (New York: NYRB, 1981), 82.

∂ Open Access. © 2022 the author(s), published by De Gruyter. (cc) BY-NC-ND This work is licensed under the Creative Commons Attribution-NonCommercial-NoDerivatives 4.0 International License.
https://doi.org/10.1515/9783110695090-009

pure seeing"[5] of the child. Benjamin's first turn toward childhood coincided, then, with a turn from the youth movement (1914–1915) and from his teacher Gustav Wyneken, an active reformer in the movement. This confluence of events is significant on two levels: some motifs used by reform pedagogy and the youth movement persist in Benjamin's works, yet he clearly mobilizes them in modified form against these very movements.

Robert Musil's *Man Without Qualities* satirizes the youth movement's cult of childhood with the character Hans Sepp, who enthusiastically reflects how

> the child was creative, it was growth personified and constantly engaged in creating itself. The child was regal by nature, born to impose its ideas, feelings, and fantasies on the world; oblivious to the ready-made world of accidentals, it made up its own world. It had its own sexuality. In destroying creative originality by stripping the child of its own world, suffocating it with the dead stuff of traditional learning, and training it for specific utilitarian functions alien to its nature, the adult world committed a barbaric sin. The child was not goal-oriented – it created through play, its work was play and tender growth; when not deliberately interfered with, it took on nothing that was not utterly absorbed into its nature; every object it touched was a living thing; the child was a world, a cosmos unto itself, in touch with the ultimate, the absolute, even though it could not express it. But the child was killed by being taught to serve worldly purposes and being chained to the vulgar routines so falsely called reality![6]

Some of Sepp's ideas appear in Benjamin's writings as well. However, closer inspection reveals crucial differences in his handling of them. For instance, Benjamin calls the nature of children's creativity into question by asking whether their fantasy is purely receptive or destructively constructive. Obvious differences also exist in Sepp's assumption that the child has no interest in the existing world – Benjamin's ideas about children's play assume the opposite. Also, for him, the child does not create itself, but remains subject to ontogenetic as well as to historical and sociological conditions. As for the child's ability to intuit "the absolute," Benjamin acknowledges as much only insofar as children possess superior mimetic gifts of reception and observation.

Benjamin was well informed on research in child psychology and education published in his time, but (with some exceptions) he criticized it sharply. Thus,

---

5 Benjamin, "One-Way Street," in *Selected Writings*, vol. 1, *1913–1926*, ed. Marcus Bullock and Michael W. Jennings (Cambridge, MA: Harvard University Press, 1996). See Heinz Brüggemann, *Walter Benjamin über Spiel, Farbe und Phantasie* (Würzburg: Königshausen + Neumann, 2007), especially Section II ("Phantasie und Farbe"). Cf. the closely related notion of the innocent eye (Ruskin) in chapter 5 of the book at hand.
6 Musil, *Man Without Qualities*, 604.

while still involved in the youth movement, he wrote to Wyneken about a research assignment the latter had given him:

> I've looked through [...] *Zeitschrift für pädagogische Psychologie* except for [volumes] 4 and 6–10 [...] at the library here, all of *Zeitschrift für angewandte Psychologie*, and *Zeitschrift für Philosophie und Pädagogik*, apart from volume 5 [...]. One gets the impression that the state of ideas in pedagogy is awful. [...] No new ideas are being produced at all; thanks to *Zeitschrift für Philosophie und Päd.*, I was made aware, in particular, of the systematic musings of the Hebartians, which obviously bear no fruit at all.[7]

His impression did not change after his abandonment of the movement. However, his criticism now shifted to the newer (reform) pedagogies themselves.[8] In "Alte vergessene Kinderbücher" (1924; "Old Forgotten Children's Books," 1996), he faults Enlightenment philanthropists for having tried to educate the young with "incomprehensible" books as "dry as dust" in order to make "creatures of nature" into "the most pious, the best, and the most sociable beings of all." But even worse, in his eyes, are the errors induced by "supposed insights into the child's psyche"[9] carried out by the newer pedagogy. Benjamin contends that these pedagogues are more interested in their own success than the child's. Their "infatuation with psychology"[10] is driven by their attempt to capture a larger audience.

> A pride in our psychological insight into the internal life of the child [...] has engendered a literature whose complacent courting of the modern public obscures the fact that it has sacrificed an ethical content which lent dignity even to the most pedantic efforts of neoclassical pedagogy. This ethical content has been replaced by a slavish dependence on the slogans of the daily press.[11]

---

[7] Benjamin to Gustav Wyneken, 19 June 1913, in Benjamin, *Gesammelte Briefe*, 1: 115. This letter has not been published in any English editions of Benjamin's correspondence.

[8] Cf. Eva Geulen, "Legislating Education. Kant, Hegel, and Benjamin on 'Pedagogical Violence,'" *Cardozo Law Review* 26, no. 3 (2005), who observes that Benjamin, in spite of the criticism he voiced, held on to some of the demands of the youth movement and pedagogical reform (e.g., "self-education" and "stress on the collective's role" [951]). Benjamin's theory culminates in paradox: "The task of education is the 'formation' of a moral will that, as absolute norm, resists by definition any and all means of its educational production. The conflict between the means and ends of education is radicalized to the point of rendering (moral) education impossible" (952). Cf. also Davide Giuriato, "Tintenbuben. Kindheit und Literatur um 1900 (Rainer Maria Rilke, Robert Walser, Walter Benjamin)," *Poetica* 42, nos. 3–4 (2010): 345–347.

[9] Walter Benjamin, "Old Forgotten Children's Books," in *Selected Writings*, vol. 1, *1913–1926*, ed. Marcus Bullock and Michael W. Jennings (Cambridge, MA: Harvard University Press, 1996), 407.

[10] Benjamin, "Old Forgotten Children's Books," 1: 408.

[11] Benjamin, "Old Forgotten Children's Books," 1: 412.

Benjamin considers the resulting image of the child to be a "thoroughly modern prejudice." Children are viewed as "esoteric, incommensurable beings" for whom a special line of products is to be devised.[12] The "cloying"[13] results are "depressingly distorted jolliness,"[14] "hellish exuberance,"[15] and a simplicity that is false because it is based on form, not on the process by which the toy is produced.[16] Unlike Musil's Hans Sepp, Benjamin does not take issue with the forced adaptation of children to the adult world undertaken by the older pedagogy. Indeed, the "remote and indigestible" impositions may even prove appropriate to the precise mindset of children and their demand for "clear, comprehensible, but not childlike books."[17] Rather, Benjamin objects to a false conception of childhood that ultimately follows a colonialist logic and "betrays what is most genuine and original" when "the child's affectionate and self-contained fantasy is understood as a psychic demand in the sense of a commodity-producing society and education [...] as a colonialist sales opportunity to distribute cultural goods,"[18] i.e., entertainment products (sold for children) and pedagogical writings (peddled to adults).[19]

In this light, toys say more about how grown-ups see children than anything else. Benjamin observes the cultic origins of many traditional toys, which served "to ward off evil spirits."[20] Now, along similar lines but to opposite effect, toys subject them to the "hideous features of commodity capital."[21] The "perceptual world of the child" hardly occupies "a fantasy realm, a fairy-tale land of pure childhood,"[22] then. Only what children seek out and create for themselves is meaningful, for this is how they engage with the adult world.

> Children are particularly fond of haunting any site where things are being visibly worked on. They are irresistibly drawn by the detritus generated by building, gardening, house-

---

12 Benjamin, "Old Forgotten Children's Books," 1: 408.
13 Benjamin, "Old Forgotten Children's Books," 1: 412.
14 Benjamin, "Old Forgotten Children's Books," 1: 407.
15 Benjamin, "Toys and Play," in *Selected Writings*, vol. 2, pt. 1, *1927–1930*, ed. Michael W. Jennings, Howard Eiland, and Gary Smith (Cambridge, MA: Harvard University Press, 1999), 119.
16 Benjamin, "Toys and Play," 2.1: 119.
17 Benjamin, "Old Forgotten Children's Books," 1: 407.
18 Walter Benjamin, "Kolonialpädagogik," *Gesammelte Schriften*, 3: 273; cf. Benjamin, "Toys and Play," 119.
19 Benjamin, "Kolonialpädagogik," 3: 273; cf. 129. Benjamin considers this "kind of children's psychology" the "exact counterpart of the celebrated 'psychology of peoples in a state of nature'" (273).
20 Benjamin, "Toys and Play," 2.1: 118.
21 Benjamin, "Toys and Play," 2.1: 119.
22 Benjamin, "Toys and Play," 2.1: 118.

work, tailoring, or carpentry. In waste products they recognize the face that the world of things turns directly and solely to them. In using these things, they do not so much imitate the works of adults as bring together, in the artifact produced in play, materials of widely differing kinds in a new, intuitive relationship. Children thus produce their own small world of things within the greater one.[23]

This famous passage from "Old Forgotten Children's Books," which also appears in "One-way Street,"[24] addresses central motifs – the wastefulness of manufacture, collecting and bricolage, the "face" of "the world of things" – that bridge Benjamin's conception of childhood with his philosophy of history (to which I will return).[25] In place of a domestication of the "child's soul" through analysis, conceptualization, and educational practice, as proposed in the "colonial pedagogy" (*Kolonialpädagogik*) he condemns, Benjamin advocates an approach that is "not psychologically but materially" oriented – which is to say not around the "child's soul" but rather around toys. [26] This approach would involve formulating a physiognomy[27] of the child's objects and activities that in its essayistic form avoids instrumentalization and – crucially – the conventional discourse on the child as a better person.

Such discourse is circulated by Karl Groos, among others. Speaking on behalf of his guild, he declares the child a "loveable"[28] object of research. Accordingly, reflections on possibly amoral conduct among children play a smaller role for developmental psychologists than would be expected, given the parallels constructed between children and figurations of the 'primitive' (see Chapter 3).[29] Criticism that already applied to Enlightenment philanthropists thus held even more for the developmental psychologists of Benjamin's day, whom he ridicules as "meek and mild educators still cling[ing] to Rousseauesque dreams" of idealized childhood. [30] Educational reformers, in particular, enshrined children as an-

---

23 Benjamin, "Old Forgotten Children's Books," 1: 408.
24 Benjamin, "One-Way Street," 1: 450.
25 Benjamin, "Old Forgotten Children's Books," 1: 408.
26 Benjamin to Siegfried Kracauer, 21 December 1927, in Benjamin, *Gesammelte Briefe*, ed. Christoph Gödde and Henri Lonitz (Frankfurt am Main: Suhrkamp, 1997) 3: 316. This letter has not been published in any English editions of Benjamin's correspondences.
27 On this point, Benjamin acts in the capacity of the collector, whom he defines as the "physiognomist of the domestic interior" (*Arcades Project*, 20); he also understands the child as a collector – see below pages 326 f.
28 Karl Groos, *Das Seelenleben des Kindes*, 2.
29 For counterexamples (such as the "wicked child"), see chapter 3.
30 Benjamin, "Old Toys," *Selected Writings*, vol. 2, pt. 1, *1927–1930*, ed. Michael W. Jennings, Howard Eiland, and Gary Smith (Cambridge, MA: Harvard University Press, 1999), 101.

gels and geniuses by nature. Musil's Hans Sepp likewise dreams of a childlike "world of ideals" and sees children's play as characterized by tenderness.[31] In striking contrast, Benjamin acknowledges the "grotesque, cruel, grim side of children's life," the "despotic and dehumanized element" that makes them "insolent and remote from the world."[32] How is one to understand this?

This chapter argues that Benjamin took the child as the model for an enchanting/disenchanting (i.e., dialectical) approach to alterity and history and in this way also as an inspiration for his *Arcades Project*. Benjamin's child functions as a utopian figure. This is, however, not in the Romantic sense of children in harmony with nature, but in view of both the "barbaric" and, above all, "primitive" tendencies they display.[33] The destructive and mimetic potential of these tendencies come together in children's play, leading dialectically to an acquisition of sovereignty in which intimacy with history and the Other, analytical destruction, and steadily new creation intertwine with one another.[34]

The Child as 'Barbarian'

If one reads the satires in *Neues Kinderspielzeug* (1913; "A New Kind of Plaything," 2012) by Mynona (Salomo Friedländer) and *Geheimes Kinderspielbuch* by Joachim Ringelnatz,[35] both of which Benjamin cited in "Old Toys," it seems that the "grim side" of children's life primarily involves the lust for destruction and the amorality associated with it. But in this mimesis of the adult world, the child exposes above all the fragility of *adults'* moral ideas (and that is certainly the main concern of Friedländer and Ringelnatz's texts). In any event, beyond this satirical and socially critical dimension, the attention Benjamin pays to the child's destructive pleasure is also tied to the disenchantment of the romantic image of childhood practiced in psychoanalysis.

According to Freud, the so-called death drive is more readily apparent in children than in adults: they act out their desire to destroy what is living, whether inwardly or outwardly directed, in a relatively open way. Benjamin repeatedly

---

[31] Musil, *Man Without Qualities*, 604.
[32] Benjamin, "Old Toys," 2.1: 101.
[33] Cf. Pan, *Primitive Renaissance*, 6–16, which, disregarding the author's primitivistic conception of the child, does not include Benjamin; on the distinction between "barbarian" and "primitive" in reference to Nietzsche, see 66–82.
[34] Regarding the concept of sovereignty used here, see page 329 and footnote 154.
[35] Cf. Brüggemann, *Walter Benjamin über Spiel, Farbe und Phantasie*, 109–111.

refers to "Beyond the Pleasure Principle," where Freud develops the concept of the death drive – such a reference is found a few months after the publication of "Old Toys," for example. And he does so in connection with child's play, whose urge toward repetition he understands, with Freud, as an expression of the death drive:

> the obscure urge to repeat things is scarcely less powerful in play, scarcely less cunning in its workings, than the sexual impulse in love. It is no accident that Freud has imagined he could detect an impulse "beyond the pleasure principle" in it.[36]

He concludes that children's play is animated by destructive desire and is meant to avoid change and achieve stasis (as he puts it, to turn "a shattering experience into habit"[37]). Indeed, destruction and repetition condition and reinforce each other: the tower must first be destroyed before it can be built again, so that the rebuilt tower can also be destroyed, and so on. At the same time, Benjamin follows Freud by identifying a culture-creating impulse at work: sublimation. He recognizes an emancipatory and self-empowering component in children's acts of destruction, which takes the form of rehearsing small victories over and over.

A third possible way of understanding Benjamin's talk of "dehumanized children" is provided by his essay on Karl Kraus, which discusses "a creature [*Unmensch*] sprung from the child and the cannibal."[38] In Benjamin's reading of Kraus, the child stands for an original purity, and the man-eater for a destruction of the mythical order upon which modern civilization rests. The two concepts (of original purity and destruction) meet up in the monstrous creature (*Unmensch*), insofar as "not purity but purification" stands "at the origin of creation."[39]

At the same time, there is something "man-eating" about children as well. Cited in "Old Toys," children's laughter at the "negative sides of life" [40] is tied to the pleasure they derive from playful imitation of destruction and here returns as the laughter of a bellicose humanity. Looking back at the First World War, Benjamin observes,

---

36 Benjamin, "Toys and Play," 2.1: 120. On Benjamin's connection to Freud here, see Doris Fittler, '*Ein Kosmos der Ähnlichkeit*': *Frühe und späte Mimesis bei Walter Benjamin* (Bielefeld: Aisthesis, 2005), 411–413.
37 Benjamin, "Toys and Play," 2.1: 120.
38 Benjamin, "Karl Kraus," in *Selected Writings*, vol. 2, part 2, *1931–1934*, ed. Michael W. Jennings, Howard Eiland, and Gary Smith (Cambridge, MA: Harvard University Press, 1999), 457.
39 Benjamin, "Karl Kraus," 2.2: 455.
40 Benjamin, "Old Toys," 2.1: 101.

the laughter of the infant carrying its foot to its mouth. This is how humankind began to nibble at itself fifteen years ago [*So begann die Menschheit vor fünfzehn Jahren von sich zu kosten*]. [...] It's the laughter of the sated infant. This humanity "devoured" everything.[41]

A problematic reading of war as the start of a necessary purification is at work here – a reading shared with other opponents of the Great War such as Benjamin's friend Ernst Bloch. Two years later, in the essay "Erfahrung und Armut" (1933; "Experience and Poverty," 1999), Benjamin returns to this constellation in the context of a "new, positive concept of barbarism." "Never has experience been contradicted more thoroughly," Benjamin writes; "strategic experience has been contravened by positional warfare; economic experience, by the inflation; physical experience, by hunger; moral experiences, by the ruling powers." The generation emerging from the war finds itself back in a landscape "in which nothing is the same except for [...] the tiny, fragile human body"[42] – a body Benjamin compares to a "newborn babe in the dirty diapers of the present."[43] Yet the "poverty of experience" that characterizes this condition is not lamented so much as longed for. "With a laugh," people participate in the ultimate downfall of a culture they have long perceived as mendacious: [44] "They have 'devoured' everything, both 'culture and people,' and they have had such a surfeit that it has exhausted them."[45] In their sleep, they dream of "completely new, lovable, and interesting creatures" that are no longer "human-like."[46]

The child then occupies a threefold position in this constellation of destruction and renewal. First, the child is the bare creature remaining after the destruction of previous humanity. This creature has not only survived the man-eating war, but is itself a man-eater by nature to the extent that it affirms and perpetuates the destruction of humankind. However, whereas the devastation of the anthropophagous order of war is instrumental, the child-creature's cannibalism manifests pure destruction: one that is only a manifestation of the death or life drive to the extent that what is at stake is its own survival. The monstrous creature (*Unmensch*), Benjamin writes,

---

[41] Benjamin, Notes on "Karl Kraus," in *Gesammelte Schriften*, 2: 1108; cf. Benjamin, "Karl Kraus," 448.
[42] Walter Benjamin, "Experience and Poverty," in *Selected Writings*, vol. 2, Pt. 2, *1931–1934*, ed. Michael W. Jennings, Howard Eiland, and Gary Smith (Cambridge, MA: Harvard University Press, 1999), 732.
[43] Benjamin, "Experience and Poverty," 2.2: 733.
[44] Benjamin, "Experience and Poverty," 2.2: 735.
[45] Benjamin, "Experience and Poverty," 2.2: 734.
[46] Benjamin, "Experience and Poverty," 2.2: 733.

has made a pact with the destructive side of nature. Just as the old conception of creaturely existence [*der alte Kreaturbegriff*] was based on love, [...] the new one, the conception of creaturely existence exemplified by the monster, is based on devouring: the cannibal purifies his relationship to fellow human beings by simultaneously satisfying the urge to eat."[47]

The question arises of how emancipation from the pressure of creaturely/monstrous drives might succeed. Benjamin's concept of a "positive barbarism" provides an answer.

The child is seen, secondly, as a "barbarian" in that barbarians are not only characterized by a "poverty of experience" but are driven by just this poverty "to begin from scratch, to make a new start."[48] That is precisely the outstanding feature of child's play as formulated by Benjamin in "Toys and Play": "a child creates the entire event anew and starts again right from the beginning."[49] This signifies more than the creaturely drive to tear down the old; it means that destruction creates the possibility for subsequent production. Such play does not simply act out a repetition compulsion; rather, the repetition is applied in such a way that hitherto unintuited prospects arise. The barbarian child creates nothing organic (these mythical concepts have also been "devoured") but is the draughtsman of "arbitrary, constructed nature,"[50] who recognizes the necessity for constant destruction of the old in order to create the possibility of a new beginning. Extended to the philosophy of language, this final point resembles Benjamin's conception of the allegorician, which will be taken up later in this chapter.[51]

Third, the child is the new being that emerges from the arbitrary constructions of the barbarian. This being no longer resembles the human; it has been "de-humanized" insofar as it requires the destruction of previous conceptions

---

47 Benjamin, Notes on "Karl Kraus," in *Gesammelte Schriften*, 2: 1106.
48 Benjamin, "Experience and Poverty," 732. Davide Giuriato also makes the connection between the child and the barbarian (*Mikrographien. Zu einer Poetologie des Schreibens in Walter Benjamins Kindheitserinnerungen [1932–1939]* [Munich: Fink, 2006], 16–17). Cf. Renate Reschke, "Barbaren, Kult und Katastrophen. Nietzsche bei Benjamin. Unzusammenhängendes im Zusammenhang gelesen," in *Aber ein Sturm weht vom Paradiese her. Texte zu Walter Benjamin*, ed. Michael Opitz and Erdmut Wizisla (Leipzig: Reclam, 1992); Manfred Schneider, *Der Barbar. Endzeitstimmung und Kulturrecycling* (Munich: Hanser, 1997), 210–215; Kevin McLaughlin, "Benjamin's Barbarism," *The Germanic Review* 81, no. 1 (2006); Sami Khatib, "Barbaric Salvage: Benjamin and the Dialectics of Destruction," *parallax* 24, no. 2 (2018).
49 Benjamin, "Toys and Play," 2.1: 120.
50 Benjamin, "Experience and Poverty," 2.2: 733.
51 Regarding the allegorician in Benjamin's work, cf., e.g., Bettine Menke, *Sprachfiguren. Name-Allegorie-Bild nach Walter Benjamin* (Munich: Fink, 1991), 161–238; Winfried Menninghaus, *Walter Benjamins Theorie der Sprachmagie* (Frankfurt am Main: Suhrkamp, 1995), 95–133.

of "the humanly and of the human."⁵² In the paralipomena to the Kraus essay, Benjamin refers to this new being, which has overcome the "mythical humanity" of old, as an "angel."⁵³ Taken together, these three aspects – the child as creature, barbarian, and angel – make it clear why the child represents a "transfiguration of creaturely existence [*Geschöpf*]"⁵⁴ as well as a "man-eater and angel" in one. Not the man-eater but the child is at the "heart of the monster [*Unmensch*]"⁵⁵ because the child offers not only ideas of the pure and primeval but also their linkage with destruction, and in this way the child already anticipates his angelic purification.⁵⁶

Thus, in light of "Experience and Poverty," the figures Benjamin invokes can be arranged in the following relation: the destruction of the mythical order, and with it mythic man, provides the precondition for the "creature" to survive and be able, as a "barbarian," to construct a new "angel," which, as a "monster," is compelled by the principle of purifying destruction that enables new production. This entanglement of destruction and production, however, is only faintly discernible in the Kraus essay or "Der destructive Charakter" (1931; "The Destructive Character," 1978). In the former, Benjamin's reflections break off at the stage of creaturely existence, at which point the issue is only survival, not that something new should be constructed out of it. However, in the "power [...] to purify," lies the "hope [...] that something might survive this age."⁵⁷ "The Destructive Character" makes it even clearer that nothing new is to be expected: "The destructive character sees no image hovering before him. He has few needs, and the least of them is to know what will replace what has been destroyed."⁵⁸ Here, too, the possibility of a new construction is only implied by the metaphor of the

---

52 Benjamin, Notes on "Karl Kraus," in *Gesammelte Schriften*, 2.2: 1112; cf. Benjamin, "Experience and Poverty," 2.2: 733.
53 Benjamin, Notes on "Karl Kraus," in *Gesammelte Schriften*, 2.2: 1106.
54 Benjamin, Notes on "Karl Kraus," in *Gesammelte Schriften*, 2.2: 1103.
55 Benjamin, Notes on "Karl Kraus," in *Gesammelte Schriften*, 2.2: 1102.
56 Cf. Winfried Menninghaus, "Walter Benjamins Diskurs der Destruktion," *Studi germanici* 29 (1991), who identifies two fundaments of destructive discourse: interruption (in the dimensions of rhetoric and poetics, anthropology, theology, the philosophy of history, and the philosophy of language) and purification (in a theological, ritualistic, technical, and aesthetic sense).
57 Benjamin, "Karl Kraus," 2.2: 455.
58 Benjamin, "The Destructive Character," in *Selected Writings*, vol. 2, pt. 2, *1931–1934*, ed. Michael W. Jennings, Howard Eiland, and Gary Smith (Cambridge, MA: Harvard University Press, 1999), 541.

path running through the rubble, which the destructive character wishes to clear.[59]

In their denial of new construction, both texts point back to an earlier essay, "Zur Kritik der Gewalt" (1921; "The Critique of Violence," 1978). Here, Benjamin elaborates the concept of a non-instrumental violence that de-poses or "suspends" (*ent-setzt*) the violence of law – which, for its part, traces back to myth[60] – but without at the same time putting anything new in its place. Structurally, it relates to the violence practiced by children in that their destruction is without purpose. Instead it is a manifestation of a drive and at the same time an emancipatory move, because it is directed against the violence of positing (*Setzung*). In the Kraus essay, this violence is very clearly carried out by the educator, but it is also found in a more general sense in the world of givens as a whole. For this reason, in "On the Critique of Violence," non-instrumental violence is not only tied to anarchy, but anarchy at the same time is connected to the child with Benjamin's reference to "childish anarchy."

Taken together, one can perhaps read those cautious references to a new construction in the two texts (i.e., "Karl Kraus" and "Destructive Character") as indications that purification is more than just annihilation in that it creates the possibility of a new beginning – without, however, a hint as to what the new might look like. The only certainty is that it would not posit (*setzen*) a new order: "First of all, for a moment at least, empty space – the place where the thing stood or the victim lived. Someone is sure to be found who needs

---

[59] This aspect is made clear by Nicolas Pethes, *Mnemographie. Poetiken der Erinnerung und Destruktion nach Walter Benjamin* (Tübingen: Niemeyer, 1999), 158–171, 367–390.

[60] Myth as "the epitome of the persistence of the spell, of the lack of freedom (fate), of the repetition compulsion (the ever-same)" (Burkhardt Lindner, "Engel und Zwerg. Benjamins geschichtsphilosophische Rätselfiguren und die Herausforderung des Mythos," in *Was nie geschrieben wurde, lesen*, ed. Lorenz Jäger and Thomas Regehly, Frankfurter Benjamin-Vorträge [Bielefeld: Aisthesis, 1992], 238). Lindner determines that Benjamin's use of the term "myth" generally retains a negative connotation, whereas attention to the mythical aspect of related phenomena, especially in later writings, proves much more positive (239, 251–254). Winfried Menninghaus has also drawn attention to this point: "In 'Fate and Character' and 'On the Critique of Violence,' Benjamin 'defines' myth almost exclusively in terms of the fateful structure of time, the compulsion of the ever-same. [...] Then, in *Berlin Childhood* and 'One-Way Street,' as well as *The Arcades Project*, it dissolves into the multiplicity of narrow mythologies, which are more fleeting and impermanent than the mythical 'totalities' of old." (Menninghaus, *Schwellenkunde. Walter Benjamins Passage des Mythos* [Frankfurt am Main: Suhrkamp, 1986], 110) "Only as Benjamin moves toward his later works are the negative accents of myth 'dialecticized' with positive ones" (111). See also Burkhardt Lindner, "Das *Passagen-Werk*, die *Berliner Kindheit* und die Archäologie des 'Jüngstvergangenen,'" in *Studien zu Benjamin*, ed. Jessica Nitsche and Nadine Werner (Berlin: Kadmos, 2016), 232–235.

this space without occupying it."[61] "Experience and Poverty" also admits interpretation along these lines in that new designs are provisional and "arbitrary"[62] or, in other words, could also have been constructed differently. Without anything "determinate" or "settled" on them, they express no duration[63] and are "improvised."[64] Similarly, the "angel" in the Kraus essay "passes into nothingness"[65] as soon as it is created.

The thought of a destruction that enables creation, which Benjamin links to the figure of the child, stands in a larger context, which will be laid out in the following section. Benjamin's discussion of the child's other side, I will argue, reflects a specific concept of liberation, namely, liberation as a gaining of sovereignty. The linchpin of this notion is the dialectical turn from mimesis as compulsion (which he deems 'primitive') to mimesis as cunning, play, and bricolage – concepts Benjamin draws from the figure of the child and applies to his philosophy of language, his philosophy of history, and his way of writing. Through them he offers a dialectical way out from the "colonial pedagogy" of the day.

The Child as 'Primitive'

"Lehre vom Ähnlichen" (1933; "Doctrine of the Similar," 1977), which Benjamin wrote in connection with the "first piece"[66] of *A Berlin Childhood*, establishes a correspondence between children's play as the ontogenetic school of the mimetic faculty and the phylogeny of humankind, which is shaped by this faculty and its transformation. The child "[playing] at being not only a shopkeeper or teacher but also a windmill and a train" is the counterpart of "ancients or even [...] primitive peoples," whose world abounds in "magical correspondences."[67] Like prehistoric humans, children obey the "compulsion to become similar and [...] to behave mimetically,"[68] which is expressed in their "transform[ation]" into the

---

61 Benjamin, "The Destructive Character," 2.2: 541. Cf. Pethes, *Mnemographie*, 373.
62 Benjamin, "Experience and Poverty," 2.2: 733.
63 Benjamin, "Experience and Poverty," 2.2: 734–735.
64 Benjamin, "Experience and Poverty," 2.2: 735.
65 Benjamin, "Karl Kraus," 2.2: 457.
66 Benjamin to Gershom Scholem, February 1933, in Benjamin and Gershom Scholem, *The Correspondence of Walter Benjamin and Gershom Scholem, 1932–1940*, trans. Gary Smith and André Lefevere (Cambridge, MA: Harvard University Press, 1992), 28.
67 Walter Benjamin, "Doctrine of the Similar," in *Selected Writings*, vol. 2, pt. 2, *1931–1934*, ed. Michael W. Jennings, Howard Eiland, and Gary Smith (Cambridge, MA: Harvard University Press, 1999), 695.
68 Benjamin, "Doctrine of the Similar," 2.2: 698.

objects and words of their play. This passage is only the best known of many passages where Benjamin draws lines of connection between children and "primitive peoples." But before these lines can be discussed in more detail, I must address the question of what Benjamin means when he speaks of the 'primitive.'[69]

According to Scholem, Benjamin's engagement with the concept of the 'primitive' can be traced back to 1916, when his foray into theories of myth led him to take interest in "animism and pre-animism."[70] His main point of reference was Karl Theodor Preuss, a well-known ethnologist of the day: "[Benjamin] often used Preuss's remarks on pre-animism. This brought us to ghosts and their role in the pre-animistic age."[71] Early twentieth-century theories of pre-animism traced the first beginnings of religion to belief in an indeterminate, omnipresent magical force rather than the soul (which E.B. Tylor and Wilhelm Wundt, among others, considered the basis of animism[72]). Preuss understood pre-animism in the same way:

> There are reliable reports that a certain power, a magical force in [natural objects], is thought to be at work, which demonstrably has nothing to do with the elements from which the so-called concept of the soul has been formed, namely Melanesian *mana*, Iroquois *orenda*, and so on.[73]

Regarding this point, Preuss conjectured that a pre-animistic age of magic, distinguished by its belief in a general magical force, preceded the age of myth, which was defined by its belief in gods.

Benjamin follows Preuss in two respects: First, he accepts the existence of a pre-mythical age. A manuscript from 1918, "Anthropologie" (Anthropology),[74] outlines a speculative historical theory in which the pre-mythical age, marked by belief in ghosts, was superseded by a mythic age marked by belief in demons.

---

69 For instance, in "Kolonialpädagogik" children are compared with "peoples in a state of nature" (273); in the *Arcades Project*, affirming the repetition of phylogeny in ontogeny, Benjamin declares that "the embryo in the womb relives the life of animals" (*Arcades Project*, 106). Two reviews ("Kulturgeschichte des Spielzeugs" and "Spielzeug und Spielen" in *Gesammelte Schriften*, 3: 116, 128) mention the cultic origins of various toys (balls, pinwheels, kites, and rattles) – a thesis already advanced by Tylor.
70 Scholem, *Walter Benjamin: The Story of A Friendship*, 40.
71 Scholem, *Walter Benjamin: The Story of A Friendship*, 40–41.
72 Arguing against E.B. Tylor, R.R. Marett coined the influential phrase "preanimistic religion" in a 1900 article: "Preanimistic Religion" (1900), in *The Threshold of Religion* (London: Methuen, 1914). His reflections were taken up (and modified) by his contemporaries in discussions of *mana* and related beliefs.
73 Preuss, *Die geistige Kultur der Naturvölker*, 19.
74 Benjamin, "Schema zur Anthropologie," in *Gesammelte Schriften*, 6: 64.

The latter, which had witnessed the emergence of law and positing language, was in turn replaced by the age of justice and revelation. In this age's overthrow of the other, one can locate the purely suspending (i.e., not positing) divine violence from "On the Critique of Violence." In contrast, the relationship between the ghostly age and the age of justice is thought to have proceeded not by revolutionary succession but through "sublation." This distinction is crucial both for understanding the importance that the pre-mythic age has for Benjamin and for central categories of his thinking such as salvation and awakening. In a dialectical turn, justice salvages elements of the ghostly into the new age, but with a decisive modification. The difference is between compulsion and freedom; a dialectical turn from mimetic compulsion and the drive to destruction toward the gaining of sovereignty, which is achieved in the passage through mimesis and expressed in analytical destruction and open-ended production.

Second, Benjamin also followed Preuss in ascribing the "primitive" with a pre-animistic belief in a mysterious magical force – a "mimetic force"[75] – pervading the world: "Mimetic genius [was] a life-determining power of the ancients"; in keeping with the parallels between phylogeny and ontogeny, "full possession of this gift" is "to be attributed to the newborn"[76] as well. Elsewhere, Benjamin also speaks of the "gift of mimesis, which was peculiar to mankind in its early times and today only works unbroken in the child."[77] For Benjamin, then, the 'primitive' world is stamped by an omnipresent "mimetic force" of whose "objective existence" (*Vorhandensein*)[78] he is convinced. He writes, "not only are [...] resemblances imported into things by virtue of chance comparisons on our part, but [...] all of them [...] are the effects of an active, mimetic force working expressly inside things."[79] Thus, here mimesis is not understood as the establishment of a relation, but substantialized (to use Cassirer's term, who Benjamin read very carefully): the mimetic force is a substance of its own that works in things and, as such, evokes similarities between them.

It would take me too far afield to explore all of the fine points of Benjamin's mimetic theory, but a few of the theory's features are important for the present discussion. For Benjamin mimetic force is the grounding for a "magical commu-

---

[75] Walter Benjamin, "On Astrology," in *Selected Writings*, vol. 2, pt. 2, *1931–1934*, ed. Michael W. Jennings, Howard Eiland, and Gary Smith (Cambridge, MA: Harvard University Press, 1999), 684. On "similarity as a primal phenomenon," see Fittler, 'Ein Kosmos der Ähnlichkeit,' 54–63. The author does not examine connections to ethnology, however.
[76] Benjamin, "On Astrology," 2.2: 684.
[77] Benjamin, further notes on "Lehre vom Ähnlichen," in *Gesammelte Schriften*, 7: 792.
[78] Benjamin, notes on "Lehre vom Ähnlichen," in *Gesammelte Schriften*, 2: 956.
[79] Benjamin, "On Astrology," 2.2: 684.

nity of material,"⁸⁰ understandable, as Doris Fittler explains, as a proto-"genetic code" causing "relational similarity" in nature: the similarity of all with all rests on a "limited supply of basic elements, features, and qualities. [...] They are simply the variants, mutations, and metamorphoses of one and the same repository."⁸¹ This "community of matter" includes human beings, onto whom mimetic force does not impress itself biologically so much as culturally, when people actively adjust to similarities or rather to processes of becoming similar (*Anähnelung*) perceived as being already at work in their surroundings.⁸² They thus possess a "mimetic faculty" encompassing both the abilities to perceive and produce similarities. In distinction to the similarity at work in the "community of material," the similarity emerging from the process of assimilation is not always already given; rather, it is the product of an active capacity for transformation, which, as such, already presumes difference. Considered against the backdrop of the ethnological discourse of the 1920s, this is precisely the difference between the participation that Lévy-Bruhl conceives as always already constituted and the association assumed by English and some German ethnologists to be the basis of 'primitive' thought (see Chapter 2).

According to Fittler, the production of likenesses represents a "response" to the "communication of matter in its magical community,"⁸³ whose "object and, at the same time, realization" is similarity.⁸⁴ Benjamin himself writes, "[t]hese natural correspondences assume decisive importance [...] only in light of the consideration that they are all, fundamentally, stimulants and awakeners of the mimetic faculty which answers them in man."⁸⁵ However, whether this can be determined as an "act of communication"⁸⁶ is questionable due to the imperative nature of the communication and the compulsory nature of the response. Benjamin speaks, after all, of a "once powerful *compulsion* to become similar and [...] to behave mimetically."⁸⁷ The "faculty" appears to be a drive rather than an ability at this juncture. Correspondingly, Benjamin also denies the originality of the human production of similarity: "We must assume in principle that processes in the sky were imitable [...] by people who lived in earlier times; indeed, that this similarity [*Nachahmbarkeit*] contained instructions for mastering

---

80 Benjamin, further notes on "Lehre vom Ähnlichen," in *Gesammelte Schriften*, 7: 795.
81 Fittler, 'Ein Kosmos der Ähnlichkeit,' 64–65, 61.
82 Benjamin, "Doctrine of the Similar," 2.2: 694; cf. Fittler, 'Ein Kosmos der Ähnlichkeit,' 70–76.
83 Benjamin, "Antithetisches über Wort und Name," in *Gesammelte Schriften*, 7: 795.
84 Fittler, 'Ein Kosmos der Ähnlichkeit,' 76, 66.
85 Benjamin, "Doctrine of the Similar," 2.2: 695.
86 Fittler, 'Ein Kosmos der Ähnlichkeit,' 77.
87 Benjamin, "Doctrine of the Similar," 2.2: 698. Emphasis added.

an already present similarity."[88] Strictly speaking, what is foregrounded here is not the production, but the handling of an already available resemblance. Likewise, when Benjamin observes that human assimilation (the process of dynamic mimesis he calls *Anähnelung*) merely mediates the similarity between things, it is clear that they do not forge new similarities, but are realizing ones that were already present.[89] However, both cases also already imply the inverse tendency. Just as stars and clouds shift in position and shape with each passing moment, the mimetic act may be only momentary, that is, exist precisely in the moment of transformation.[90] The question, then, is how, in the course of its phylogenetic and ontogenetic transformation – the focus of "Doctrine of Similarity" and "On the Mimetic Faculty" – the mimetic faculty shifts from mimetic compulsion to a mimetically-inspired production of the new.

But first let us consider the connections Benjamin traces – in line with the developmental psychologists of his time – between the child and the figure of the "primitive." He generates an abundance of such links, both structural and motif based, in *Berlin Childhood*.[91] The places children seek out, which adults have forgotten, often represent a "wilderness"[92] where one finds tribal sorcerers,[93] masquerades,[94] demons,[95] sacred animals,[96] ghosts and spirits,[97] and goddesses and temples.[98] In the chapter, "Das Karussell" ("The Carousel"), Benja-

---

[88] Benjamin, "Doctrine of the Similar," 2.2: 695.
[89] Benjamin, notes on "Lehre vom Ähnlichen," in *Gesammelte Schriften*, 2: 956.
[90] Cf. Werner Hamacher, "The Word *Wolke* – If It Is One," in *Benjamin's Ground: New Readings of Walter Benjamin*, ed. Rainer Nägele (Detroit: Wayne State, 1988).
[91] Scholem writes, "Benjamin's predilection for the imaginative world of associations [...] was also evident in his marked interest in the writings of insane persons. [...] What primarily fascinated him about them was the architectonic (today one would call it the structural) element of their world systems and the fantastic tables often associated therewith, tables of coordinates that are no longer variable, as they are with children, but are marked by the onset of a grim rigidity. His interest was not pathologic-psychological but metaphysical in nature" (*Walter Benjamin: The Story of A Friendship*, 82). And elsewhere, "The 'world systems' of the mentally deranged [...] provided him with material for the most profound philosophical reflections on [...] the nature of the associations that nourish the thinking and imagination of the mentally sound and unsound alike" (Scholem, "Walter Benjamin," 175).
[92] Benjamin, *Berlin Childhood around 1900*, in *Selected Writings*, vol. 3, *1935–1938*, ed. Howard Eiland and Michael W. Jennings (Cambridge, MA; Harvard University Press, 2002), 350, 352, 354.
[93] Benjamin, *Berlin Childhood*, 3: 365, 375.
[94] Benjamin, *Berlin Childhood*, 3: 375–376.
[95] Benjamin, *Berlin Childhood*, 3: 365, 375.
[96] Benjamin, *Berlin Childhood*, 3: 366.
[97] Benjamin, *Berlin Childhood*, 3: 369, 375, 376, 399.
[98] Benjamin, *Berlin Childhood*, 3: 366, 375, 403.

min even speaks of a child who travels through the "jungle" surrounded by "natives" and credits him (his younger self) with knowledge of the "eternal return of all things," which brings together the distant past ("thousands of years ago") and recent times ("just now"). [99]

Furthermore, the child displays magical thinking. The boy believes in supernatural beings and practices magical rites; thinking figurally, he revives lexicalized metaphors and back-translates unknown words and names into images. Accordingly, the phrase "waging war" (*Krieg führen*, literally, "leading war") evokes the idea of a man "leading a rhinoceros or a dromedary"[100]; the salutation *gnädige Frau* (an archaic phrase meaning "gracious woman") is taken to refer to his mother's needlework and becomes *Näh-Frau* ("sew-woman"); "Steglitz" – and the aunt who lives in the neighborhood of this name – turns into *Stieglitz*[101] ("goldfinch"); and *Kupferstich* ("copperplate," in the sense of an engraving) becomes *Kopfversteck* ("head-hiding place").[102] The child thinks he possesses the whole in possessing a part (Peacock Island by means of a peacock feather[103]), and he takes similarity as an indication that unrelated things belong together, e.g., the waiting areas for hackneys are provinces of "my back yard" because "the trees were similarly rooted" in both places. The most incidental phenomena are not trivial, but point to connections yet to be discovered (e.g., "everything in the courtyard became a sign [...] to me").[104] The child's animism is also evident. In "Wintermorgen" ("Winter Morning"), "the flame" that "barely had room to move" in the narrow oven "peeps out" at him.[105] In "Schmetterlingsjagd" ("Butterfly Hunt"), "Wind and scents, foliage and sun" "govern the flight of the butterflies."[106] The butterfly is also credited with emotions, in keeping with the exchange of identity that takes place between the animal and the hunter:

> Between us, now, the old law of the hunt took hold: the more I strove to conform, in all the fibers of my being, to the animal – the more butterfly-like I became in my heart and soul – the more this butterfly itself, in everything it did, took on the color of human volition.[107]

---

99 Benjamin, *Berlin Childhood*, 3: 386.
100 Benjamin, *Berlin Childhood*, 3: 348.
101 Benjamin, *Berlin Childhood*, 3: 358.
102 Benjamin, *Berlin Childhood*, 3: 390.
103 Benjamin, *Berlin Childhood*, 3: 367.
104 Benjamin, *Berlin Childhood*, 3: 345; cf. 356.
105 Benjamin, *Berlin Childhood*, 3: 357.
106 Benjamin, *Berlin Childhood*, 3: 350.
107 Benjamin, *Berlin Childhood*, 3: 351.

In *Berlin Childhood* the repeated transformation of the child into a thing or animal is even more prevalent than such anthropomorphosis. Thus, in "Die Farben" ("Colors"), Benjamin writes, "I took on the colors of the landscape [...]. I traveled in [soap bubbles] through the room"[108]; in "The Mummerehlen," words exert a "compulsion" on the child to make himself "similar to dwelling places, furniture, clothes"[109]; and "Verstecke" ("Hiding Places") describes how the child adapts to his surroundings to become first a "ghost," then an "idol," a "door," and finally a "sorcerer."[110] These transformations make it plain that demarcations of identity are not yet clearly drawn for the child. Piaget coined the term "realism" for this state: for the child, subjective phenomena are just as "real" as objective ones; the distinction between the self and the external world is still blurry. *Berlin Childhood* illustrates as much in the fluid boundaries between dream, fantasy, and reality as well as in the intersections between voicing a wish and its fulfillment. These metamorphoses may be seen in light of Lévy-Bruhl and Piaget's notion of participation. The child can transform into its animate or inanimate counterpart for the same reason that wind and sun command the butterfly and that butterflies and flowers communicate with each other: they participate with each other with the help of "spirits" and "demons" whose traces the child gets wind of in "Unordentliches Kind" ("Untidy Child") from *One Way Street*, and which can enter him as in "Butterfly Hunt" and "Hiding Places." In view of Benjamin's terminology, however, it seems most appropriate to attribute the child's acts of assimilation to his "mimetic faculty," with which he reacts to the "mimetic force" at work in things and in himself.[111] This force is the common feature that makes them always already related to one another.

Magical thinking also serves as *Berlin Childhood*'s aesthetic principle, not only on the motivic level but also structurally. In the book, the child's thinking is determined by associations based on similarities or simultaneities. For example, the seaside resort Westerland and Athens turn into colonies of "Blumeshof

---

[108] Benjamin, *Berlin Childhood*, 3: 380.
[109] Benjamin, *Berlin Childhood*, 3: 391.
[110] Benjamin, *Berlin Childhood*, 3: 375.
[111] Thus, as Gérard Raulet has shown, the "pertinent passage" from the "Doctrine of the Similar," in which "the gift of seeing similarity" is determined to be a "weak rudiment of the formerly formidable compulsion to become and behave similarly," reappears "text-identically in a preliminary study of *Berlin Childhood* under the title 'Zur Lampe' and confirms the close connection between the phylogenetic speculations and the ontogenetic reflections on childhood" ("Mimesis. Über anthropologische Motive bei Walter Benjamin – Ansätze zu einer anthropologischen kritischen Theorie," *Deutsche Zeitschrift für Philosophie* 64, no. 4 [2016]: 585).

12" when the narrator's grandmother, who lives at this address, sends postcards from those places.[112] Many parts of the work *structurally* follow the same principle. For instance, the section "Zwei Blechkapellen" ("Two Brass Bands") jumps from the description of a stroll down the Lästerallee to an account of diversions on Rousseau Island, an artificial island in a Berlin park.[113] The only connection between the scenes is that brass bands are performing at both locations. The first paragraph of "Tiergarten" ("Zoo"), which quickly moves from the child purposely getting lost in the metropolis to traces of ink left on blotting paper to the zoo, is tied together by a labyrinth motif.

The same applies to the archaic principle of repetition, which Benjamin calls the "great law that presides over the rules and rhythms of the entire world of play." Specifically, "for a child repetition is the soul of play, [...] nothing gives him greater pleasure than to 'Do it again!'"[114] *Berlin Childhood* confirms as much in the endless delight the child experiences when playing with a stocking, fascinated by its metamorphoses. Transferred to the structural level, this means that certain themes repeat without developing in *Berlin Childhood*, circling around the fascination of remembered childhood. It is only logical then that the author had trouble settling on a final sequence of this work's contents (and editors still disagree on what it should be). The principle of repetition is evident in the writing process as well, with Benjamin repeatedly rewriting the various parts of the book – not necessarily by adding new material but revising what was already there.[115] This process corresponds to the principle of 'starting over again' that governs children's games.

In this manner, Benjamin assimilates with the child he is recalling. And as with the child and butterfly, a double transformation takes place: he affirms the child's perspective and, at the same time, the child he recalls becomes an adult insofar as he is always already shaped by his adult self. This intertwining comes out in the narrative perspective, which oscillates between a child-like first-person voice that simply recounts what takes place and an adult first-person voice that comments on it. This is evident, for example, in the use of personifications that translate the child's animism into language. "Blumeshof 12" tells how the old-fashioned furnishings from the 1870s elicit ambivalent feelings, then jumps to the threatening goings-on on the landing and stairs (where an elf or imp [*Alb*] casts a spell on the child) before switching again to declare

---

112 Benjamin, *Berlin Childhood*, 3: 369.
113 Benjamin, *Berlin Childhood*, 3: 383–384.
114 Benjamin, "Toys and Play," 2.1: 120.
115 On the composition of *Berlin Childhood*, see Giuriato, *Mikrographien*.

the events to be from a dream. "Winter Morning" omits such retrospective distancing. The flame "was peeping out at me," one reads, but the phenomenon is not explained as a child's (mis)perception.[116] And when, in other parts of *Berlin Childhood,* the first-person voice retreats behind an omniscient narrator and his main character, "the child," this move is by no means accompanied by a distancing from the childlike perspective: "The child who stands behind the doorway curtain *himself* becomes [...] a ghost."[117]

With these rotating perspectives, a distancing from the law of participation is indicated once again. In contrast to the idea of participation, the notion of assimilation assumes that a successful separation has already taken place, so that the focus here is not on an always-already-given participation, but on the becoming similar (*Anähnelung*) of the child to the world of animals and things surrounding him. Also addressed here is the perspective of the one who only *remembers* magical thinking and therefore cannot think of participation as anything other than assimilation.

A Dialectical Turn

However, not only must participation be distinguished from the process of dynamic mimesis Benjamin calls *Anähnelung*; but the child's performance of the latter must also be differentiated from the practice of participation observed by ethnologists. The difference comes out in the concept of play, specifically its distance from compulsion on the one hand and illusion on the other. Benjamin is here concerned with the dialectic turn from mimesis as a sign of powerlessness to mimesis as an instrument of self-empowerment. In many scenes of *Berlin Childhood*, a "magic spell" threatens to strike the child, creating conditions that evoke the pre-animistic age of the ghostly. Ghostly entities appear and threaten the child's autonomy. As in the above-mentioned episode of the *Alp* (imp), they most often surface in dreams (cf. *Alptraum*, Eng. "nightmare"), an indication of their pre-mythic origin.[118] Scholem recalls that in the period when his friend was thinking about the ghostly age, he often spoke of children's dreams in which ghosts carried out their mischief.

In "Über das Grauen" (On Horror, ca. 1920–1922), Benjamin ties the appearance of ghosts to "immersion [...] in the alien,"[119] which he understands as a pri-

---

[116] Benjamin, *Berlin Childhood*, 3: 357.
[117] Benjamin, *Berlin Childhood*, 3: 375. Emphasis added.
[118] Benjamin, *Berlin Childhood*, 3: 354.
[119] Benjamin, "Über das Grauen I," in *Gesammelte Schriften*, 6: 76.

meval expression of the mimetic faculty. In such a state, the human mind is not present to itself any longer; at the same time, the body – the dwelling-place and expression of the mind (or spirit) – is equally empty and functionless.[120] What remains is the bare, mindless body – the "sub-corporeality" (*Unter-Leiblichkeit*) ascribed to the ghostly in the above-mentioned manuscript, "Anthropologie." The ghostly, for Benjamin, is this same "depotentized body" confronting the human being as his uncanny double. By immersing oneself in the other, one becomes/summons ghosts. In this light, 'primitive mimesis' can be seen as having a tendency toward a state of identity; or else, the "ghostly" aspect of the pre-mythical age does not allow demarcations between self and other to emerge in the first place. In contrast, the liberation from ghosts and their kin aims to constitute the self by traversing the other and thereby doing more justice to both. This liberation is what is at stake for the child. Or rather Benjamin portrays the child as always already existing in a state of liberation, with play remaining "always liberating." [121]

This mode of being is particularly pronounced in "Butterfly Hunt." "Powerless" before the interrelations of nature, the child assimilates himself to the butterfly, thus placing his "human existence" at risk.[122] But despite the danger, this procedure is the only way to learn the "laws" of the "foreign language" of nature. The butterfly's behavior will only become predictable to the child through this acquired knowledge, which will thus allow him to capture the insect. The powerlessness stressed at the beginning stands counter to "confidence" at the end: belief in one's own abilities. "Hiding Places" also shows the child's assimilation accompanied by his initial state of powerlessness. The child complies with the compulsion of similarity by camouflaging himself in the "material world." In this case, the process is not willed so much as it occurs through the mediation of a "demon." Accordingly, the narrator recalls his apprehension that he could remain trapped in the metamorphosis: "Whoever discovered me could hold me petrified as an idol under the table, could weave me as a ghost for all time into the curtain, confine me for life within the heavy door."[123] Ultimately, however, the opposite happens. The child initiates a "struggle with the demon," "anticipating its arrival with a cry of self-liberation," and the event is

---

120 The following takes up Fittler's reading ('Ein Kosmos der Ähnlichkeit,' 371–379), but instead of viewing immersion in horror as the "mystically or pathologically heightened synonym of mimetic adaptation" (374), I consider it the primitivistic version of becoming similar (*Anähnelung*).
121 Benjamin, "Old Toys," 2.1: 100.
122 Benjamin, *Berlin Childhood*, 3: 360.
123 Benjamin, *Berlin Childhood*, 3: 375.

reworked into a process of personal empowerment that the child is "never tired of." The struggle concludes with the transformation of "magical experience" into "science." The boy who has grown so intimate with the apartment by means of assimilating himself to it is soon able to "disenchant" the space as an "engineer."[124]

That for Benjamin mimesis has a dimension of self-empowerment is also shown in his attributing it (in a text on Brecht) not to empathy but to astonishment, conceived since antiquity (as in Greek *thaumazein*) as a spur to the search for knowledge.[125] Hence the assimilative process stands in the service of successfully completing that search and thereby empowering the subject. Its flip side is the hunter's "lust for blood," which leaves "destruction [...] and violence"[126] in its wake. The destruction at work here is incorporating because it is built on mimesis. Its proximity to a "human devouring" destructive pleasure (discussed at the beginning of this chapter) is striking. It is confirmed in a radio speech on "Children's Literature" when Benjamin remarks how books are "deformed and destroyed"[127]; children do not read empathetically so much as they "devour" them. By intensively engaging with what is read, they "increase [themselves]"; the process relates intimately "to their growth and their sense of power."[128]

Playing is thus always already liberation to the extent that assimilation changes from a compulsion born of powerless necessity into a trick played by children in standing up to their environments. Because the child recognized the functionings of this environment through his assimilation of it, he gained power over it and himself by experiencing himself, the cognizant subject, as distinct from what he cognizes. The child savors the pleasure afforded by this victory with each repetition of the game he plays.

As a liberation from a "spell," the mimetic process of making oneself similar to something has an affinity with cunning and ruse; indeed, it can be understood as a strategy of cunning, since it works not through force but by fooling the op-

---

124 Benjamin, *Berlin Childhood*, 3: 375–376.
125 Benjamin, "What is Epic Theatre? [First Version]," in *Understanding Brecht*, trans. Anna Bostock (London: Verso, 1998), 11.
126 Benjamin, *Berlin Childhood*, 3: 351.
127 Benjamin, "Children's Literature," in *Selected Writings*, vol. 2, pt. 1, *1927–1930*, ed. Michael W. Jennings, Howard Eiland, and Gary Smith (Cambridge, MA: Harvard University Press, 1999), 255.
128 Benjamin, "Children's Literature," 2.1: 256. This concept is superficially similar to that of Robert Müller's man-eating colonizer, personified in the engineer Brandlberger (see chapter 7). However, while Müller's perspective instrumentalizes what has been determined as foreign, Benjamin is concerned precisely with doing away with the determination of such categories.

ponent.[129] In his reflections on the fairy tale, Benjamin repeatedly emphasizes the cunning of the fairy-tale hero as a strategy for overcoming myth – superior even to "divine violence" in "On the Critique of Violence" and to the expiation offered by the tragic hero.

> The fairy tale tells us of the earliest arrangements that mankind made to shake off the nightmare which myth had placed upon its chest. [...] The wisest thing – so the fairy tale taught mankind in olden times, and teaches children to this day – is to meet the forces of the mythical world with cunning and with high spirits.[130]

Benjamin attributes this strategy to the child from the start. His interest in fairy tales concerns the "complicity [of nature] with liberated man,"[131] exemplified by the relationship between animals and children in these stories. Accordingly, the essay on Robert Walser credits the fairy tale figures with "childlike nobility,"[132] and the one on Kafka refers to cunning as a "childish [...] means of rescue."[133]

The essay on Kafka also makes it clear why cunning represents the "most prudent" strategy for fighting myth: it is not deployed against the violence that is already in effect in myth, but against its lures, the false promise of redemption from the amorphous, primeval existence. The childish cunning of the fairy-tale hero both resists the enticements of myth and functions as an alternative means for exiting the ghostly realm. Its superiority follows from the fact that its structure is not indebted to mythical violence; this structure draws on pre-mythical mimesis instead of suspending what myth posits or atoning for mythical guilt.[134] Only cunning (and not divine violence or tragic expiation) follows the dialectic of enchantment and disenchantment – a feature to which Benjamin returns again and again in his affirmation of the "liberating magic which the fairy tale has at its disposal."[135]

---

129 Cf. Fittler's discussion of cunning in children's games (*'Ein Kosmos der Ähnlichkeit,'* 362–370); Fittler, however, does not identify its function as an alternative to myth or way out from the pre-mythical sphere.
130 Walter Benjamin, "The Storyteller: Observations on the Works of Nikolai Leskov," in *Selected Writings*, vol. 3, *1935–1938*, ed. Howard Eiland and Michael W. Jennings (Cambridge, MA: Harvard University Press, 2002), 157.
131 Benjamin, "The Storyteller," 3: 157.
132 Walter Benjamin, "Robert Walser," in *Selected Writings*, 2.1: 259.
133 Walter Benjamin, "Franz Kafka," in *Selected Writings*, 2.2: 799.
134 Along similar lines, Menninghaus points out that the act of interruption, for Benjamin, does not mean freedom so much as it "suspends the opposition between freedom and fate" ("Walter Benjamins Diskurs der Destruktion," 302).
135 Benjamin, "The Storyteller," 3: 157.

Thinking of emancipative assimilation as cunning does not entail that the undertaking is done in pretense. For this might not at all be the case. Rather, the main feature of this cunning is that children *determine* the assimilation with their ability to independently begin or at least end it. In the process of becoming-similar, they become the *sovereign* recipients of the foreign, and to that extent their assimilation is simultaneously a process of liberation. And only as such – which is to say as release from compulsion – can the process be understood as play. But this means that it can only be understood as play *retroactively:* at the point, that is, when it has exhibited its emancipative quality.

The child's sovereignty finds expression in the way the material of play is handled, which, to use Lévi-Strauss's term, may be described as "bricolage."[136]

> In our own time the "bricoleur" is still someone who works with his hands and uses devious means compared to those of a craftsman. The characteristic feature of mythical thought is that it expresses itself by means of a heterogeneous repertoire which, even if it is extensive, is nevertheless limited. It has to use this repertoire, however, whatever the task in hand because it has nothing else at its disposal. Mythical thought is therefore a kind of intellectual "bricolage."[137]

In Benjamin's work, on the other hand, bricolage is attributed to a process of thinking that liberates itself from the spell of myth. The bricolage at work here follows the following formula: "the signified changes into the signifying and vice versa."[138] For the bricoleur does not use his materials in the established sense but reorients past purposes as the means to a new end or sees new ends in past purposes.[139]

Benjamin repeatedly describes the child as a collector of fragments and scraps, assembling a new world of things from what he has collected. Such activity is significant on three registers: First, the world of things turns and "faces"

---

[136] Giorgio Agamben has made this point in reference to toys (*Infancy and History: Essays on the Destruction of Experience*, trans. Liz Heron [London: Verso, 1993], 72).

[137] Lévi-Strauss, *The Savage Mind*, 16–17.

[138] Lévi-Strauss, *The Savage Mind*, 21.

[139] The concept of bricolage has forerunners in the classifications proposed by contemporaries of Benjamin, whose works he knew – for instance, Lev Vygotsky, who observes that children think first in "an *unorganized congeries*, or 'heap'" and then in "complexes" (*Thought and Language* [1936], trans. Alex Kozulin [Cambridge, MA, London: The MIT Press, 1986], 110, 112). In either case, the child creates new signs that are still too close to concrete *realia* to be concepts but are also more than isolated impressions. William Stern also identifies a principle of children's language in his psychological studies, affirming that "speech-invention" does not arise "from nothing" but uses the linguistic "material" already given (*Psychology of Early Childhood*, 159).

the child.[140] The collector's non-instrumental approach to things enables them to tell their stories and show their potential for transformation. Accordingly, Benjamin speaks of the collector as a "magician"[141] and assigns him to the age of myth in multiple senses, e. g., as subject to the law of fate[142] or (in contrast to the destructive character) as a preserver and guardian.[143] Second, thanks to his access to the "magic" of things, the collector "disenchants"[144] what he gathers through bricolage (that is, the liberating mimesis described above). This is why Benjamin describes children's discoveries as victories: collecting immerses them in the world of objects. In turn, they detach these objects from their contexts and place them in new ones. This amounts, third, to a "renewal of existence."[145] The child has several means to achieve this end; bricolage, as Benjamin writes, is just one of them. First of all the "old world" must be discovered in its magic so that it may be dismembered into fragments, newly assembled, and thereby "renew[ed]." On this score, the point of contact with – but also the difference from – the figure of the barbarian is manifest: such renewal presupposes not only destruction but first and foremost an intimate adaptation to things; it does not aim at restoring identity or even at something entirely novel, but at transformation. This transformed material is defined neither by Benjamin nor the child, but remains variable and non-positing, as both a "creature[] of [...] blissful caprice" [146] and an opposition against being "bound by sense."[147]

What ethnology designates as bricolage could in the (Benjaminian) philosophy of language be called allegory. The bricoleur corresponds to the allegorician in that both "detach[] things from their context" and allow "meaning" to emerge from the inherent "profundity" of those things.[148] The procedure of bricolage is in linked opposition with that of mimesis, with each only being able to unfold its sensibility and productivity by passing through the other. Much in the same way, allegory is contrasted yet paired with the Romantic symbol. The mutual passage of each through the other is captured by Benjamin in the concept of gestural language, which is both motivated by its object and posited by the speaker. In "Pro-

---

140 Benjamin, *Einbahnstrasse*, in *Gesammelte Schriften*, 4: 93 ("Baustelle"); and "Alte vergessene Kinderbücher," 3: 16.
141 Benjamin, "Lob der Puppe," in *Gesammelte Schriften*, 3: 217; and "Pariser Passagen I," V: 1027.
142 Benjamin, *Arcades Project*, 207.
143 Benjamin, notes on "Der destruktive Charakter," in *Gesammelte Schriften*, 4: 1000.
144 Benjamin, "One-Way Street," 1: 466.
145 Benjamin, "Unpacking My Library," in *Selected Writings*, 2.2: 487.
146 Benjamin, *Berlin Childhood*, 3: 349
147 Benjamin, "A Glimpse into the World of Children's Books," in *Selected Writings*, 1: 435.
148 Benjamin, *Arcades Project*, 211.

gram for a Proletarian Children's Theater," he describes the gesture as an interlocking of reception and creation and considers the process physiological: "the receptive innervation of the eye muscles [passes] into the creative innervation of the hand. What characterizes every child's gesture is exactly proportioned to receptive innervation."[149] I will return to this point below.

Bricolage accounts for the opposition of sovereign children's play not to compulsion but to illusion. In a footnote to the second version of "The Work of Art in the Age of Mechanical Reproduction," Benjamin observes that mimesis has two sides, illusion (*Schein*) and play. He understands the "space for play" (*Spiel-Raum*) as a stage for "experimenting procedures."[150] His examples for this stem from film. Within this new medium, actors no longer imitate; their main concern is no longer creating illusion. Rather, their natural behavior is broken into moments that are then reassembled. The director and film technology experiment, as it were, with the material that actors place at their disposal. In this context, children can be understood as directors of their own play material – which includes, potentially, their own metamorphosis – rather than as beings who transform themselves into that material. Benjamin accordingly describes children as directors or "theater producers" of the stories they tell, not as "actors" in them.[151]

This approach produces yet another connection to contemporary developmental psychology, whose representatives tend to explain the child's animism and transformations as associational processes, which is thus related to Benjamin's notion of assimilation (as opposed to participation). At the same time, they must determine how these childhood behaviors relate to illusion. Do children deceive themselves concerning the reality of their productions? Or do they know that simple illusion is at play here? Or do they find themselves somewhere in between, in conscious self-deception? In such discussions, developmental psychologists consistently raise the question of play – as the site where illusion, whatever its status is, can be legitimately engaged in (see Chapters 3 and 5). Now Benjamin's position on this question is that the cultivation of

---

[149] Walter Benjamin, "Program for a Proletarian Children's Theater," in *Selected Writings*, vol. 2, pt. 1, *1927–1930*, ed. Michael W. Jennings, Howard Eiland, and Gary Smith (Cambridge, MA: Harvard University Press, 1999), 204.

[150] Walter Benjamin, "The Work of Art in the Age of Mechanical Reproduction: Second Version," in *The Work of Art in the Age of its Technological, and Other Writings on Media*, ed. Michael W. Jennings, Brigid Doherty, and Thomas Y. Levin, trans. Edmund Jephcott, Rodney Livingstone, Howard Eiland, et al. (Cambridge, MA, London: The Belknap Press of Harvard University Press, 2008), 49, 48.

[151] Benjamin, "A Glimpse into the World of Children's Books," in *Selected Writings*, 1: 435.

similarities during childhood can *not* be understood in direct relation to the *illusion* of play, but rather to the *sovereignty* children gain over that illusion and to their sovereign handling of the play material.

In distinction to developmental psychologists, Benjamin proposes that a dialectical shift takes place in the child's magical thinking, after which the question of illusion, regardless of whether it has been believed in or not, becomes obsolete. Benjamin's insight is that children's sovereignty is decided only by the question of whether or not they have acted as directors (even of believed-in illusion). If this is the case, play can actually be play, that is, liberation. And as such, it is always located in the sphere of non-illusion. Accordingly, Benjamin stresses that children do not identify with the hero when they read or hear fairy tales. On the contrary, their narcissism – "childish superiority"[152] – stands front and center. For "children are able to manipulate fairy stories [*schaltet mit Märchenstoffen*] with the same ease and lack of inhibition [*so souverän und unbefangen*] that they display with pieces of cloth and building blocks." Instead of immersing themselves in the fairy-tale world, they draw material from it for their own designs: "They build their world out of motifs from the fairy tale."[153] This reconfirms their sovereignty.[154]

The Sovereign Child

Numerous texts show how central the idea of sovereignty is for Benjamin's understanding of the figure of the child and children's play. *Berlin Childhood* describes the child's "power to supervise the game," the animistically transformative "doings of [the] fingers."[155] A bicycle's handlebars, "which seemed to move of [their] own accord," ultimately are mastered, giving the child dominion over the terrain through which he now can travel.[156] In "The Carousel," the child sits "enthroned, as faithful monarch, above a world that belongs to him."[157]

---

152 Benjamin, "Kolonialpädagogik," 273.
153 Benjamin, "Old Forgotten Children's Books," 1: 408.
154 With this term I do not refer to the political concept of the *Trauerspiel* book – transcendant sovereignty that has no need for the other. Instead, the sovereign here possesses autonomy and self-determination in an ongoing process of encounter with the other. Cf. Geulen: "The transformation [into a moral and educated subject] is an act of self-empowerment that lacks a preceding subject [...] and cannot be traced to any positing authority" ("Legislating Education," 953).
155 Benjamin, *Berlin Childhood*, 3: 364.
156 Benjamin, *Berlin Childhood*, 3: 368.
157 Benjamin, *Berlin Childhood*, 3: 385.

Benjamin's theory of the sovereign playing child is expanded upon in two lesser known texts. The first is "Grünende Anfangsgründe" (Blossoming Elements, 1931), a review of Tom Seidmann-Freud's play-primers, which Benjamin appreciates for the ample space they grant to the "power of command, which is so decisive for the play of children": "At every point, care has been taken to preserve the sovereignty of the individual at play."[158] Benjamin's attention falls on letting children act on their own initiative in the process of mimetically learning numbers and letters; they are being called on to understand that material as a means to invent stories, "nonsense, mischief and absurdities" – indeed, to undertake barbaric "clearing work."[159] Consequently, letters and numbers do not appear as powerful "idols" eliciting "dread," but as building blocks for sovereign play. The same view is expounded in "Programm eines proletarischen Kindertheaters" (1929; "Program for a Proletarian Children's Theater," 1999), where Benjamin writes that "the child inhabits his world like a dictator. [...] Almost every childlike gesture is a command and a signal."[160] These signals bring "improvisation" and "variation"[161] to fruition in an anarchistic "carnival."[162] To reiterate, this sovereignty arises not from mere positing, but from a union of sensitive reception and self-confident production, and it opens up play spaces for the productive handling of the given material.

Benjamin's remarks on the "despotism and dehumanized element" of children, with which I began my discussion, must then be understood in terms of the "dictatorial" behavior they use to demonstrate their sovereignty. The behavior is embedded in the dialectical turn from magic to disenchantment, evoked again and again at central points of Benjamin's texts on children and later taken up in his philosophy of history: As a counterpart to the fairy tale's above-cited "liberating magic," Benjamin speaks in "Grünende Anfangsgründe"

---

158 Benjamin, "Grünende Anfangsgründe," in *Gesammelte Schriften*, 3: 312.
159 Benjamin, "Grünende Anfangsgründe," 3: 313.
160 Benjamin, "Program for a Proletarian Children's Theater," 2.1: 204. For a thorough discussion of this text and its contexts, see Karin Burk, *Kindertheater als Möglichkeitsraum. Untersuchungen zu Walter Benjamins "Programm eines proletarischen Kindertheaters"* (Bielefeld: Aisthesis, 2015), which connects to my own reflections (Gess, "Walter Benjamin und 'die Primitiven'"). See also Hans-Thies Lehmann, "Eine unterbrochene Darstellung. Zu Walter Benjamins Idee des Kindertheaters," in *Szenarien von Theater und Wissenschaft*, ed. Christel Weiler and Hans-Thies Lehmann (Berlin: Theater der Zeit, 2003); the author examines Benjamin's references to commands, signals, dictatorship, gestures, and the "dehumanized" child, identifying "not a communistic so much as an anarchistic conception" at work, reflecting a "Nietzschean" and "Surrealist" sensibility (181).
161 Benjamin, "Program for a Proletarian Children's Theater," 2.1: 204.
162 Benjamin, "Program for a Proletarian Children's Theater," 2.1: 205.

of the child's "enchanting-disenchanting play" (*bezaubernd-entzauberndes Spiel*).[163] In *Berlin Childhood*, the child, as an engineer, "disenchants" the apartment that he has previously mimetically made himself similar to. Finally, as a collecting bricoleur caught in a forest of dreams, the child continuously "disenchants" his spoils.[164]

Throughout, the compulsion to mimetically assimilate things leads to liberation from those things, and mimesis thus becomes sovereignly employed cunning and play (replacing compulsion and illusion). In this process, the child as bricoleur sovereignly disassembles the things that have become intimately familiar to him and puts them together in a new way. The destructive element of play proves constructive insofar as it aims at *analysis* (etymologically, something like "breaking up") that enables new construction. At the same time such destruction is not carried out blindly, but requires great intimacy with the things it dismantles.

In this way, children's play brings together two fundamental principles and gives them a dialectical turn: destruction and mimesis. Initially, Benjamin associates both of them with figures of the foreign and the uncivilized, 'barbarians' and 'primitives' respectively. However, he recognizes that they represent two opposing approaches to the other: elimination of alterity and self-renunciation. The "colonial pedagogy" that he criticizes has forged a false unity between the two approaches, whereby seeming convergence with the other serves only to open a new market, which ultimately leads to the former's extinction. With the principle of sovereignty, Benjamin is aiming at another possibility of mediation, amounting to the dialectical turn by which the two laws interact: in mimesis, the tendency toward liberation is underscored – a liberation containing a destructive moment without, however, being attached to a deposing structure of *Ent-setzung* negatively bound to mythic violence. With cunning, mimesis emancipates itself from compulsion, making the other or the past into material at its disposal. The temporary new construction of bricolage is then simultaneously a transformation of the other into one's own and communication of one's own with the other. It creates and preserves the sovereignty of the one, without thus disregarding the otherness of the other. At the same time it makes clear that the self can only be gained through a descent into the strange and that inversely the strange only constitutes itself in view of the self.

In this light, the child appears as the dialectically turned "primitive." In terms of Benjamin's early anthropological model, the child is the figure who suc-

---

[163] Benjamin, "Grünende Anfangsgründe," 314.
[164] Benjamin, "One-Way Street," 1: 466.

ceeds in dialectically elevating the pre-mythic age of the ghostly into the postmythic age of justice – and this with a circumvention of mythic violence. For even though Benjamin speaks of "despotism" and "dictatorship," he is not referring to the mythic violence of positing, which founds and maintains categories and classes. Instead, the child's despotism emerges from destructive-productive mimesis and results from a cunning, not suspending (*ent-setzend*), form of liberation. To be sure, it also acts destructively in bricolage, but such destruction is grounded in mimetic rapprochement and oriented toward a new non-positing construction. In this way, Benjamin derives an emancipating element from the child's "despotism": an element fundamentally different from the regressive reflections of many of his contemporaries who embraced – and prescribed – 'primitive' violence as a salutary force.

Toward the Child's Language of Gesture

"Imitation may be a magical act; at the same time, however, the imitator also disenchants nature by bringing it closer to language," Benjamin writes, and observes further that this process takes millennia.[165] Over this course of time, language becomes the archive of "nonsensuous similarities,"[166] thus replacing connections between things originally perceived by the senses. In the end, it is no longer nature or humankind, but language that works magic by establishing relations between things and standing in mimetic contact with them.

But how is the phrase "nonsensuous similarity" to be understood? When Benjamin affirms, in "Doctrine of the Similar" and "On the Mimetic Faculty," that a similarity exists between language and the writer's intended meaning as well as the writer's unconscious, he is pointing to the double, receptive-productive imprint of language (writing, in this context). In the course of human development, the "mimetic faculty" has made language an "archive of nonsensuous similarities" – not just with what used to exist, but also with images from the unconscious of previous writers. Benjamin calls this the "magical aspect" of language. Such magic is not a matter of conjuring through language so much as its precondition: language, as it relates to the world of objects, is a medium where the objects' "essences" meet.[167] Thus, the focus shifts from the production of language bound to the mimetic faculty to a likewise bound reception of lan-

---

[165] Benjamin, notes on "Lehre vom Ähnlichen," in *Gesammelte Schriften*, 2: 956.
[166] Benjamin, "Doctrine of the Similar," 2.2: 697.
[167] Benjamin, "Doctrine of the Similar," 2.2: 697.

guage: reading, which now attends to the mimetic aspect of language also assumes a "magical"[168] meaning.

Phylogeny repeats itself in the child, who creates language anew. On the other hand, the child is born into a language that already exists. This language has the archival character noted above and at the same time always already communicates a linguistically mediated relation to the objects. Thus, like Groos before him, Benjamin assigns the child an intermediate position between his notion of the 'primitive' as an archaic, indigenous origin and the modern European adult. In contrast to the latter, the child still inhabits a universe of "magical correspondences" and demonstrates the corresponding mimetic faculty. In contrast to the former, these qualities are strongly tied to language,[169] which interposes itself, as it were, between the child and the similarities he recognizes. On the one hand, he discovers the connection among entities through words (e.g., in *Berlin Childhood*, between the aunt and the goldfinch). In this regard, the child is the ideal researcher in the linguistic archive of similarities. On the other hand, language intervenes between the child and the object to which he makes himself similar. Accordingly, Benjamin describes his transformation into a butterfly as mastering the rules of a foreign language.[170] Here, the focus does not bear on the similarity between the thing and its name so much as on the similarity between subject and object, which is achieved with the help of the appropriation of a foreign language.

Ultimately, however, the subject (the "translator"), not the object, finds expression in this language. Benjamin's discussion of verbal misunderstanding makes as much clear: the child hears an unknown word, assimilates its sound to familiar words, and creates a new meaning by assimilating himself to it. *Kupferstich* (copperplate) becomes *Kopfversteck* (head-hiding place) when the child sticks his head out from under the chair. The separation, not correspondence, between an object and its name is the precondition for this event.[171] By the same token, the process no longer depends on a correspondence between subject and object, but on the child's correspondence with a word that only has a referent in this very correspondence. In this way, words become masks – "mummery" – that the child puts on (as *Mummerehlen*,[172] a word produced by misunderstanding, suggests). But at the same time, the child, or rather the process of

---

168 Benjamin, "Doctrine of the Similar," 2.2: 698.
169 See Anja Lemke, *Gedächtnisräume des Selbst. Walter Benjamins 'Berliner Kindheit um neunzehnhundert'* (Würzburg: Königshausen + Neumann, 2008), 64.
170 Benjamin, *Berlin Childhood*, 3: 351.
171 Cf. Benjamin, "Doctrine of the Similar," 2.2: 697.
172 See Giuriato, *Mikrographien*, 188–190.

his receptive-productive appropriation of words, is expressed in these masks. In contrast to the modern adult, the child not only reads in the archive of similarities, but also brings forth new words and new similarities.

Scholars have already discussed in detail how the Benjaminian child employs language mimetically. Yet, less attention has been paid to the gestural aspect of that language, which is indispensable to children's mimetic use of language. In "Program for a Proletarian Children's Theater," Benjamin writes,

> For the true observer [...] every childhood action and gesture becomes a signal. Not so much a signal of the unconscious [...] (as the psychologists like to think), but a signal from another world, in which the child lives and commands. [...] The child inhabits his world like a dictator. For this reason, the "theory of signals" is no mere figure of speech. Almost every childlike gesture is a command and a signal in a world which only a few unusually perceptive men [...] have glimpsed.[173]

In this passage and others like it attending to signals, commands, and orders, a language seems to be called upon that is completely unlike the language of nonsensuous similarities. Such language is deictic, that is, it performs a demonstrative pointing-out that defines a situation or constitutes it in the first place, as opposed to mimetic reference to an object or the mimetic expression of a speaker.

The nature of deixis can be elucidated in the same way that "nonsensuous similarity" was above: with the gesture that Benjamin declares essential to children's signaling activity.[174] My thesis is that the seemingly opposing ideas of a mimetic and a deictic language come together in the concept of a language of gestures, which forms a centerpiece of Benjamin's later theory of language as a whole.[175] Benjamin develops the concept in reference to the figure of the child because for him childhood repeats the early stages of human evolution and linguistic development.[176]

Benjamin's writings on children do not provide the focus for the following, however; and neither do his discussions of Brecht and Kafka, which are usually

---

[173] Benjamin, "Program for a Proletarian Children's Theater," 2.1: 203–204.
[174] Benjamin, "Program for a Proletarian Children's Theater," 2.1: 203–204.
[175] I consider this more substantial than the definition of gesture proposed by Carrie Asman ("Die Rückbindung des Zeichens an den Körper. Benjamins Begriff der Geste in der Vermittlung von Brecht und Kafka," *The Other Brecht II. The Brecht-Yearbook* 18 [1993]: 107, 115), who stresses the oscillation between its mimetic and semiotic dimensions – a quality displayed by language in general for Benjamin.
[176] Like many of his contemporaries, Benjamin follows Ernst Haeckel's claim that ontogeny repeats phylogeny. However, he departs from the model when he identifies differences between the ways early humankind and children understand the relationship between language and the world.

enlisted to shed light on his theory of gesture. Instead, attention will be given to "Problems in the Sociology of Language: An Overview," an essay that has received relatively little scholarly notice,[177] even though Benjamin deemed it a belated forerunner to "Doctrine of the Similar" and "On the Mimetic Faculty."[178] Despite its title, this text is much more than a work commissioned by the Institute for Social Research. It presents Benjamin's own philosophy of language and, as such, should certainly be classified among his texts on language theory. Yet, in contrast to the latter, this essay enables a more nuanced categorization of the author's reflections in the context of linguistic anthropology.[179] A detailed comparison between "Problems in the Sociology of Language" and the texts it references will make it possible to elaborate Benjamin's theory of gesture in such a way that expands (and, in some points, corrects) the prevailing view of it among scholars.

A Theory of Gestures in the "Problems in the Sociology of Language"

Descriptive Vocal Gestures

By the author's own account, "Problems in the Sociology of Language" leads up to where "Doctrine of the Similar" begins. The text revolves around the "origin of language itself."[180] Benjamin begins by noting the "stimulating effect" of "variants of onomatopoeic theory" proposed by Lucien Lévy-Bruhl and Ernst Cassirer. For Benjamin, these variants lie in the understanding of onomatopoeia as a "descriptive vocal gesture." Accordingly, Benjamin refers to Lévy-Bruhl's talk

---

[177] One of the few studies to examine this text in detail is by Günter Karl Pressler, *Vom mimetischen Ursprung der Sprache. Walter Benjamins Sammelreferat Probleme der Sprachsoziologie im Kontext seiner Sprachtheorie* (Frankfurt am Main: Peter Lang, 1992). See also Anja Lemke, "Zur späteren Sprachphilosophie," in *Benjamin-Handbuch*, ed. Burkhardt Lindner (Stuttgart: Metzler, 2006).

[178] Benjamin to Werner Kraft, 30 January 1936, in *The Correspondence of Walter Benjamin, 1910–1940*, ed. Gershom Scholem and Theodor W. Adorno, trans. Manfred R. Jacobson and Evelyn M. Jabobson (Chicago: University of Chicago Press, 2012), 521.

[179] Benjamin himself points out that it is not enough to enlist sociology; child psychology, depth psychology, ethnology, and psychopathology must also be consulted when pursuing "the question of the origin of language" ("Problems in the Sociology of Language: An Overview," in *Selected Writings* 3: 68).

[180] Benjamin, "Problems in the Sociology of Language," 3: 69.

of the "graphic character"[181] of language and claims that language's origins lie in the "language of the hand."[182]

Indeed, Lévy-Bruhl never indicates that language emerges from onomatopoeia but rather from gestures reproducing the behaviors/manners of their objects.[183] As he writes, "[i]f verbal language, therefore, describes and delineates in detail positions, motions, distances, forms, and contours, it is because the language of gestures uses exactly the same means of expression."[184] This results in the "pictorial concepts"[185] that Benjamin cites, which, instead of generalizing, are suited to particularities and therefore innumerable. Significantly, Benjamin contends that such linguistic depiction explains "the magical use of words"[186] for Lévy-Bruhl. This claim is based on an incomplete reading, however, inasmuch as for Lévy-Bruhl language's pictoriality is related to but not responsible for its magic. Instead, the magic of language derives from the mystical participation at work in established, handed-down initiation rites.[187]

Benjamin's slanted reading practice is even more pronounced in his reading of Cassirer, whose remarks on mythical thinking and its relation to language are not as similar to Lévy-Bruhl's as Benjamin makes out. Benjamin describes Lévy-Bruhl's "pictorial concepts" as having a "concreteness," but this does not correspond to the "concentration and compression"[188] that Cassirer ascribes to mythical concepts and "primitive linguistic concepts."[189] Cassirer's interest bears on the moment of "self-predication,"[190] when the sacred detaches from the profane and the mythical/linguistic concept emerges. This is identified with the object not on the basis of gestural depiction, as the connection to Lévy-Bruhl suggests. Instead it derives from the spontaneous expression of affect in sound; word and phenomenon merge due to the violence the latter exerts on the experiencing subject.[191] This process, not any form of likeness, provides the basis for Cassirer's

---

181 Benjamin, "Problems in the Sociology of Language," 3: 70.
182 Benjamin, "Problems in the Sociology of Language," 3: 73. Cf. "Reflections on Humboldt," where, against Humboldt's claim that the word is "the most important component of language," Benjamin suggests comparing "the word to the index finger on the hand of language" (in Selected Writings, 1: 424).
183 Cf. Lévy-Bruhl, How Natives Think, 136–152.
184 Lévy-Bruhl, How Natives Think, 140, translation slightly modified.
185 Benjamin, "Problems in the Sociology of Language," 3: 71.
186 Benjamin, "Problems in the Sociology of Language," 3: 73.
187 Lévy-Bruhl quoted in Benjamin, "Problems in the Sociology of Language," 3: 71.
188 Benjamin, "Problems in the Sociology of Language," 3: 71.
189 Benjamin, "Problems in the Sociology of Language," 3: 70.
190 Cassirer, Language and Myth, 77.
191 Cassirer, Language and Myth, 58.

magical equation of word and thing. Benjamin suggests otherwise when he declares that Cassirer considers the "linguistic magic of the primitives" to be rooted in "complexes" that are supposed to correspond to Lévy-Bruhl's "pictorial concepts."[192]

From Benjamin's treatment of Lévy-Bruhl and Cassirer's writings, therefore, one can draw two conclusions: First, Benjamin overestimates both authors' interest in linguistic magic. Second, there is a skewed, in the case of Lévy-Bruhl, and, in the case of Cassirer, false attribution of linguistic magic to the pictoriality (*Abbildlichkeit*) of language.

Physio-logic and Expressive Movement

In his readings of various linguistic anthropologists, Benjamin shows a particular interest in the premise of an original language of gesture. As I have noted, his starting point for this is the work of Lévy-Bruhl, whom he defends against critics by invoking the "simpler and more sober considerations" of the Russian linguist and ethnographer Nikolai Marr that "primeval man, who did not possess any articulated language, was happy if he could point to or draw attention to an object, and to do this he had a particularly well-adapted tool, the hand."[193]

Contrary to Benjamin's suggestion, however, Marr does not assume that this deictic language of hand gesture is the basis of spoken language. On the contrary, for him the raw material of spoken language, natural animal sounds, exists in parallel to gestural language, and the prerequisite for the formation of spoken language is ultimately represented by the use of tools. For, as he sees it, precisely such a "tool refined by special art" is at work in articulated language.[194] Benjamin's reading obscures (if not eliminates) this difference. He also claims that, according to Marr, the use of tools "liberated the *hand* for the tasks of language."[195]

---

192 The quotation that Benjamin appends makes it clear that he means that the "complex" advanced by Karl Theodor Preuss is the precondition for magical thinking. However, Preuss's "complex" cannot be equated with Lévy-Bruhl's "mystical" intuition or Cassirer's "predication" as easily as Benjamin claims. Lévy-Bruhl posits *participation*, whereas Preuss has an undifferentiated state in mind. Evidently, Benjamin still conflates participation as identity at this point (cf. "Problems in the Sociology of Language," 3: 73). Cassirer enlists Preuss only to show that mythical thinking must first undergo "the process of separation and liberation" (Cassirer, *Language and Myth*, 97–98) on the level of the individual (this is "self-predication").
193 Benjamin, "Problems in the Sociology of Language," 3: 73–74.
194 Nikolaus Marr, "Über die Entstehung der Sprache," *Unter dem Banner des Marxismus* 1, no. 3 (1926): 558–599, here: 593.
195 Benjamin, "Problems in the Sociology of Language," 3: 81. Emphasis added.

Yet Marr's point is precisely the opposite: that the mouth could now take over the essential tasks of language.[196]

Likewise, Benjamin takes up the deictic function of language as described by the developmental psychologist and linguistic theorist Karl Bühler, who illustrates that function with repeated references to the pointing gesture of the index finger.[197] Wrested from context, the passages Benjamin quotes in "Problems of the Sociology of Language" seem to support his claims that nouns emerged from demonstratives. For example, he quotes Bühler as follows:

> Within the broad development of human language, we can imagine that single-class systems of deictic utterances were the first stage. But then came the need to include what was absent, and that meant severing the direct link of utterance to situation [...]. The liberation of linguistic expression from the field of showing – from the *demonstratio ad oculos* – had begun.[198]

In fact, Bühler held the exact opposite view: "deictic words and naming words are two different word classes that must be clearly separated; there is no justification for assuming that [...] the one emerged from the other." He argues *against* the "myth of the deictic source of representative language."[199] Benjamin's tendentious readings of Lévy-Bruhl, Marr, and Bühler imply that he subscribed to the very myth Bühler wished to refute (and in so doing departed from the notion of language originating in imitative depictions): that naming was originally a matter of pointing-and-showing.

Nonetheless, that conclusion must also be modified. For Benjamin advances his thesis that language originated in gesture to a different end. Accordingly, he voices enthusiasm about the arguments made by Richard Paget, who "understands [language] as gesticulation of the speech organs." This definition is not as "surprising"[200] as Benjamin thinks since it was common at the time to trace articulated language back to gestures performed by the body and/or the mouth.[201] In contrast to scholars working in the Cratylist tradition, conceiving the relation-

---

196 Marr, "Über die Entstehung der Sprache," 592–593.
197 Bühler, *Theory of Language*, 94–95, 100, 112.
198 Benjamin, "Problems in the Sociology of Language," 3: 79; and Bühler, *Theory of Language*, 418.
199 Bühler, *Theory of Language*, 101.
200 Benjamin, "Problems in the Sociology of Language," 3: 83.
201 Pethes points out the connection to "oral gesture" developed by Nietzsche in *Human, All Too Human* ("Die Transgression der Codierung," in *Gestik. Figuren des Körpers in Text und Bild*, ed. Margreth Egidi, Oliver Schneider, and Matthias Schöning [Tübingen: Narr, 2000], 303).

ship between vocal gestures and objects in terms of imitation (e.g., Clara and William Stern[202]), Paget focuses on physiology:

> If the mouth, tongue and lips be moved as in eating, this constitutes a gesture sign meaning "eat"; if, while making this sign, we blow air through the vocal cavities, we automatically produce the whispered sounds mnyam-mnyam (mnyum), or mnia-mnia (mnya) – words which probably would be almost universally understood, and which actually occur as a children's word for food in Russian, as well as in English. Similarly, the action of sucking liquid in small quantities into the mouth, if "blown" as before, produces the whispered words sip, sap, according to the exact position of the tip of the tongue behind the lower teeth.[203]

From this, Benjamin extracts that the "gesture" of slurping up liquid brought forth the word "soup" and that the inaudible "gesture" of smiling produced the utterance of "ha ha."[204]

Enlisting Paget for his own purposes, Benjamin understands these gestures (slurping, smiling) as "expressive movements" (*Ausdrucksbewegungen*)[205] along the lines proposed by Wilhelm Wundt, who explains gesture physiologically as the involuntary discharge of an inner tension. For Wundt, the movements at issue externalize an inner state and serve as a declaration (*Kundgabe*)[206] of emotion or to communicate wishes.[207] The same can be argued of Paget: the gesture of smiling serves to express emotion and that of slurping soup signifies the fulfillment of a wish insofar as it exerts influence first in a palpable way on the object and later in the form of an appeal (soup!) on the listener. Semiotically speaking, both are indexical signs, connected to the referent not by a similarity available to the senses, but by physiologically motivated contiguity. Both therefore have indicative (not imitative) characters.

Theories of language's gestural origins appealed to Benjamin in part because they offered an alternative to the narrow conception of mimesis in onomatopoeic theory, which could be "called a mimetic theory in the narrower sense, [...] supplemented by a mimetic theory in a far wider sense."[208] The physiological correspondences between the referent, oral gesture, and speech sound avoid the

---

202 Stern and Stern, *Die Kindersprache*, 355–357.
203 Richard Paget, *Human Speech: Some Observations, Experiments, and Conclusions as to the Nature, Origin, Purpose and Possible Improvement of Human Speech* (London: Kegan, 1930), 136–137.
204 Benjamin, "Problems in the Sociology of Language," 3: 84.
205 Benjamin, "Problems in the Sociology of Language," 3: 73.
206 Wundt, *Elements of Folk Psychology*, 58.
207 Wundt, *Elements of Folk Psychology*, 90–91.
208 Benjamin, "Problems in the Sociology of Language," 3: 84.

issue of oral imitation without lapsing into arbitrariness, offering an indexical model of the sign. In this way, Benjamin found confirmation that nonsensuous similarities are in effect between the referent and the sounds of speech, as proposed in "Doctrine of the Similar" and "On the Mimetic Faculty."

Moreover, the idea of expressive movement enabled him to understand physiologically motivated gestures as a form of declaration (*Kundgabe*). At the same time, he was able to stabilize his concept of a nonsensuous similarity between the speaker and speech sounds. This feature is most evident perhaps in the way children assimilate to the words they themselves have made up in the process of misunderstanding – as illustrated by the example of "copperplate" becoming "head hiding-place" discussed above. To be recalled as well are the images of the unconscious archived in handwriting that are discussed in "On the Mimetic Faculty" and the "way of meaning" introduced in "The Task of the Translator."[209] In all of these cases, emphasis is laid on how the speaker or writer *themselves*, not their signified meanings, enter into language.[210]

Gestural Language as Motivated Positing

Benjamin lauds Heinz Werner for presenting the "most advanced"[211] of the theories he surveys and also voices appreciation for Rudolf Leonhard's work. Both authors focus not on the enunciation of the subject, but on that of the object and simultaneously on the expression of language itself. To that end, they develop an understanding of language as motivated positing that becomes important for Benjamin's theory.

Werner's *Grundfragen der Sprachphysiognomik* (Foundational Questions of Linguistic Physiognomy, 1932), like Benjamin's "On Language as Such and on the Language of Man," starts from the hypothesis that everything human beings encounter communicates an expression to them. Indeed, Werner holds that language itself – as an "objective, particular world of objects"[212] – possesses this ex-

---

[209] Benjamin, "The Task of the Translator," in *Selected Writings*, vol. 1, *1913–1926*, ed. Marcus Bullock and Michael W. Jennings (Cambridge, MA: Harvard University Press, 1996), 257.
[210] See Menninghaus, *Walter Benjamins Theorie der Sprachmagie*, e.g., "Nevertheless, the more significant and, indeed, sounder aspect of Benjamin's theory of mimesis in language and writing lies in his reflection on [...] larger linguistic figures [...], which do not concern the relationship of language and writing to 'meaning'[...] so much as 'naming' (on the part of the 'speaker' or 'writer')" (66).
[211] Benjamin, "Problems in the Sociology of Language," 3: 85.
[212] Heinz Werner, *Grundfragen der Sprachphysiognomik* (Leipzig: Barth, 1932), 10.

pressive dimension, too. His monograph is shaped by the paradoxical task of ascribing to language an inherent expressiveness *of its own* and, at the same time, allowing such expression to coincide again and again with what it designates. He cites the example of an experimental subject referring to the word "wood" (*Holz*) as "something coarse, rough, crude. One gets stuck on it when one sweeps one's eyes over it."[213]

Werner justifies his approach by emphasizing that "all language in the sphere of expression has an image-relationship [*Bildbeziehung*] to reality," but at the same time he rejects the claim that expressive language depicts things.[214] The expression of language he is pursuing is not an attribute based on referential convention or imitation of its object. To counter such views, Werner appeals to language's "ideality."[215] Invoking Plato's *Cratylus*, he stresses the "moment of reconfiguring" performed by the "linguistic creator," who neither enlists arbitrary sounds to designate things nor makes an acoustic copy of them. Instead – as asserted by Johann Gottfried Herder more than a century earlier – the speaker chooses sounds that are motivated by his particular perspective on the objects he is naming: "This sonic material, which the creator of language forms, is not an imprint of reality, but a tool with which the characters of things are designated, aspects of the essence of things are brought out."[216]

For Werner, the representational function of language serves its declarative function: "[The speaker] does not want to produce the things themselves, but to declare something about things," which, at the same time, is a declaration of his own point of view on them. Such a perspective is not arbitrary, but a motivated *Setzung* or positing. Language communicates information not only about the speaker's subjectivity but also about the nature of the object spoken of, its "essential aspects" (albeit from a specific and personal standpoint). In this context, "expression" does not refer to the usual declarative function of language so much as to its secondary, representational function. Werner is not concerned with the speaker or with the speaker's declaration of his or her inner state; instead, he focuses on how language gives shape to an actual "aspect of [the thing's] being."[217] This ontological feature of the object is tied to its existence in language, but it is thought of not as an invention so much as a *discovery*. In this light, what language proclaims is something authentically linguistic

---

213 Werner, *Grundfragen der Sprachphysiognomik*, 35.
214 Werner, *Grundfragen der Sprachphysiognomik*, 12; cf. 44.
215 Werner, *Grundfragen der Sprachphysiognomik*, 44.
216 Werner, *Grundfragen der Sprachphysiognomik*, 15.
217 Werner, *Grundfragen der Sprachphysiognomik*, 15.

and, at the same time, a feature belonging to the thing itself. Werner's claim is that language as an "objective, special world of its own" does not express the subjectivity of the speaker; rather, it expresses itself as a form of knowledge as well as the object's essence.[218]

Leonhard also sets out to solve the riddle of language's expressive dimension, and in doing so he abstracts, even more than Werner, from the speech situation and investigates language as a "phenomenon *sui generis* with an existence according to its own laws."[219] Yet in the process he scarcely attends to how objects motivate words. Instead, his interest is directed to the physiognomic associations that words elicit – whether or not these associations coincide with the object signified (most of his examples involve cases where they do). He shows, again even more forcefully than Werner, the "constitutive" dimension of language: "The word constitutes not only itself, but also [...] the idea, the idea precisely assigned to reality."[220]

Both theories interest Benjamin because they approach an aspect of language's origins in gesture that his own theory is trying to elaborate as well. However, Werner and Leonhard are not concerned with likenesses or the physiological relationship between words and things. They are interested in an expression of language *as such*, which is also an expression of the essence of things. As Leonhard's text exhibits, these considerations ultimately lead to an insight into language as symbolic form. Only in and as language can the things of the world be known. Their relationship is *inversely* motivated, then, insofar as language posits their existence in the first place.

Leonhard also calls the constitutive power of language its "magic."[221] Bühler does the same and understands positing language as a magical appeal to objects to take shape in conformity with language. In a passage quoted by Benjamin, he writes, "[n]aming the things by their 'true' name becomes a powerful (a benign or baleful) means for the speaker to appeal to the world of things itself."[222] Here naming and appealing merge in relation to the world of things. Bühler observes such a behavior in children as well: "under the influence of high affective tension [...] the world is transmuted before the eyes of the child much as the theo-

---

[218] The proximity to the considerations in Benjamin's early language essay ("On Language as Such") is obvious here. What is theologically justified in Benjamin's work, however, tends to amount to epistemological optimism in Werner's.
[219] Rudolf Leonhard, *Das Wort* (Berlin: Graetz, 1932), 5.
[220] Leonhard, *Das Wort*, 5.
[221] Leonhard, *Das Wort*, 5.
[222] Bühler, *Theory of Language*, 244.

rists of the magical attitude of mind think." At the same time, however, Bühler notes another attitude at work, namely,

> [t]he completely unmagical *experimental attitude* of the child, by virtue of which the newcomer in this life matures gradually, in step with the successful results of his struggles when he "encounters resistant matter" [...], maturing to become a master of the techniques required by life. The child has no trouble switching from one attitude to the other, and, for example, quite tranquilly puts the piece of wood that a moment before "was" a sobbing and pacified foster-child into the stove. It is not by any stretch of the imagination the foster-child that then burns before its eyes, but the common piece of wood.[223]

Both considerations resurface in Benjamin's work. Passages on the child's perception of the magic of the name occur throughout *Berlin Childhood*, as I have already remarked (*Steglitz* and *Stieglitz*, *gnädige Frau* and *Näh-Frau*, and so on). Similarly, the experimental bearing that Bühler discusses corresponds, in "Program for a Proletarian Children's Theater," to "improvisation [...] [which] is the framework from which the signals, the signifying gestures, emerge."[224] But unlike Bühler, Benjamin does not oppose the magical stance to an experimental one free of magic. Instead, he sees the two as dialectically mediated by gesture. Gesture is at once both magic and liberation from magic.

Benjamin ends his overview with a lengthy excerpt from a study on aphasia by Kurt Goldstein, who takes issue with the instrumental conception of language. For him, language is instead a "manifestation, a revelation of our innermost being and of the psychic bond linking us to ourselves and to our fellow human beings."[225] With this conclusion, Benjamin underscores once again the relational character of linguistic positing, insofar as the speaker and situation of his or her speech return to the forefront. At such moments, language is no longer thought of as the positing of something. Instead positing language is thought of as the innermost essence of humanity and human community.

This finding corresponds to hints, in "Doctrine of the Similar" and "On the Mimetic Faculty," regarding what or whom language might resemble. But in contrast to Werner and Leonhard, Benjamin makes virtually no effort to elaborate. He has no interest in interpreting the images glimpsed in handwriting, for in-

---

223 Bühler, *Theory of Language*, 245.
224 Benjamin, "Program for a Proletarian Children's Theater," 2.1: 204.
225 Benjamin, "Problems in the Sociology of Language," 3: 86. Kurt Goldstein, "L'analyse de l'aphasie et l'étude de l'essence du langage," in Ernst Cassirer, Leo Jordan, Henri Delacroix et al., *Psychologie du langage* (Paris: F. Alcan, 1933).

stance, even in texts concerned with graphology.²²⁶ Instead it is merely noted that they *are* visible. *That* the nonsensuous similarity exists is more important than *what* it means. The images refer primarily to their own existence *as* images instead of to whatever they may depict.

In this sense, Giorgio Agamben has called gesture a "communication of a communicability": "It has precisely nothing to say because what it shows is the being-in-language of human beings as pure mediality."²²⁷ For Benjamin, Agamben continues, this notion entertains a relationship with the "expressionless" (a point of difference with Werner) and therefore with the process of "showing": "Gesture is what remains expressionless in every expression. In this sense the gesture may be essentially deictic."²²⁸ This distinction, between language as the positing *of* something and positing language *as such* as the innermost essence of humanity, should be viewed in light of Benjamin's earlier distinction between expression *through* and *in* language. It is not with the aid of language so much as *in* language that human nature, individual and collective, comes out. This is not the case because language can be traced back to God (as Benjamin affirms in his earlier essay, "On Language as Such"), but because, as his reference to Marr makes clear, linguistic positing is a manmade tool for disclosing a world that is historically and sociologically constituted.

Creative Innervation of the Hand

In discussions of the motivated nature of language, the physiological perspective and the theory of positing tend to be set in opposition. The former holds that a physiologically motivated connection between language and its objects is at work in indexical expressive movements. The latter holds that the link between language and its objects is posited, that is, positing language outlines the con-

---

226 Benjamin, "Der Mensch in der Handschrift" and "Zur Graphologie" in *Gesammelte Schriften*, 3: 137 and 6: 185, respectively.
227 Giorgio Agamben, "Notes on Gesture," in *Means Without End*, trans. Vicenzo Binetti and Cesare Catarino (Minneapolis: University of Minnesota Press, 2000), 58.
228 This passage is not found in the English translation cited in footnote 227 and has been translated by Susan Solomon from the German edition: Giorgio Agamben, "Noten zur Geste," *Postmoderne und Politik*, ed. Jutta Georg-Lauer (Tübingen: edition discord, 1992), 105–106. Werner Hamacher also interprets gesture in Benjamin along these lines, pointing to the caesura constuted in the process: "The decision, a pure caesura in the language of predications, laying-bare what simply says without saying *something*, lies in what Benjamin calls gesture" ("Die Geste im Namen. Benjamin und Kafka," in *Entferntes Verstehen. Studien zu Philosophie und Literatur von Kant bis Celan* [Frankfurt am Main: Suhrkamp, 1998], 318).

tours of the object in the first place. Benjamin elaborates a model in which the two perspectives are mediated through one another.

The final pages of "Problems of the Sociology of Language" include a quote in which Mallarmé calls a dancer a "metaphor" that "may give expression to one aspect of the elementary forms of our existence: sword, goblet, flower, and others."[229] In this context, dance is not an expressive movement in the sense of a declaration of affect or desire, but rather of the "expression" of something external. And Benjamin must be using "expression" deliberately here, because the French original speaks of "a *metaphor summarizing* one of the elementary aspects of our form" (une *métaphore résumant* un des aspects élémentaires de notre forme).[230]

What is one to make of this disparity? A passage from "Program for a Proletarian Children's Theater" is instructive. Here, Benjamin emphasizes that the gestural signals of the child are to be "applied to materials." In this context, Benjamin follows art historian Konrad Fiedler in understanding gesture as a "seeing with the hand" and as a physiological process, whereby "the receptive innervation of the eye muscles [is transferred] into the creative innervation of the hand. What characterizes every child's gesture is that creative innervation is exactly proportioned to receptive innervation."[231] "Innervation" is a physiological term that refers both to the neurological disposition of an organ and the process by which stimuli reach it. For example, Freud writes,

> all our psychical activity starts from stimuli (whether internal or external) and ends in innervations. Accordingly, we shall ascribe a sensory and a motor end to the [psychic] apparatus. At the sensory end there lies a system which receives perceptions; at the motor end there lies another, which opens the gateway to motor activity.[232]

At the same time, Benjamin distinguishes between "receptive" and "creative" innervation. This reflects his thoughts on the development of the mimetic faculty

---

229 Benjamin, "Problems in the Sociology of Language," 3: 84.
230 Stéphane Mallarmé, "Ballets," trans. Evlyn Gould, *Performing Arts Journal* 15, no. 1 (1993): 107. Emphasis added.
231 Benjamin, "Program for a Proletarian Children's Theater," 2.1: 204.
232 Freud, *The Interpretation of Dreams*, 539. On Benjamin's use of the term *innervation*, cf. Miriam Hansen, "Benjamin and Cinema: Not a One-Way Street," in *Benjamin's Ghosts: Interventions in Contemporary Literary and Cultural Theory*, ed. Gerhard Richter (Stanford: Stanford University Press, 2002). Hansen devotes more attention to Freud and stresses that Benjamin views innervation as a "two-way process, that is, not only a conversion of mental, affective energy into somatic, motoric form but also the possibility of reconverting, and recovering, split-off psychic energy through motoric stimulation" (50).

from gaze to bodily gesture and finally to articulation, which he formulates in a draft on the mimetic faculty and which, against this background, can be understood as a development from receiving to creating, from reception to production.[233] Even more importantly, Benjamin's above quoted passage on innervation corrects the common understanding of signals. The gestural signal does not apply to objects from without, and therefore it is neither arbitrary nor simply instrumentalized by the subject. Instead it develops out of them, so to speak. It receives, imitates, and reshapes in one process.

The same holds for the dancer mentioned above. Like the child's gesture, Benjamin understands dance as a transfer of visual perception into bodily gesture. Its way of bringing objects into expression is first based on the physio-logic of innervation and then on the transfer of neurological stimuli from one organ to another. In this process, innervation is motivated at least as much by the received object as it is by the creative subject. In other words, dance does not appear only as the reference to an object, but as the object's producer. At the same time, it expresses the creative innervation of the subject that is constituted in the process.[234] Indexical mimesis and positing deixis are mediated through one another in a dance-like gestural language.[235]

Benjamin's notes on "Doctrine of the Similar" make clear that his model of the "creative innervation of the hand" goes hand in hand with the paradoxical idea of a simultaneous liquidation and establishment of magic. On the one hand, the relocation of the mimetic faculty from the eye through the body to the lips implies the "overcoming of myth," that is, the overcoming of magical compul-

---

233 Benjamin, Notes on "Zum mimetischen Vermögen," in *Gesammelte Schriften*, 2: 958.
234 At no point does Agamben examine Benjamin's discussion of dance. However, he offers his own thoughts on dance in keeping with the quality described: "If dance is gesture, it is so [...] because it is nothing more than the endurance of and the exhibition of the media character of corporal movements. *The gesture is the exhibition of a mediality; it is the process of making a means visible as such*" ("Notes on Gesture," 58). Likewise, Agamben invokes Mallarmé to make his point, albeit from different passages – for instance, "The body takes possession of itself again and again: its dance is the analysis, the sequencing of all of the tendencies toward movement that it discovers in itself" ("Noten zur Geste," 107). As above, this passage is not found in the English translation cited in footnote 227 and has been translated by Susan Solomon from the German edition just cited. The "expression" of an external element is missing here; the focus is instead self-referential showing. The interpretation I have proposed combines both perspectives.
235 This view contradicts the one proposed by Jürgen Habermas ("Walter Benjamin: Consciousness-Raising or Rescuing Critique [1972]," in *Philosophical-Political Profiles*, trans. Frederick G. Lawrence [Cambridge, MA: MIT Press, 1983]), who disregards the deictic component of Benjamin's theory of language. Cf. Anja Lemke ("Zur späteren Sprachphilosophie," 652).

sion. Obviously, the dialectical turn of the mimetic faculty from a compulsive to a sovereign behavior, used with cunning, is here linked to the transformation from being spellbound by the object to determining the object, i.e., from being looked at to bodily and lip gestures to spoken language. On the other hand, he argues that the dancer's "mimetic mode of behavior" stands in a dialectical relationship with the "dynamic side" of dance, namely, the magical "transfer of energy" to its respective objects.[236] In other words, Benjamin claims that the formation of the world of things (by means of the magical transfer of energy) takes place in the gestures of dance, which are at the same time mimetically derived from that same world of things. Here, in contrast to the passage above, "magic" serves creative production, not the compulsion to become similar.

"On the Mimetic Faculty" and "Doctrine of the Similar" allow for the same finding. In the first essay, Benjamin writes that with the transfer of the mimetic faculty to language, magic has been "liquidated." But in "Doctrine of the Similar" he calls the mimetic aspect of language as well as the reading of it "magical." Clearly, then, two different notions of magic are at work, or, more likely, a dialectical turn of magic is in evidence.[237] Magic is overcome insofar as people are no longer bound in a compulsive relationship to similarity. Instead, they come out of reception into production and from assimilation to the creation of something new but similar, through which they acquire a fleeting sovereignty. The gesture remains "magical"; however, it is just as motivated by the object as it is (physiologically) by the subject, especially insofar as it has the force of symbolic formation at its disposal.

In light of the quotation from Mallarmé, the model of the "creative innervation of the hand" can be related to Benjamin's theories on the nonsensuous similarity of writing, in which the gestures of the hand have left their traces behind. Indeed, Mallarmé goes on to say (although Benjamin does not quote the passage) that the dancer does not dance, but writes:

> *She does not dance,* suggesting, by way of prodigious abbreviations and expansions, with a corporal writing that would necessitate paragraphs of dramatic dialogue as well as prosaic description, to be expressed, in the rewriting: poem disengaged from all of the scribe's apparatus.[238]

Indeed, given Mallarmé's reference to "metaphor," the "creative innervation of the hand" may also be grasped in semiotic terms as an interactionist "meta-

---

236 Benjamin, notes on "Zum mimetischen Vermögen," in *Gesammelte Schriften*, 2: 957.
237 Cf. Menninghaus, *Walter Benjamins Theorie der Sprachmagie*, 75–77.
238 Mallarmé, "Ballets," 107.

phor," which gives expression to what is taken from the world, while at the same time transforming it and making it the material for one's own new production. A note Benjamin made (12 October 1928) when writing this essay confirms as much: "Upon close inspection, the metaphor ultimately becomes the only possible manifestation of the thing. The path to reach it: impassioned play with things. On that same path children reach the heart."[239]

Defining "creative innervation of the hand" as a metaphor points to a middle ground between the conflicting interpretations of Benjamin's theory of gesture: One, the allegorical interpretation, understands gesture as pure positing.[240] The other is bound to the model of the Romantic symbol and sees in gesture a possible remnant of immediate or at least motivated language.[241] The physiological conception of expressive motion and the notion of language as positing come together in the idea of language/writing as metaphor, i.e., as transference, which resonates with Benjamin's early model of "translation" in "On Language as Such." There he writes, "For conception and spontaneity together, which are found in this unique union only in the linguistic realm, language has its own

---

[239] Benjamin, "Verstreute Notizen. 12. Oktober 1928," in *Gesammelte Schriften*, 6: 417. The quotation is taken from a discussion with Bloch and Rethel about how things point to the social relations they are a part of, i.e., the way in which they are, as it were, metaphors for the social. Thus, the perspective on language as a product and expression of human community, already mentioned and to be discussed in more detail below, is echoed here.

[240] Pethes connects the gesture with allegorical emblems: "The essential quality of gesture is that its meaning vacillates, which [...] occurs in allegorical emblems: the gesture is material that only takes on meaning when it [...] is inscribed. This deferral of meaning makes physical bearing in Kafka's works into the self-referential model of representation in the figurative sense of 'gesture': 'gestural texts' expose the staged character of the meaning they offer" (Pethes, *Mnemographie*, 119). Rainer Nägele also considers the gesture to have the structure of an emblem and reads it as the caesura and dismemberment of bodily wholeness ("Von der Ästhetik zur Poetik der Zäsur," in *Lesarten der Moderne. Essays* [Eggingen: Isele, 1998], 110–120).

[241] This concerns, e.g., Habermas's reading of gesture as immediate expression. However, the deconstructive interpretations of Agamben and Hamacher, who understand gesture as the showing of showing, could also be placed here inasmuch as they assume that the problems of difference between sign and referent have been suspended by the sign's self-referentiality. Cornelia Zumbusch identifies a middle ground in Benjamin's conception of the dialectical image, which she traces back to the "true symbol" invoked in the author's early works (*Wissenschaft in Bildern. Symbol und dialektisches Bild in Aby Warburgs Mnemosyne-Atlas und Walter Benjamins Passagen-Werk* [Berlin: Akademie, 2004], 14). This conception agrees with that proposed by Aby Warburg: "With the symbol and dialectical image, Warburg and Benjamin aim for a third form between the magical symbol and the purely arbitrary sign. Warburg's symbol and Benjamin's dialectical image [...] bridge the common distinction between symbol and allegory" (20). On the ambivalence of the Benjaminian symbol, cf. Menke, *Sprachfiguren*, 432–433.

word: translation."²⁴² The assimilation to language, that I observed earlier in the examples of the child's handling of language, and the appealing signals of the child dictator thus become legible as two sides of the same gestural language, in which reception and production are each dialectically mediated through the other.

The Politics of Gestural Language

How does this definition of gesture relate to statements on the same in Benjamin's writings on Kafka and Brecht? ²⁴³ In the former, Benjamin refers to gesture as a matter of "bodily innervation" or even "reflex." However, it does not react to just any arbitrary object, but rather to a threatening "nightmare" (*Alb*) that must be combatted. Here, the gesture is marked by an "ambiguity before a decision": it can be either a "reflex of liberation" or "of submission."²⁴⁴

Benjamin made this note when planning to revise his Kafka essay at the beginning of 1935. It refers to a passage that treats how gestures of power precipitate into frameworks of social roles (employee and boss, sinner and clergyman). In contrast to the liberating, dialectical turn from the subject's assimilation to the creative transformation of the object presented earlier in this chapter, here Benjamin regards the gesture as an ambivalent expression of a socially predetermined power structure. Likewise, in his essays on Brecht, the gesture relates to the "devastations of our social order," the "one-eyed monster whose name is 'class society.'"²⁴⁵ Such structures of power become visible in the gesture because it results from the interruption of an action, at which point the events sol-

---

242 Benjamin, "On Language as Such and the Language of Man," in *Selected Writings*, vol. 1, *1913–1926*, ed. Marcus Bullock and Michael W. Jennings (Cambridge, MA: Harvard University Press, 1996), 69. Cf. Menninghaus, *Walter Benjamins Theorie der Sprachmagie*, 35–37.
243 On Benjamin's theory of gesture in his writings on Kafka and Brecht, cf. Asman, "Die Rückbindung des Zeichens an den Körper"; Hamacher, "Die Geste im Namen"; Samuel Weber, "Citability – of Gesture" and "Violence and Gesture. Agamben Reading Benjamin Reading Kafka Reading Cervantes," in *Benjamin's -abilities* (Cambridge, MA: Harvard University Press, 2008); Nägele, "Von der Ästhetik zur Poetik"; Nikolaus Müller-Schöll, "Nachahmbarkeit. Zur Theorie des Gestischen als eines Theaters der Spur," in *Das Theater des 'konstruktiven Defaitismus.' Lektüren zur Theorie eines Theaters der A-Identität bei Walter Benjamin, Bertolt Brecht und Heiner Müller* (Frankfurt am Main: Stroemfeld, 2002).
244 Benjamin, notes related to Benjamin's writings on Franz Kafka, in *Gesammelte Schriften*, 2: 1261.
245 Benjamin, "What is Epic Theatre? [First Version]," 5.

idify into "real conditions"²⁴⁶ of society. Benjamin also discusses how the gesture preserves social reality in "Problems of the Sociology of Language," where, apropos of Marr's theory of language, he writes, "[t]he essential element in the life of language [...] appears to be the link between its evolution and certain social and economic groupings which underlie the groupings of social strata and tribes."²⁴⁷

Social power relations then are sedimented in gestures. According to Benjamin's reading of Kafka and Brecht, the task is to make the reader or spectator aware of this fact. Gestures do not only result from the interruption of an action, but also need this interruption in order to become visible as a language of the social. In keeping with Brecht's vision of the theater, Benjamin writes that the actor "must be able to space his gestures as the compositor produces spaced type."²⁴⁸ Spectators should not become familiar with actors but be "distanced" from them. For only in this way can their astonishment at the seemingly familiar be roused and thereby also their interest in knowledge.²⁴⁹ Elsewhere, he observes that Kafka offers no interpretation of gestures. Instead Kafka makes them the object of endless consideration by wresting them from their normal contexts and withholding any explanation of them. Benjamin presents Kafka's works also as theater in which "the author trie[s] to derive such a meaning from them in ever-changing contexts and experimental groupings."²⁵⁰ Here as elsewhere, Benjamin is not interested in the production of gestures, but in their analytical reception. This occurs along the same lines as the reading of gesture discussed above. There it was about a magical reading attending to the physiognomic dimension of language, which was initially defined as its gestural dimension. In the essay on Kafka, magical reading is not directed toward a supposedly immediate expression of the object or subject that has been sedimented in gestures, and within this the receptive-productive language itself, but rather toward social and historical elements sedimented in gestures.

The essay on Brecht also mentions "setting up an experiment"²⁵¹ regarding the interaction with gestural conditions, but with this Benjamin envisions more than a search for sociological sediment. He quotes the playwright, "[i]t can happen this way, but it can also happen quite a different way." ²⁵² The experimental arrangement promotes the realization of "freedom" through engagement

---

246 Benjamin, "What is Epic Theatre? [First Version]," 4.
247 Benjamin, "Problems in the Sociology of Language," 3: 75.
248 Benjamin, "What is Epic Theatre? [First Version]," 11.
249 Benjamin, "What is Epic Theatre? [First Version]," 11.
250 Benjamin, "Franz Kafka," 801.
251 Benjamin, "What is Epic Theatre? [First Version]," 4.
252 Benjamin, "What is Epic Theatre? [First Version]," 8.

with gesture. As with Kafka, the gesture is ambivalent in Benjamin's reading of Brecht: "Twice Galy Gay is summoned to a wall, the first time to change his clothes, the second time to be shot."[253] Within the same gesture there is space for developmental play. What in one case means domination might in another context signify revolt. Making the gesture "quotable" – a Brechtian prescription to the actor that Benjamin endorses – aims for precisely this end. It involves not only alienation (*Verfremdung*) but also displacement into new contexts, a bricolage of gestures. The productive force that the theater displays, enacts, and rouses by means of this procedure recalls the productive force (which is likewise sedimented as the social element in gesture) of the social collective that created it. The subject that comes into language here is not that of individual psychology; it is a collective subject and inscribed with a specific historical and social index.

In the essay on Brecht, Benjamin calls the gesture the "mother of the dialectic."[254] And this is not only because it mediates between the moment of its own occurrence and the play's temporal flow. Like Brecht, he calls for the act of showing to be shown. Two gestures stand at issue: First is the theater's gestural reference to gestures in which both the social reality and linguistic-creative force of the historical collective are sedimented. That is the dialectic of the first gesture. The dialectic of the second gesture, that is, the theater's gestural referencing, lies in how its imitation of the first gesture simultaneously manifests its own freedom because it is not only imitating, but bricolaging and thereby creating space for interpretation. This second gesture is the one performed by the child dictator and director, mediating reception and creation as well as innervated nature and creative subject through one another. Yet this child is not necessarily the singular subject of individual psychology. On the contrary, Benjamin embeds this figure in a "children's collective"[255] and in the proletarian children's theater, where "the themes and symbols of class struggle [...] have a place."[256]

A gesture – whether that of the child director-dictator or that of the actor in epic theater – is subject to a double dialectic, namely both the uncovering of the sedimented social reality and the creative force of the historical collective, as well as the simultaneous imitation and re-creation of the gesture. With this double dialectical structure, Benjamin can read the child's gesture as "the *secret signal* of what is to come."[257] This is meant also in the sense that child actors

---

253 Benjamin, "What is Epic Theatre? [First Version]," 12.
254 Benjamin, "What is Epic Theatre? [First Version]," 12.
255 Benjamin, "Program for a Proletarian Children's Theater," 2.1: 203.
256 Benjamin, "Program for a Proletarian Children's Theater," 2.1: 205.
257 Benjamin, "Program for a Proletarian Children's Theater," 2.1: 206.

who experience the "wild liberation of the [...] imagination" through the proletarian children's theater will not bear the burden of an unlived childhood later in life: "Through play, their childhood has been fulfilled. They carry no superfluous baggage around with them, in the form of overemotional childhood memories that might prevent them later on from taking action in an unsentimental way."[258] From undisturbed immersion in play there ultimately emerges an adult who does not perceive disenchantment as a deficiency but as a prerequisite for creating a different society.

The *Arcades Project*: The Child as Historiographical Model

On a semiotic register, the dialectical shift of 'primitive thinking' from mimesis as compulsion to liberating mimesis and mimesis as cunning, as well as from mimesis as illusion to mimesis as the play of bricolage, corresponds to a gestural language that is not spellbound by the world of things in mirroring imitation and as such, in turn, has a banishing effect on the world and users of language. Nor does this gestural language relate to objects by a mere arbitrary positing or instrumentalization by its users. Rather, in nonsensuous manner, it resembles the realm of objects and its users simultaneously, inasmuch as it is animated by the metaphoric model of the "creative innervation of the hand." Gestural language relates to things, but only by means of people, or more specifically, through the stimulation of their sensory systems by things and the creative transformation of these stimuli into artistic products. Gestural language always already implies appropriation, manipulation, transfer, translation, and transformation, through which human beings gain sovereignty without ignoring or colonizing the object in the process.

The sovereign child and his or her activities – cunning, destruction/bricolage, and gestural language – are models for the dialectical "detachment from an epoch" pursued by Benjamin in the early *Arcades Project* (until 1929).[259] Such detachment, which Benjamin also calls "awakening," is not a rupture so much as

---

[258] Benjamin, "Program for a Proletarian Children's Theater," 2.1: 205.
[259] Benjamin, *Arcades Project*, 173. Benjamin indicates in a note to "Theses on the Philosophy of History," a late work, that children still play a central role – "as representatives of paradise" (*Gesammelte Schriften*, 1: 1243). In a letter to Adorno, Benjamin declares children to be "a kind of corrective to society" (7 May 1940, in Adorno and Benjamin, *Complete Correspondence*, 330). Cf. Lindner, "Das *Passagen-Werk*," 236–242. Adorno's criticism that Benjamin does not proceed dialectically enough in the *Arcades Project* would therefore require qualification (Adorno to Benjamin, 2–4 August 1935, in Adorno and Benjamin, *Complete Correspondence*, 105).

the simultaneous engagement with and overcoming of the past, salvaging and new configuration, through which the past maintains or acquires its relevance for the present. In the early drafts of the *Arcades Project*, children (unlike the Surrealists) do not number among the figures caught in the realm of dreams.[260] Instead, the figure of the child exemplifies the strived for dialectical shift from enchantment to disenchantment. The recollections of childhood that Benjamin invokes in the early *Arcades Project* are not simply the memory of an enchanted world of things – he also calls the "dream figure" of the nineteenth century its "child's side"[261] because the child perceives the world as enchanted and because he himself was a child at that time[262]; but they are also the memory of this shift, this turn from enchantment to disenchantment. Benjamin models his project's detachment from the epoch of the nineteenth century after the child's "technique[s]."[263]

First of all, cunning: By immersing oneself in its dream side (for example, through involuntary memory or a childlike perception), one gets to know one's past epoch so well that one can interpret it. In so doing, one can free oneself from its mythical timelessness and its appearance as nature. The child one once was and one's own children play a vital role in this process:

> The fact that we were children during this time belongs together with its objective image. [...] The dream waits secretly for the awakening: the sleeper [...] waits for the second when he will cunningly wrest himself from its clutches. So, too, the dreaming collective, whose children provide the happy occasion for its own awakening.

Children recall to adults the dreamworld of their own childhoods, or rather: they make them aware of the world of their childhood *as* a dreamworld, from which they can now finally awaken. From their children and their own childhood experience, it becomes clear that the objects remembered from the past have a "symbolic character"[264] and therefore have the potential to be interpreted. This potential is decisive to the present adult's understanding. Benjamin differentiates this cunning detachment from a violent one.

---

260 Benjamin, *Arcades Project*, 13.
261 Benjamin, *Gesammelte Schriften*, 5: 1006.
262 Benjamin, *Gesammelte Schriften*, 5: 1006, 1024.
263 Benjamin, *Gesammelte Schriften*, 5: 1002.
264 Benjamin, *Arcades Project*, 390.

> The genuine liberation from an epoch [...] has the structure of awakening in [that] it is entirely ruled by cunning. Only with cunning, not without it, can we work free of the realm of dream. But there is also a false liberation; its sign is violence.[265]

The violent detachment would in fact be bound right back (whether negatively in its de-posing suspension or positively in the new positing it carries out) to the myth whose spell it is seeking liberation from.

That cunning implies a demarcation from mythic violence is also indicated by Benjamin's referral to the *Arcades Project* as a "féerie,"[266] where he thus associates it with the fairy tale, which, as shown above, circumvents the violence of myth with cunning. Likewise, in another key passage, he affirms that the "most radical expression" of the "dialectical schematism" underlying the transition from dream to waking is found in Chinese "fairy tales."[267] Immediately after this statement, he presents his project's program:

> The new, dialectical method of doing history as the art of experiencing the present as waking world, a world to which the dream we name the past refers in truth. To pass through and carry out *what has been* in remembering the dream! [268]

In other words, the "difficulty of this dialectical technique"[269] of awakening that he noted earlier may be resolved with help from the fairy tale and its cunning hero. The cunning fairy-tale hero of the early *Arcades Project*, however, is Benjamin himself, insofar as he only gets involved with the dreamworld of the nineteenth century in order to be able to first interpret it and then understand the present. And it is his text that, through its procedures, shields itself both against the spell of what has been and against its own positings.

Second, destruction and bricolage: In a letter to Adorno, Benjamin relates his use of the term *féerie* to the text's form: "This subtitle suggests the rhapsodic character of the presentation."[270] Later, he speaks of "rhapsodic naiveté" (implicitly referring to childhood) and "romantic form."[271] In music, rhapsody is characterized by the absence of a fixed form and the loose connection of motifs and themes often taken from a profane realm. Literary montage, the method Benjamin uses in the *Arcades Project*, takes this principle to an extreme. Thus,

---

265 Benjamin, *Arcades Project*, 173.
266 Benjamin, *Arcades Project*, 389.
267 Benjamin, *Arcades Project*, 884.
268 Benjamin, *Arcades Project*, 389.
269 Benjamin, *Arcades Project*, 834.
270 Benjamin to Adorno, 31 May 1935, in *Complete Correspondence*, 88.
271 Benjamin to Adorno, 31 May 1935, in *Complete Correspondence*, 89.

the text gathers fragments – "rags, [...] refuse"²⁷² – from the nineteenth century in order to draw attention to and situate them in new relationships. The prime exponents of this method are the collector and the child. The collector has "withdrawn [the object] from [its] functional context" and takes no instrumental interest in it. Under his physiognomic gaze, the world, by means of this object, rearranges itself and invites unconventional interpretations; the perspective shifts from collecting to bricolage. The same process is at work among children. One of Benjamin's earliest notes for the *Arcades Project* reads, "Game in which children have to form a brief sentence out of given words. This game is seemingly played by the goods on display: binoculars and flower seeds, screws and musical scores, makeup and stuffed vipers, fur coats and revolvers."²⁷³ Tellingly, Benjamin describes this novel "assembly" of things/words as "construct[ing] an alarm clock"²⁷⁴ that serves the aim of awakening from enchantment to disenchantment.

Third, gesture: In the same note where Benjamin identifies his method as montage, he stresses that he does not want to describe or "*say* anything," only "show." The *Arcades Project* is to display a collection of texts. He intends to "purloin no valuables, appropriate no ingenious formulations," but simply display "the rags, the refuse."²⁷⁵ In other words, this means that the gestural method of the *Arcades Project* is based on the citation of passages: "This work has to develop to the highest height the art of citing without quotation marks."²⁷⁶ Citation represents an intensive involvement with the source (intensified by the fact that the cited materials often receive no commentary) as well as a sovereign intervention into its original context, especially when this is made unrecognizable ("without quotation marks").²⁷⁷ Citation performs a gesture that Benjamin discusses in relation to Brecht's epic theater:

> Interruption is one of the fundamental methods of all form-giving. [...] It is [...] the origin of the quotation. Quoting a text implies interrupting its context. It will be readily understood, therefore, that epic theatre, which depends on interruption, is quotable in a very specific sense.²⁷⁸

---

272 Benjamin, *Arcades Project*, 460.
273 Benjamin, *Arcades Project*, 540.
274 Benjamin, *Arcades Project*, 883.
275 Benjamin, *Arcades Project*, 460.
276 Benjamin, *Arcades Project*, 458.
277 Pethes deems this the destruction of destruction inasmuch as the trace of quotation is erased (Pethes, *Mnemographie*, 403).
278 Walter Benjamin, "What is Epic Theatre? [Second Version]," in *Understanding Brecht*, trans. Anna Bostock (London: Verso, 1998), 19. On interruption as destruction, see Menninghaus, "Wal-

Just as the epic theater suspends the situations in which characters find themselves, Benjamin interrupts his material by taking passages out of context and incorporating them as citations into the *Arcades Project*. He seeks out what has been overlooked ("the rags, the refuse") and shows them – in this sense his project proceeds gesturally. At the same time, as in the situations in Brecht's theater, the citations themselves become recognizable as gestures in which a receptive-productive approach to the nineteenth century has sedimented. With this procedure, Benjamin brings to light the "dream side" of the texts and the world from which they come, i.e., an interpretation is to be gained from them that had been previously obscured.

Benjamin therefore connects quotation with the hope of knowledge ("awakening of a not-yet-conscious knowledge of what has been"[279]). The essay on Kraus, which was written at the same time, hints at this too. Quotation is said to "summon [...] the word by its name" inasmuch as it "wrenches it destructively from its context" and "calls it back to its origin."[280] Here, the term origin is used in the same sense as in the Brecht essay, where it refers to the "real conditions" to which the "astonished" spectator is awakened by the alienation effect.

In the early *Arcades Project*, the destruction of textual context is attended by the hope of a dissolution of mythic timelessness, tied to a "now of recognizability"[281] that would facilitate the reader's insight into "what has been" and, at the same time, promote critical understanding of the present. For Benjamin, "writing history" means "citing history."[282] The bricolage of quotations follows the pattern of the "forceful impact"[283] of cinematic images that trains the viewer to become aware of and reflect on stimuli. Benjamin compares this pattern to the procedure of gesture in his discussion of the epic theater. Thus, the finger's signifying gesture is preserved and the 'showing is shown' to the extent that the citational quality of the citations is exhibited through their alienating composition. Just as Brecht juxtaposes situations, Benjamin places contrasting quotations side by side. Such bricolage does not create a new whole – a new historical order or definitive interpretation of history – but serves to highlight individual quotes and their reciprocal alienation, challenging the reader to think.

---

ter Benjamins Diskurs der Destruktion." On the poetics of destruction in *The Arcades Project*, see Pethes, *Mnemographie*, 391–437.
279 Benjamin, *Arcades Project*, 907.
280 Benjamin, "Karl Kraus," 2.2: 454.
281 Benjamin, *Arcades Project*, 486.
282 Benjamin, *Arcades Project*, 458. For a thorough discussion of the role of quotation in Benjamin, see Menke, *Sprachfiguren*, 371–393.
283 Benjamin, "What is Epic Theatre? [Second Version]," 21.

These final remarks indicate how central the figure of the child (with its procedures of cunning, destruction/bricolage, and gesture elaborated above) is for Benjamin's philosophy of history and thought as a whole, specifically as a model for Benjamin's approach to the nineteenth century. Here, immersion in what has been forms the precondition for detaching from it, and the fragmentation and temporary montage of its parts can only succeed on the basis of intimate knowledge. The two laws governing child's play – destruction and mimesis – proceed dialectically and determine the process of the materialist historian. Also at stake is a sovereignty that no longer stands in the mythic spell of what has been (as, for example, Benjamin observes of the Surrealists). Instead, it can be gained by means of an intensive passage through that past and its subsequent reflection.

If Benjamin's works present the child as a sublation of the 'primitive,' and if this dialectical turn is, in his view, missing from the efforts of the Collège de Sociologie to renew the sacred and mythical present,[284] then the same expectation holds for his vision for the materialist historian: he must proceed like the child. In this way, the *Arcades Project* may also be understood as an ethnology-in-reverse that seeks out a foreign perspective in order to defamiliarize one's own culture. Indeed, the epigraph for "Paris, the Capital of the Nineteenth Century," which outlines the book he never completed, is taken from a visitor to the city, Nguyen-Trong-Hiep. Describing the French metropolis, the Vietnamese traveler observes with curiosity, "[o]ne goes for a walk; the *grandes dames* go for a walk; behind them stroll the *petites dames*."[285] European customs seem strange and incomprehensible to foreign eyes. The enchantment experienced through the childlike gaze is joined by the alienation of the everyday, which challenges readers not to dream, but to interpret their own culture.

---

[284] On Benjamin's relationship to this group, see Moebius, *Die Zauberlehrlinge*, 370–375.
[285] Benjamin, *Arcades Project*, 3.

Epilogue

This book has traced the scientific, aesthetic and literary discourse on 'primitive thinking,' which exercised a decisive influence on how human beings, history, culture, and art were conceived in the early twentieth century. To this end, I have examined the paradigm of the 'primitive' in theories of art, language, and metaphor, as well as in texts representing the human sciences: ethnology, developmental psychology, and psychopathology. The concept of 'primitive thinking' served not only to buttress each field's claims of scientific validity but also to shed light on putative origins by pairing indigenous cultures with the figure of the child and the mentally ill. All three functioned as figurations of humanity's first beginnings, representing different aspects of 'primitive thinking,' e.g., myth and community, play and illusion, delusion and protest. Aesthetic theories of the period took up these aspects to develop their own accounts of the essence and purpose of art. In particular, art scholars as well as artists believed they could solve the riddle of creativity by understanding its workings as a survival of 'primitive thinking'.

At the same time, this book has historicized and contextualized these theories, revealing the questions and processes by which they were governed. At frequent junctures and on multiple registers, their proximity to literary operations comes to light. This is why in this study the 'primitive' is defined not only as a paradigm and figure of thought but also as a scientific reverie or *poème*. In fact, the deconstruction of the concept of 'primitive thinking' began within the field of ethnology itself, notably by Claude Lévi-Strauss, who both critically traced the emergence of primitivist discourse and at the same time perpetuated it in his praise of the "savage mind."[1] The convergence of scientific texts with literature, however, can also be read as a resistance to their usual form and methodology and, in this respect, as an opportunity to *create* the alterity postulated in the texts, yet – as a result of imperialist, pedagogical, or psychiatric colonialization – hardly still in existence at the time or at any rate only marginally appreciated by many scholars. While this resistance holds only in part for texts

---

[1] See, for example, Lévi-Strauss, *The Elementary Structures of Kinship, Totemism, The Savage Mind*, as well as Lienhardt, "Modes of Thought." Regarding the persistence of the paradigm of the 'primitive' in the work even of its critics, see Hsu, "Rethinking the Concept 'Primitive'"; Fabian, *Time and the Other*; Kuper, *The Invention of Primitive Society*; and Derrida on Lévi-Strauss, *Of Grammatology*, trans. Gayatri Chakravorty Spivak (Baltimore: Johns Hopkins University Press, 1997), 112–115; for a critique of Derrida's critique, cf. Därmann, *Fremde Monde der Vernunft*, chapter 8.

from the human sciences and art studies, it thoroughly applies to the literary turn explicitly carried out in writings by erstwhile scientists such as Gottfried Benn and Robert Musil.

The discourse of 'primitive thinking' stands in the context of an ambivalent search for origins that seeks to secure its own beginnings in the 'primitive,' but nevertheless feels compelled to demarcate itself from the latter to stabilize its own identity. At the same time, this discourse also expresses a longing for the archaic, in which the 'primitive' functions as a utopian alternative to modern society. Yet, as I have shown above, the 'primitive' served not only as a story of origins and critical utopia but also provided an image of the present, a signature of a "disenchanted" modernity that, in Max Weber's phrase, was nevertheless experienced mythically, exemplifying how, as Alfred Döblin puts it, "Prometheanism" turns back into "primitivism."[2] In view of this diagnosis, writers such as Robert Musil and Walter Benjamin in their treatments of 'primitive thinking' sought to sketch the concept of a critical re-enchantment: instead of ferrying readers off into enchanted worlds of the past, they placed them in a skeptical distance from the "other conditions" (Musil) and "féeries" (Benjamin) of modernity.

This is also what Theodor W. Adorno expected from Benjamin's *Arcades Project*, which he once described as the most important philosophical undertaking of the epoch.[3] Adorno was finely attuned to the "archaizing tendency" of associating myth with a yearning for the enchanted world of nineteenth-century commodities and a classless society of prehistory. (This occurs in the works of Ernst Bloch, as shown in the Introduction.) Instead, he holds, myth must be exposed as the "alienated character of the commodity itself" and repeatedly reminds Benjamin in his letters of the 1930s of his own (i.e., Benjamin's) conviction that the 'primitive archaic' is indeed the condition of the newest, thus comprising "objective constellations in which the condition of society finds itself represented."[4] Indeed, Benjamin's early notes on the *Arcades Project* make it clear that he immersed himself in the nineteenth-century dreamworld precisely in order to awaken from it and thus bring about "the dissolution of 'mythology' into the space of history."[5] At the same time, Adorno recognized his correspondent's desire to salvage procedures attributed to figurations of the 'primitive' (mimetic assimilation, for instance) and put them in the service of demystifica-

---

[2] Alfred Döblin, "Prometheus und das Primitive (1938)," in *Schriften zur Politik und Gesellschaft* (Freiburg: Walter, 1972), 364. Cf. Introduction, 18–19.
[3] Adorno to Benjamin, 20 May 1935, in *Complete Correspondence*, 84.
[4] Adorno to Benjamin, 2–4 August 1935, in *Complete Correspondence*, 110.
[5] Benjamin, *Arcades Project*, 458.

tion.⁶ Benjamin famously speaks of the "axe of reason" with which the nineteenth century is to be "cleared of the undergrowth of delusion and myth."⁷

Taking distance from the "féeries" of modernity in this way begins with textual operations. Literary primitivism turned features considered central to 'primitive thinking' into formal innovations like associative narration, literal treatments of metaphor, or figures of participation. This holds for authors who approach the phantasm of the 'primitive' in an affirmative manner, such as Robert Müller and Gottfried Benn, as well as for writers who, despite their fascination, are critical of it. In addition, the critical impulse also generated its own innovative methods of writing. Examples include Robert Musil's essayistic style, which interrupts linguistic mimesis of the 'primitive,' and Walter Benjamin's use of bricolage and gesture in his montages of citations.

The literary texts treated in this book thus never resort to 'primitive thinking' as mere imitation. In the best case, literature under the sign of the 'primitive' means not just adaptation, but also critical engagement with 'primitive thinking' and its discourse. The only re-enchantment these texts promise is one from which, as Benjamin demanded, it is necessary to "awaken."

---

6 Adorno and Horkheimer also pursue this angle in *Dialectic of Enlightenment* (trans. Edmund Jephcott [Stanford: Stanford University Press, 2002]) when they insist on the proximity between word and thing and subject and object in the context of magic (7). The work of art inherits this dynamic and, in "renunciation of external effects," shows "the appearance of the whole in the particular" and potentially affords insight superior to "conceptual knowledge" (14).
7 Benjamin, *Arcades Project*, 456–457.

Bibliography

Achebe, Chinua. "An Image of Africa: Racism in Conrad's *Heart of Darkness*." In *Heart of Darkness: An Authoritative Text, Backgrounds and Sources, Criticism*. Ed. Robert Kimbrough. New York: Norton, 1988. 251–261.
Adler, Guido. "Antrittsvorlesung an der Universität Wien, Musik und Musikwissenschaft." *Jahrbuch der Musikbibliothek Peters* 5 (1898): 28–39.
Adorno, Theodor W., and Walter Benjamin. *The Complete Correspondence, 1928–1940*. Translated by Nicholas Walker. Ed. Henri Lonitz. Cambridge, MA: Harvard University Press, 2000. [*Briefwechsel 1928–1940*. Ed. Henri Lonitz. Frankfurt am Main: Suhrkamp, 1994.]
Adorno, Theodor W., and Max Horkheimer. *Dialectic of Enlightenment*. Translated by Edmund Jephcott. Stanford: Stanford University Press, 2002. [*Dialektik der Aufklärung*. In Adorno, vol. 3 of *Gesammelte Schriften*. Vol. 3. Ed. Rolf Tiedemann. Frankfurt am Main: Suhrkamp, 1997.]
Aebli, Hans. "Einführung." In Jean Piaget, *Das Weltbild des Kindes*. Stuttgart: Klett-Cotta, 1978. 8–12.
Agamben, Giorgio. *Infancy and History: Essays on the Destruction of Experience*. Translated by Liz Heron. London: Verso, 1993.
Agamben, Giorgio. "Noten zur Geste." In *Postmoderne und Politik*. Ed. Jutta Georg-Lauer, 97–108. Tübingen: edition discord, 1992.
Agamben, Giorgio. "Notes on Gesture." In *Means Without End*. Translated by Vicenzo Binetti and Cesare Catarino. Minneapolis: University of Minnesota Press, 2000. 49–62.
Albers, Irene. "Mimesis and Alterity: Michel Leiris's Ethnography and Poetics of Spirit Possession." *French Studies* 62.3 (2008): 271–289.
Albers, Irene. *Der diskrete Charme der Anthropologie. Michel Leiris' ethnologische Poetik*. Konstanz: UVK, 2018.
Albers, Irene, and Stephan Moebius. "Nachwort." In Denis Hollier, *Das Collège de Sociologie, 1937–1939*. Frankfurt am Main: Suhrkamp, 2012. 757–828.
Anacker, Regine. *Aspekte einer Anthropologie der Kunst in Gottfried Benns Werk*. Würzburg: Königshausen + Neumann, 2004.
Anz, Thomas. *Literatur der Existenz. Literarische Psychopathographie und ihre soziale Bedeutung im Frühexpressionismus*. Stuttgart: Metzler, 1977.
Anz, Thomas. *Literatur und Lust. Glück und Unglück beim Lesen*. Munich: DTV, 2002.
Anz, Thomas. "Schizophrenie als epochale Symptomatik. Pathologie und Poetologie um 1910." In *Epochen/Krankheiten. Konstellationen von Literatur und Pathologie*. Ed. Frank Degler and Christian Kohlroß. St. Ingbert: Röhrig Universitätsverlag, 2006. 113–130.
Asman, Carrie. "Die Rückbindung des Zeichens an den Körper. Benjamins Begriff der Geste in der Vermittlung von Brecht und Kafka." *The Other Brecht II. The Brecht-Yearbook* 18 (1993). 105–119.
Aue, Maximilian. "'Pandämonium verschiedener Formen des Wahns?' Vom Wahnsinn und seinen Grenzen in Musils *Mann ohne Eigenschaften*." In *Literatur und Kultur im Österreich der zwanziger Jahre. Vorschläge zu einem transdisziplinären Epochenprofil*. Ed. Primus-Heinz Kucher. Bielefeld: Aisthesis, 2007. 135–144.

Auerbach, Erich. "Figura." In Auerbach, *Scenes from the Drama of European Literature*. Manchester: Manchester University Press, 1984. 11–71. ["Figura." In Auerbach, *Neue Dantestudien*. Istanbul: Basimeva, 1944. 11–71]

Bachelard, Gaston. *The Psychoanalysis of Fire*. Translated by Alan C. M. Ross. Boston: Beacon, 1968. [*La psychanalyse du feu*. Paris: Gallimard, 2000.]

Bachelard, Gaston. "Scientific Objectivity and Psychoanalysis." In *The Formation of the Scientific Mind*. Translated by Mary McAllester Jones. Manchester: Clinamen, 2002. 237–250.

Bachelard, Gaston. *The Formation of the Scientific Mind*. Manchester: Clinamen, 2002. [*La formation de l'esprit scientifique*. Paris: J. Vrin, 1996.]

Balint, Michael. *Therapeutische Aspekte der Regression. Die Theorie der Grundstörung*. Stuttgart: Klett-Cotta, 1970.

Basu, Priyanka. "Die 'Anfänge' der Kunst und die Kunst der Naturvölker: Kunstwissenschaft um 1900." In *Image match. Visueller Transfer, "Imagescapes" und Intervisualität in globalen Bildkulturen*. Ed. Martina Baleva, Ingeborg Reichle, and Oliver Lerone Schultz. Paderborn: Fink, 2012. 109–129.

Bataille, Georges. *Visions of Excess: Selected Writings, 1927–1939*. Ed. Allan Stoekl. Minneapolis: University of Minnesota Press, 1985.

Beard, Philip H. "Clarisse und Moosbrugger vs. Ulrich/Agathe: Der 'andere Zustand' aus neuer Sicht." *Modern Austrian Literature* 9.3 (1976): 114–130.

Behrens, Heike, and Werner Deutsch. "Die Tagebücher von Clara und William Stern." In *Theorien und Methoden psychologiegeschichtlicher Forschung*. Ed. Helmut E. Lück and Rudolf Miller. Göttingen: Hogrefe, 1991. 66–76.

Beil, Ulrich Johannes, Michael Gamper, and Karl Wagner, eds. *Medien, Technik, Wissenschaft: Wissensübertragung bei Robert Musil und in seiner Zeit*. Zurich: Chronos, 2011.

Benjamin, Walter. *Selected Writings*. Vol. 1, *1913–1926*. Ed. Marcus Bullock and Michael W. Jennings. Cambridge, MA: Harvard University Press, 1996.

Benjamin, Walter. "On Language as Such and the Language of Man." In Benjamin, *Selected Writings*, 1: 62–74. ["Über Sprache überhaupt und über die Sprache des Menschen." In Benjamin, *Gesammelte Schriften*, 2.1: 140–156.]

Benjamin, Walter. "The Task of the Translator." In Benjamin, *Selected Writings*, 1: 257–263. ["Die Aufgabe des Übersetzers." In Benjamin, *Gesammelte Schriften*, 4.1: 9–22.]

Benjamin, Walter. "Old Forgotten Children's Books." In Benjamin, *Selected Writings*, 1: 406–413. ["Alte vergessene Kinderbücher." In Benjamin, *Gesammelte Schriften*, 3: 14–22.]

Benjamin, Walter. "Reflections on Humboldt." In Benjamin, *Selected Writings*, 1: 424–425. ["Reflexionen zu Humboldt." In Benjamin, *Gesammelte Schriften*, 4: 26.]

Benjamin, Walter. "A Glimpse into the World of Children's Books." In Benjamin, *Selected Writings*, 1: 435–443. ["Aussicht ins Kinderbuch." In Benjamin, *Gesammelte Schriften*, 4.2: 609–614.]

Benjamin, Walter. "One-Way Street." In Benjamin, *Selected Writings*, 1: 444–488. ["Einbahnstrasse." In Benjamin, *Gesammelte Schriften*, 4.1: 85–150.]

Benjamin, Walter. "What is Epic Theatre? [First Version]." In *Understanding Brecht*. Translated by Anna Bostock. London: Verso, 1998. 1–14. ["Was ist das epische Theater?" (1) In Benjamin, *Gesammelte Schriften*, 2.2: 519–531.]

Benjamin, Walter. "What is Epic Theatre? [Second Version]." In *Understanding Brecht*. Translated by Anna Bostock. London: Verso, 1998. 15–22. ["Was ist das epische Theater?" (2) In Benjamin, *Gesammelte Schriften*, 2.2: 532–538.]
Benjamin, Walter. *The Arcades Project*. Translated by Howard Eiland and Kevin McLaughlin. Cambridge, MA: Harvard University Press, 1999. [*Das Passagen-Werk*. In Bejamin, *Gesammelte Schriften*, vol. 5.1 and 5.2.]
Benjamin, Walter. *Selected Writings*. Vol. 2, Part 1, *1927–1930*. Ed. Michael W. Jennings, Howard Eiland, and Gary Smith. Cambridge, MA: Harvard University Press, 1999.
Benjamin, Walter. "Old Toys." In Benjamin, *Selected Writings*, 2.1: 98–102. ["Altes Spielzeug." In Benjamin, *Gesammelte Schriften*, 4.1: 511–514.]
Benjamin, Walter. "Toys and Play." In Benjamin, *Selected Writings*, 2.1: 117–121. ["Spielzeug und Spielen." In Benjamin, *Gesammelte Schriften*, 3: 127–131.]
Benjamin, Walter. "Kulturgeschichte des Spielzeugs." In Benjamin, *Gesammelte Schriften*, 3: 113–116.
Benjamin, Walter. "Verstreute Notizen Juni bis Oktober 1928." In Benjamin, *Gesammelte Schriften*, 6: 415–417.
Benjamin, Walter. "Der Mensch in der Handschrift." In Benjamin, *Gesammelte Schriften*, 3: 135–138.
Benjamin, Walter. "Program for a Proletarian Children's Theater." In Benjamin, *Selected Writings*, 2.1: 201–206. ["Programm eines proletarischen Kindertheaters." In Benjamin, *Gesammelte Schriften*, 2.2: 763–768.]
Benjamin, Walter. "Children's Literature." In Benjamin, *Selected Writings*, 2.1: 250–256. ["Kinderliteratur." In *Gesammelte Schriften*, 7.1: 250–256.]
Benjamin, Walter. "Robert Walser." In Benjamin, *Selected Writings*, 2.1: 257–261. ["Robert Walser." In Benjamin, *Gesammelte Schriften*, 2.1: 324–327.]
Benjamin, Walter. *Selected Writings*. Vol. 2, Part 2, *1931–1934*. Ed. Michael W. Jennings, Howard Eiland, and Gary Smith. Cambridge, MA: Harvard University Press, 1999. 486–493.
Benjamin, Walter. "Grünende Anfangsgründe." In Benjamin, *Gesammelte Schriften*, 3: 311–314.
Benjamin, Walter. "Karl Kraus." In Benjamin, *Selected Writings*, 2.2: 433–458. ["Karl Kraus." In Benjamin, *Gesammelte Schriften*, 2.1: 334–367.]
Benjamin, Walter. "Unpacking my Library." In Benjamin, *Selected Writings*, 2.2: 486–493. ["Ich packe meine Bibliothek aus." In Benjamin, *Gesammelte Schriften*, 4: 389–390.]
Benjamin, Walter. "The Destructive Character." In Benjamin, *Selected Writings*, 2.2: 541–542. ["Der destruktive Charakter." In Benjamin, *Gesammelte Schriften*, 4.1: 396–397.]
Benjamin, Walter. "On Astrology." In Benjamin, *Selected Writings*, 2.2: 684–685. ["Zur Astrologie." In Benjamin, *Gesammelte Schriften*, 6: 192–193.]
Benjamin, Walter. "Doctrine of the Similar." In Benjamin, *Selected Writings*, 2.2: 694–699. ["Lehre vom Ähnlichen." In Benjamin, *Gesammelte Schriften*, 2.1: 204–209.]
Benjamin, Walter. "Experience and Poverty." In Benjamin, *Selected Writings*, 2.2: 731–736. ["Erfahrung und Armut." In Benjamin, *Gesammelte Schriften*, 2.1: 213–218.]
Benjamin, Walter. "Franz Kafka." In Benjamin, *Selected Writings*, 2.2: 794–818. ["Franz Kafka." In Benjamin, *Gesammelte Schriften*, 2.1: 409–437.]
Benjamin, Walter. *Selected Writings*. Vol. 3, *1935–1938*. Ed. Howard Eiland and Michael W. Jennings. Cambridge, MA: Harvard University Press, 2002.

Benjamin, Walter. "Problems in the Sociology of Language: An Overview." In Benjamin, *Selected Writings*, 3: 68–93. ["Probleme der Sprachsoziologie." In Benjamin, *Gesammelte Schriften*, 3: 452–479.]
Benjamin, Walter. "The Storyteller: Observations on the Works of Nikolai Leskov." In Benjamin, *Selected Writings*, 3: 143–166. ["Der Erzähler." In Benjamin, *Gesammelte Schriften*, 2.1: 438–464.]
Benjamin, Walter. "Berlin Childhood around 1900." In Benjamin, *Selected Writings*, 3: 344–413. ["Berliner Kindheit um Neunzehnhundert." In Benjamin, *Gesammelte Schriften*, 4.1: 237–306.]
Benjamin, Walter. "Eine kommunistische Pädagogik." In Benjamin, *Gesammelte Schriften*, 3: 206–208.
Benjamin, Walter. "Mimetic Faculty." In Benjamin, *Selected Writings*, 2.2: 720–722. ["Über das mimetische Vermögen." In Benjamin, *Gesammelte Schriften*, 2.1: 210–212.]
Benjamin, Walter. "Theses on the Philosophy of History." In *Illuminations*. Translated by Harry Zohn. Ed. Hannah Arendt. New York: Schocken Books, 1969. 253–264. ["Über den Begriff der Geschichte." In Benjamin, *Gesammelte Schriften*, 1.2: 691–706.]
Benjamin, Walter. *Gesammelte Schriften*. Ed. Rolf Tiedemann and Hermann Schweppenhäuser. 14 vols. Frankfurt am Main: Suhrkamp, 1991.
Benjamin, Walter. "Kolonialpädagogik." In Benjamin, *Gesammelte Schriften*, 3: 272–274.
Benjamin, Walter. "Lob der Puppe." In Benjamin, *Gesammelte Schriften*, 3: 213–217.
Benjamin, Walter. "Pariser Passagen I." In Benjamin, *Gesammelte Schriften*, 5.2: 991–1040.
Benjamin, Walter. "Schema zur Anthropologie." In Benjamin, *Gesammelte Schriften*, 6: 64.
Benjamin, Walter. "Über das Grauen." In Benjamin, *Gesammelte Schriften*, 6: 75–76.
Benjamin, Walter. "Zur Graphologie." In Benjamin, *Gesammelte Schriften*, 6: 185.
Benjamin, Walter. *Gesammelte Briefe*. 6 vols. Ed. Christoph Gödde and Henri Lonitz. Frankfurt am Main: Suhrkamp, 1997.
Benjamin, Walter. *The Correspondence of Walter Benjamin, 1910–1940*. Ed. Gershom Scholem and Theodor W. Adorno. Translated by Manfred R. Jacobson and Evelyn M. Jabobson. Chicago: University of Chicago Press, 2012.
Benn, Gottfried. "Thalassal Regression." Translated by Edgar Lohner and Cid Corman, *Quarterly Review of Literature*, 7 (1953): 290–297. ["Regressiv [1927]." In Benn, *Sämtliche Werke*. Vol. I, *Gesammelte Gedichte*, 126.]
Benn, Gottfried. *Gesammelte Werke in vier Bänden (GW)*. Ed. Dieter Wellershoff. Wiesbaden: Limes, 1958–1994.
Benn, Gottfried. "Akademie-Rede." In *Gesammelte Werke in vier Bänden*. Vol. 1, *Essays, Reden, Vorträge*. 431–440.
Benn, Gottfried. *Gesammelte Werke in der Fassung der Erstdrucke (GWED)*. Ed. Brunde Hillebrand. Frankfurt am Main: Fischer Taschenbuch, 1982–1990.
Benn, Gottfried. *Sämtliche Werke. Stuttgarter Ausgabe (SW)*. Ed. Gerhard Schuster (vols. 1–5) and Holger Hof (vols. 6–7.2). Stuttgart: Klett-Cotta, 1986–2003.
Benn, Gottfried. "Problematik des Dichterischen." In *Sämtliche Werke*. Vol. 3, *Prosa (1910–1932)*. 232–247.
Benn, Gottfried. "Der Aufbau der Persönlichkeit. Grundriss einer Geologie des Ich." In *Sämtliche Werke*. Vol. 3, *Prosa (1910–1932)*. 263–277.
Benn, Gottfried. "Dorische Welt." In *Sämtliche Werke*. Vol. 4, *Prosa 2 (1933–1945)*. 124–153.

Benn, Gottfried. "Unter der Großhirnrinde. Briefe vom Meer." In *Sämtliche Werke. Stuttgarter Ausgabe.* Vol. 7.1, *Szenen und andere Schriften.* Ed. Holger Hof. Stuttgart: Klett-Cotta, 2003. 354–363.
Benn, Gottfried. *Künstlerische Prosa. In der Fassung der Sämtlichen Werke – Stuttgarter Ausgabe (KT).* Ed. Holger Hof. Stuttgart: Klett-Cotta, 2006.
Bergengruen, Maximilian. "Moosbrugger oder die Möglichkeiten der Paranoia. Psychiatrie und Mystik in Musils 'Der Mann ohne Eigenschaften.'" *Zeitschrift für deutsche Philologie* 135.4 (2006): 545–568.
Biese, Alfred. *Die Philosophie des Metaphorischen. In Grundlinien dargestellt.* Leipzig: L. Voss, 1893.
Bilang, Karla. *Bild und Gegenbild. Das Ursprüngliche in der Kunst des 20. Jahrhunderts.* Stuttgart: Kohlhammer, 1990.
Binswanger, Ludwig. "Über Phänomenologie." *Zeitschrift für die gesamte Neurologie und Psychiatrie* 82 (1923): 10–45.
Bleuler, Eugen. "Das autistische Denken." *Jahrbuch für psychoanalytische und psychopathologische Forschungen* 4, no. 1 (1912): 1–39.
Bleuler, Eugen. *Textbook of Psychiatry.* Translated by A. A. Brill. New York: Macmillan, 1924. [*Lehrbuch der Psychiatrie.* Berlin: Springer, 1923.]
Bloch, Ernst. *Werkausgabe.* Vol. 9, *Literarische Aufsätze.* Frankfurt am Main: Suhrkamp, 1965.
Bloch, Ernst. "Die Felstaube, das Neandertal und der wirkliche Mensch [1929]." In Bloch, *Werkausgabe,* 9: 462–469.
Bloch, Ernst. "Loch Ness, die Seeschlange und Dacqués Urweltsage [1934]." In Bloch, *Werkausgabe,* 9: 470–475.
Bloch, Ernst. *Heritage of Our Times.* Translated by Neville and Stephen Plaice. Berkeley, CA: University of California Press, 1991. [*Erbschaft dieser Zeit.* Frankfurt am Main: Suhrkamp, 1992.]
Bloch, Ernst. "Songs of Remoteness." In Bloch, *Heritage of Our Times,* 179–183.
Bloch, Ernst. "Summary Transition: Non-Contemporaneity and Obligation to its Dialectic." In Bloch, *Heritage of Our Times,* 97–148.
Bloch, Ernst. "Final Form: Romantic Hook-Formation." In Bloch, *Heritage of Our Times,* 149–185.
Bloch, Ernst. "Philosophies of Unrest, Process, Dionysus." In Bloch, *Heritage of Our Times,* 299–331.
Blome, Eva. *Reinheit und Vermischung. Literarisch-kulturelle Entwürfe von "Rasse" und Sexualität (1900–1930).* Cologne: Böhlau, 2011.
Blumenberg, Hans. *Paradigms for a Metaphorology.* Translated by Robert Savage. Ithaca: Cornell University Press, 2010. [*Paradigmen zu einer Metaphorologie.* Frankfurt am Main: Suhrkamp, 1998.]
Boas, Franz. *The Mind of Primitive Man.* New York: MacMillan, 1961.
Boas, George. *The Cult of Childhood.* London: Warburg, 1966.
Böhme, Hartmut. *Fetischismus und Kultur. Eine andere Theorie der Moderne.* Hamburg: Reinbek, 2006.
Bohrer, Karl Heinz. "Nachwort." In André Breton. *Nadja.* Translated by Bernd Schwibs. Frankfurt am Main: Suhrkamp, 2002. 141–155.
Bonacchi, Silvia. *Die Gestalt der Dichtung. Der Einfluss der Gestalttheorie auf das Werk Robert Musils.* Bern: Peter Lang, 1998.

Borges, Jorge Luis. "Die Fahrt ins Land ohne Tod, de Alfred Döblin [1938]." In *Textos cautivos. Ensayos y reseñas en "El Hogar."* Barcelona: Tusquets, 1986. 196–197.
Bormuth, Matthias. "Max Weber im Lichte Nietzsches." In *Wissenschaft als Beruf: Mit zeitgenössischen Resonanzen und einem Gespräch mit Dieter Henrich.* Berlin: Matthes and Seitz, 2018. 7–36.
Bovet, Pierre. *The Fighting Instinct.* Translated by J.Y.Y. Greig. New York: Dodd, Mead and Company, 1923. [*L'instinct combatif: psychologie, éducation.* Neuchâtel: Delachaux et Niestlé, 1917.]
Bracken, Christopher. *Magical Criticism: The Recourse of Savage Philosophy.* Chicago: University of Chicago Press, 2007.
Brandstetter, Gabriele, and Sybille Peters, eds. *De figura: Rhetorik – Bewegung – Gestalt.* Munich: Fink, 2002.
Braun, Wilhelm. "Moosbrugger Dances." *The Germanic Review* 35 (1960): 214–230.
Brehm, Alfred Edmund. *Brehm's Life of Animals.* Chicago: Marquis, 1895. [*Illustrirtes Thierleben.* Hildburghausen: Verlag des Bibliogr. Instituts, 1864, 1869.]
Breton, André. "Manifesto of Surrealism." (1924). Translated by Richard Seaver and Helen Lane. In André Breton. *Manifestos of Surrealism.* Ann Arbor: University of Michigan Press, 1990. 1–48.
Breton, André. *Nadja.* Translated by Richard Howard. New York: Grove, 1960. [*Nadja.* Paris: Gallimard, 1928.]
Breuer, Josef, and Sigmund Freud. *Studies on Hysteria.* New York, Washington: Nervous and Mental Disease Publishing, 1936. [*Studien über Hysterie.* Frankfurt am Main: Fischer Taschenbuch, 1991.]
Bronfen, Elisabeth. *Over Her Dead Body: Death, Femininity and the Aesthetic.* Manchester: Manchester University Press, 1992.
Brüggemann, Heinz. *Walter Benjamin über Spiel, Farbe und Phantasie.* Würzburg: Königshausen + Neumann, 2007.
Bücher, Karl. "Arbeit und Rhythmus." In *Abhandlungen der philologisch-historischen Classe der königlich sächsischen Gesellschaft der Wissenschaften* 17 (1897): 1–130.
Bühler, Charlotte. *Das Märchen und die Phantasie des Kindes.* Leipzig: Barth, 1918.
Bühler, Karl. *The Mental Development of the Child: A Summary of Modern Psychological Theory.* Translated by Oscar Oeser. London: Routledge, 2002. [*Die geistige Entwicklung des Kindes.* Jena: Gustav Fischer, 1921.]
Bühler, Karl. *Theory of Language: The Representational Function of Language.* Translated by Donald Fraser Goodwin. Amsterdam: John Benjamins, 2011. [*Sprachtheorie. Die Darstellungsfunktion von Sprache.* Stuttgart: Gustav Fischer, 1965 (1934).]
Büssgen, Antje. *Glaubensverlust und Kunstautonomie. Über die ästhetische Erziehung des Menschen bei Friedrich Schiller und Gottfried Benn.* Heidelberg: Winter, 2006.
Burk, Karin. *Kindertheater als Möglichkeitsraum. Untersuchungen zu Walter Benjamins "Programm eines proletarischen Kindertheaters."* Bielefeld: Aisthesis, 2015.
Buschendorf, Bernhard. "Zur Begründung der Kulturwissenschaft. Der Symbolbegriff bei Friedrich Theodor Vischer, Aby Warburg und Edgar Wind." In *Edgar Wind. Kunsthistoriker und Philosoph.* Ed. Horst Bredekamp, Bernhard Buschendorf, Freia Hartung, and John Michael Krois. Berlin: De Gruyter, 1998. 227–248.
Caillois, Roger. *Man, Play, and Games.* Translated by Mayer Barash. Champaign, IL: University of Illinois Press, 2001. [*Les jeux et les hommes.* Paris: Gallimard, 1958.]

Cassirer, Ernst. *Language and Myth.* Translated by Susanne K. Langer. New York: Dover, 1953.
Cassirer, Ernst. *Mythical Thought.* Vol. 2, *The Philosophy of Symbolic Forms.* Translated by Ralph Manheim. New Haven, CT: Yale University Press, 1955. [*Philosophie der symbolischen Formen. Zweiter Teil: Das mythische Denken.* Hamburg: Meiner, 2002.]
Cavanaugh, John C. "Cognitive Developmental Psychology before Preyer: Biographical and Educational Records." In *Contributions to a History of Developmental Psychology. International William T. Preyer Symposium.* Ed. Georg Eckardt, Wolfgang G. Bringmann and Lothar Sprung. Berlin: Mouton, 1985. 187–208.
Cheng, Joyce. "Primitivismen." In *Neolithische Kindheit. Kunst in einer falschen Gegenwart.* Ed. Anselm Franke and Tom Holert. Zurich: diaphanes, 2018. 185–198.
Clifford, James. "Introduction: Partial Truths." In *Writing Culture: The Poetics and Politics of Ethnography.* Ed. Clifford, James and George E. Marcus. Berkeley: University of California Press, 1986. 1–26.
Clifford, James. *The Predicament of Culture. Twentieth-Century Ethnography, Literature, and Art.* Cambridge, MA: Harvard University Press, 1988.
Clifford, James. "On Ethnographic Authority." In Clifford, *The Predicament of Culture,* 21–54.
Clifford, James. "On Ethnographic Self-Fashioning: Conrad and Malinowski." In Clifford, *The Predicament of Culture,* 92–113.
Clifford, James, and George E. Marcus, eds. *Writing Culture: The Poetics and Politics of Ethnography.* Berkeley: University of California Press, 1986.
Cohen, Hermann. *Die dichterische Phantasie und der Mechanismus des Bewusstseins.* Berlin: Ferd. Dümmler's Verlagsbuchhandlung, 1869.
Conrad, Joseph. *Heart of Darkness.* Ed. Owen Knowles and Allan H. Simmons. Cambridge: Cambridge University Press, 2018.
Cooper, Frederick. *Colonialism in Question: Theory, Knowledge, History.* Berkeley: University of California Press, 2005.
Crapanzano, Vincent. "The Moment of Prestidigitation: Magic, Illusion, and Mana in the Thought of Emile Durkheim and Marcel Mauss." In *Prehistories of the Future: The Primitivist Project and the Culture of Modernism.* Ed. Elazar Barkan and Ronald Bush. Stanford: Stanford University Press, 1995. 95–113.
Dacqué, Edgar. *Urwelt, Sage und Menschheit. Eine naturhistorisch-metaphysische Studie.* Munich: De Gruyter, 1924.
Därmann, Iris. *Fremde Monde der Vernunft. Die ethnologische Provokation der Philosophie.* Paderborn: Fink, 2005.
de Man, Paul. "Aesthetic Formalization in Kleist." In *The Rhetoric of Romanticism.* New York: Columbia University Press, 1984. 263–290.
Derrida, Jacques. *Of Grammatology.* Translated by Gayatri Chakravorty Spivak. Baltimore: Johns Hopkins University Press, 1997. [*De la grammatologie.* Paris: Éditions de Minuit, 1967.]
Dietrich, Stephan. *Poetik der Paradoxie. Zu Robert Müllers fiktionaler Prosa.* Siegen: Carl Böschen, 1997.
Dilthey, Wilhelm. *The Imagination of the Poet: Elements for a Poetics.* In *Selected Works.* Vol. 5, *Poetry and Experience.* Ed. Rudolf A. Makkreel and Frithjof Rodi. Princeton, NJ: Princeton University Press, 1985. 29–173. ["Die Einbildungskraft des Dichters. Bausteine für eine Poetik." In *Die Geistige Welt. Einleitung in die Philosophie des Lebens.* In *Gesammelte Werke.* Vol. 6. Ed. Karlfried Gründer. Leipzig: Teubner, 1924. 103–241.]

Döblin, Alfred. *Amazonas. Romantrilogie*. Ed. Werner Stauffacher. Munich: Deutscher Taschenbuch Verlag, 1991.
Döblin, Alfred. "Prometheus und das Primitive (1938)." In *Schriften zur Politik und Gesellschaft*. Freiburg: Walter, 1972. 346–367.
Durkheim, Emile. *The Elementary Forms of the Religious Life*. Translated by Joseph Ward Swain. London: Allen & Unwin, 1915. [*Les formes élémentaires de la vie religieuse*. Paris: Alcan, 1912.]
Durkheim, Emile, and Marcel Mauss. *Primitive Classification*. Translated by Rodney Needham. London: Cohen & West, 1963. ["De quelques formes primitives de classification." In *Année sociologique* (1901–1903).]
Eckardt, Georg. "Preyer's Road to Child Psychology." In Eckardt et al., *Contributions to a History of Developmental Psychology: International William T. Preyer Symposium*. Berlin: Mouton, 1985. 177–186.
Eckardt, Georg, Wolfgang G. Bringmann, and Lothar Sprung, eds. *Contributions to a History of Developmental Psychology: International William T. Preyer Symposium*. Berlin: Mouton, 1985.
Einstein, Carl. *Negro Sculpture*. Translated by Patrick Healy. Amsterdam: November Editions, 2016. [*Negerplastik*. Leipzig: Verlag der weissen Bücher, 1915.]
Einstein, Carl. *Bebuquin, or the Dilettantes of the Miracle*. Amsterdam: November Editions, 2017. [*Bebuquin oder die Dilettanten des Wunders*. Berlin: Die Aktion, 1912.]
Einstein, Carl."Drei Negerlieder. Nachdichtung." *Die Aktion 6* (1916): Sp. 651 and *Die Aktion 7* (1917): Sp. 324.
Einstein, Carl. *Afrikanische Legenden*. Berlin: Rowohlt, 1925.
Einstein, Carl. "Berliner Vortrag über den Surrealismus." In *Carl Einstein. Existenz und Ästhetik. Einführung mit einem Anhang unveröffentlichter Nachlasstexte*. Ed. Sibylle Penkert. Wiesbaden: Steiner, 1970. 51–61.
Einstein, Carl. *Die Fabrikation der Fiktionen*. Ed. Sibylle Penkert. Reinbek bei Hamburg: Rowohlt, 1973.
Etherington, Ben. *Literary Primitivism*. Stanford: Stanford University Press, 2017.
Evans-Pritchard, Edward E. *Theories of Primitive Religion*. Oxford: At the Clarendon Press, 1965.
Fabian, Johannes. *Time and the Other: How Anthropology Makes Its Object*. New York: Columbia University Press, 2002.
Fanta, Walter. "Die Spur der Clarisse in Musils Nachlass." *Musil-Forum* 27 (2001/2002): 242–286.
Ferenczi, Sándor. *Thalassa. A Theory of Genitality*. Translated by Henry Alden Bunker. New York: Norton, 1968.
Fiedler, Konrad. "Ursprung der künstlerischen Tätigkeit." In *Schriften zur Kunst. Text nach der Ausgabe München 1913/14 mit weiteren Texten aus Zeitschriften und dem Nachlass*. 2 vols., 2nd ed. Munich: Fink, 1991. Vol. 1: 111–220.
Fischer, Bernhard. "'Stil' und 'Züchtung' – Gottfried Benns Kunsttheorie und das Jahr 1933." *Internationales Archiv für Sozialgeschichte der deutschen Literatur* 12.1 (1987): 190–212.
Fittler, Doris. *'Ein Kosmos der Ähnlichkeit.' Frühe und späte Mimesis bei Walter Benjamin*. Bielefeld: Aisthesis, 2005.
Flechsig, Paul. *Gehirn und Seele. Rede, gehalten am 31. October 1894 in der Universitätskirche zu Leipzig*. 2nd ed. Leipzig: Veit, 1896.

Fleck, Ludwik. *Genesis and Development of a Scientific Fact.* Chicago: University of Chicago Press, 1979.
Foucault, Michel. *The Order of Things: An Archaeology of the Human Sciences.* Translated by Alan Sheridan. New York: Vintage, 1994. [*Les mots et les choses.* Paris: Gallimard, 1966.]
France, Anatole. *The Garden of Epicurus.* London: Bodley Head, 1908. [*Le Jardin d'Épicure.* Paris: Calmann Lévy, 1895.]
Frank, Michael C. "Überlebsel. Das Primitive in Anthropologie und Evolutionstheorie des 19. Jahrhunderts." In *Literarischer Primitivismus.* Ed. Nicola Gess. Berlin: De Gruyter, 2013. 159–188.
Franke, Anselm, and Tom Holert, eds. *Neolithische Kindheit. Kunst in einer falschen Gegenwart, ca. 1930.* Zurich: HKW, 2018.
Franke, Anselm, and Tom Holert. "Einführung." In *Neolithische Kindheit. Kunst in einer falschen Gegenwart, ca. 1930.* Ed. Anselm Franke and Tom Holert. Zurich: HKW, 2018. 10–16.
Frazer, James George. *The Golden Bough: A Study in Magic and Religion.* New York: Penguin, 1996.
Frazer, James George. "Preface." In Bronislaw Malinowski. *Argonauts of the Western Pacific.* London: George Routledge & Sons, 1922. viii–xv.
Freud, Sigmund. *A Phylogenetic Fantasy: Overview of the Transference Neuroses.* Ed. Ilse Grubrich-Simitis. Cambridge, MA: Harvard University Press, 1987. [*Übersicht der Übertragungsneurosen. Ein bisher unbekanntes Manuskript.* Ed. Ilse Grubrich-Simitis. Frankfurt am Main: S. Fischer, 1985.]
Freud, Sigmund. *Introductory Lectures on Psychoanalysis.* Translated by James Strachey. New York: Norton, 1989. [*Studienausgabe.* Vol. 1, *Vorlesungen zur Einführung in die Psychoanalyse.* Frankfurt am Main: Fischer Taschenbuch Verlag, 2000.]
Freud, Sigmund. *The Interpretation of Dreams.* Translated by James Strachey. New York: Basic Books, 2010. [*Studienausgabe.* Vol. 2, *Die Traumdeutung.* Frankfurt am Main: Fischer Taschenbuch Verlag, 2000.]
Freud, Sigmund. *Beyond the Pleasure Principle.* Translated by James Strachey. New York: Norton, 1961. ["Jenseits des Lustprinzips." In *Studienausgabe.* Vol. 3, *Psychologie des Unbewussten.* Frankfurt am Main: Fischer Taschenbuch Verlag, 2000. 213–272.]
Freud, Sigmund. "Formulations on the Two Principles of Mental Functioning." Translated by James Strachey. In *The Freud Reader.* Ed. Peter Gay. New York: W.W Norton, 1989. 301–306. ["Formulierungen über die zwei Prinzipien des psychischen Geschehens." In *Studienausgabe.* Vol. 3, *Psychologie des Unbewussten.* Frankfurt am Main: Fischer Taschenbuch Verlag, 2000. 13–24.]
Freud, Sigmund. *Totem and Taboo: Some Points of Agreement between the Mental Lives of Savages and Neurotics.* Translated by James Strachey. London: Routledge, 2001. ["Totem und Tabu. Einige Übereinstimmungen im Seelenleben der Wilden und der Neurotiker." In *Studienausgabe.* Vol. 9, *Fragen der Gesellschaft. Ursprünge der Religion.* Frankfurt am Main: Fischer Taschenbuch Verlag, 2000. 287–444.]
Freud, Sigmund. "Creative Writing and Day-Dreaming." In *The Freud Reader.* Ed. Peter Gay. New York: Norton, 1995. 436–442. ["Der Dichter und das Phantasieren." In *Studienausgabe.* Vol. 10, *Bildende Kunst und Literatur.* Frankfurt am Main: Fischer Taschenbuch Verlag, 2000. 169–180.]

Frobenius, Leo. *Der Ursprung der Kultur*. Berlin: Bornträger, 1898.
Frobenius, Leo. *Kulturgeschichte Afrikas. Prolegomena zu einer historischen Gestaltlehre*. Zurich: Phaidon, 1933.
Fry, Roger. "Children's Drawings." *The Burlington Magazine for Connoisseurs* 30, no. 171 (1917): 225–227.
Gardian, Christoph. *Sprachvisionen. Poetik und Mediologie der inneren Bilder bei Robert Müller und Gottfried Benn*. Zurich: Chronos, 2014.
Gauguin, Paul. *Gauguin's Letters from the South Seas*. 1923. New York: Dover Publications, 1992.
Geertz, Armin W. "'Can We Move Beyond Primitivism?' On Recovering the Indigenes of Indigenous Religions in the Academic Study of Religion." In *Beyond Primitivism: Indigenous Religious Traditions and Modernity*. Ed. Jacob K. Olupona. New York, 2004. 37–70.
Geertz, Clifford C. "I-Witnessing. Malinowski's Children." In *Works and Lives: The Anthropologist as Author*. Stanford: Stanford University Press, 1988. 73–101.
Geertz, Clifford C. "Thick Description: Towards an Interpretative Theory of Culture." In *The Interpretation of Cultures*. New York: Basic Books, 2000. 3–32.
Geisenhanslüke, Achim. *Das Schibboleth der Psychoanalyse*. Bielefeld: transcript, 2008.
Geissler, Peter. *Mythos Regression*. Gießen: Psychosozial-Verlag, 2001.
Genette, Gérard. *Mimologics*. Translated by Thaïs E. Morgan. Lincoln: University of Nebraska Press, 1995. [*Mimologiques: voyage en Cratylie*. Paris: Seuil, 1976.]
Georg, Eugen. *Verschollene Kulturen. Das Menschheitserlebnis. Ablauf und Deutungsversuch*. Leipzig: Voigtländer, 1930.
Gerber, Gustav. *Sprache als Kunst*. Hildesheim: Olms, 1961.
Gess, Heinz. *Vom Faschismus zum Neuen Denken. C.G. Jungs Theorie im Wandel der Zeit*. Lüneburg: Klampen, 1994.
Gess, Nicola. "Musikalische Mörder. Krieg, Musik und Mord bei Hermann Hesse." In *Literatur und Musik in der klassischen Moderne*. Ed. Joachim Grage. Würzburg: Ergon, 2006. 189–206.
Gess, Nicola. "Kunst und Krieg. Zu Thomas Manns, Hermann Hesses und Ernst Blochs künstlerischer Verarbeitung des Ersten Weltkriegs." In *Imaginäre Welten im Widerstreit. Krieg und Geschichte in der Literatur seit 1900*. Ed. Lars Koch and Marianne Vogel. Würzburg: Königshausen + Neumann, 2007. 30–57.
Gess, Nicola. "Magisches Denken im Kinderspiel. Literatur und Entwicklungspsychologie im frühen 20. Jahrhundert." In *Literatur als Spiel. Evolutionsbiologische, ästhetische und pädagogische Aspekte. Beiträge zum Deutschen Germanistentag 2007*. Ed. Thomas Anz and Heinrich Kaulen. Berlin, New York: De Gruyter, 2009. 295–314.
Gess, Nicola. "'So ist damit der Blitz zur Schlange geworden.' Anthropologie und Metapherntheorie um 1900." *Deutsche Vierteljahresschrift für Literaturwissenschaft und Geistesgeschichte* 83.4 (2009): 643–666.
Gess, Nicola. "Walter Benjamin und 'die Primitiven.' Reflexionen im Umkreis der Berliner Kindheit." *Text+Kritik. Zeitschrift für Literatur. Walter Benjamin* 31–32 (2009): 31–44.
Gess, Nicola. "Gaining Sovereignty. The Figure of the Child in Benjamin's Writing." *Modern Language Notes* 125.3 (2010): 682–709.

Gess, Nicola. "Intermedialität reconsidered. Vom Paragone bei Hoffmann bis zum Inneren Monolog bei Schnitzler." *Poetica. Zeitschrift für Sprach- und Literaturwissenschaft* 42.1–2 (2010): 139–168.
Gess, Nicola. "Expeditionen im Mann ohne Eigenschaften. Zum Primitivismus bei Robert Musil." In *Robert Musil und die Fremdheit der Kultur. Musil-Studien 2010*. Ed. Norbert Christian Wolf and Rosmarie Zeller. Munich: Fink, 2011. 5–22.
Gess, Nicola. "'Schöpferische Innervation der Hand.' Zur Gestensprache in Benjamins *Probleme der Sprachsoziologie*." In *Benjamin und die Anthropologie*. Ed. Carolin Duttlinger, Ben Morgan, and Anthony Phelan. Freiburg: Rombach, 2012. 181–198.
Gess, Nicola. "Sie sind, was wir waren. Literarische Reflexionen einer biologischen Träumerei von Schiller bis Benn." *Jahrbuch der deutschen Schillergesellschaft* 56 (2012): 107–125.
Gess, Nicola. *Primitives Denken. Wilde, Kinder und Wahnsinnige in der Literarischen Moderne (Müller, Musil, Benn, Benjamin)*. Munich: Fink, 2013.
Gess, Nicola. "Vom Täuschen und Zerstören. Spiel und Kunst aus der Perspektive der Entwicklungspsychologie um 1900." In *'Sich selbst aufs Spiel setzen.' Spiel als Technik und Medium von Subjektivierung*. Ed. Christian Moser and Regina Strätling. Paderborn, Fink, 2016. 119–134.
Gess, Nicola. "Böse Kinder. Zu einer literarischen und psychologischen Figur um 1900 (Lombroso, Wulffen) 1950 (Golding, March) und 2000 (Hustvedt, Shriver)." In *Kindheit und Literatur. Konzepte – Poetik – Wissen*. Ed. Davide Giuriato, Philipp Hubmann, and Mareike Schildmann. Freiburg i. Br.: Rombach, 2018. 285–308.
Gess, Nicola, ed. *Literarischer Primitivismus*. Berlin: De Gruyter, 2013.
Gess, Nicola, and Sandra Janssen, eds. *Wissens-Ordnungen. Zu einer historischen Epistemologie der Literatur*. Berlin: De Gruyter, 2013.
Geulen, Eva. "Legislating Education. Kant, Hegel, and Benjamin on 'Pedagogical Violence.'" *Cardozo Law Review* 26.3 (2005): 943–956.
Giddens, Anthony. *Conversations with Anthony Giddens: Making Sense of Modernity*. Stanford: Stanford University Press, 1998.
Gisi, Lucas Marco. "Die Biologisierung der Utopie als Apokalypse. Der neue Mensch in Robert Müllers *Tropen*." In *Utopie und Apokalypse in der Moderne*. Ed. Reto Sorg and Stefan Bodo Würffel. Munich: Fink, 2010. 215–228.
Gisi, Lucas Marco. "Die Genese des modernen Primitivismus als wissenschaftliche Methode." In *Literarischer Primitivismus*. Ed. Nicola Gess. Berlin: De Gruyter, 2013. 141–158.
Giuriato, Davide. *Mikrographien. Zu einer Poetologie des Schreibens in Walter Benjamins Kindheitserinnerungen (1932–1939)*. Munich: Fink, 2006.
Giuriato, Davide. "Tintenbuben. Kindheit und Literatur um 1900 (Rainer Maria Rilke, Robert Walser, Walter Benjamin)." *Poetica* 42.3–4 (2010): 325–351.
Gockel, Bettina. *Die Pathologisierung des Künstlers. Künstlerlegenden der Moderne*. Berlin: Akademie Verlag, 2010.
Goebel, Eckart. *Konstellation und Existenz. Kritik der Geschichte um 1930. Studien zu Heidegger, Benjamin, Jahnn und Musil*. Tübingen: Stauffenburg, 1996.
Goldstein, Kurt. "L'analyse de l'aphasie et l'étude de l'essence du langage." In Ernst Cassirer, Leo Jordan, Henri Delacroix, et al. *Psychologie du langage*. Paris: F. Alcan, 1933. 430–496.
Goldwater, Robert. *Primitivism in Modern Painting*. New York: Harper, 1938.

Götze, Carl. *Das Kind als Künstler. Ausstellung von freien Kinderzeichnungen in der Kunsthalle zu Hamburg*. Hamburg: Kunsthalle zu Hamburg, 1898.
Gould, Stephen Jay. *Ontogeny and Phylogeny*. Cambridge, MA: Harvard University Press, 1977.
Greenblatt, Stephen. *Marvelous Possessions*. Oxford: Oxford University Press, 1991.
Grill, Genese. "The 'Other' Musil: Robert Musil and Mysticism." In *A Companion to the Works of Robert Musil*. Ed. Philipp Payne, Graham Bartram, and Galin Tihanov. Rochester: Bowdell and Brewer, 2007. 333–354.
Grill, Genese. "Word Magic." In *The World as Metaphor*. Rochester, NY: Camden House, 2012.
Groos, Karl. *The Play of Animals*. Translated by Elizabeth A. Baldwin. New York: D. Appleton, 1898. [*Die Spiele der Tiere*. Jena: Gustav Fischer, 1896.]
Groos, Karl. *The Play of Man*. Translated by Elizabeth A. Baldwin. New York: D. Appleton, 1913. [*Die Spiele der Menschen*. Jena: G. Fischer, 1899.]
Groos, Karl. *Der ästhetische Genuss*. Giessen: Ricker, 1902.
Groos, Karl. *Das Seelenleben des Kindes*. Berlin: Reuter & Reichard, 1904.
Grosse, Ernst. *Die Literatur-Wissenschaft. Ihr Ziel und ihr Weg*. Diss. Halle-Wittenberg, 1887.
Grosse, Ernst. *The Beginnings of Art*. New York: D. Appleton and Company, 1897. [*Die Anfänge der Kunst*. Leipzig: Mohr, 1894.]
Gruber, Gernot. "Das 'Archaische' in der Musikkultur der Wiener Moderne. Eine Skizze." In *Kunst, Kontext, Kultur. Manfred Wagner 38 Jahre Kultur- und Geistesgeschichte an der Angewandten*. Ed. Gloria Withalm, Anna Spohn, and Gerald Bast. Berlin: De Gruyter, 2012. 132–146.
Gruber, Howard E., and J. Jacques Vonèche. "Introduction." In *The Essential Piaget*. Ed. Howard E. Gruber and J. Jacques Vonèche. New York: Basic Books, 1977. xix–xlii.
Grubrich-Simitis, Ilse. "Metapsychology and Metabiology." In Sigmund Freud, *A Phylogenetic Fantasy*. Ed. Ilse Grubrich-Simitis. Cambridge, MA: Harvard University Press, 1987. 73–107. [Grubrich-Simitis, Ilse. "Metapsychologie und Metabiologie." In Sigmund Freud. *Übersicht der Übertragungsneurosen. Ein bisher unbekanntes Manuskript*. Ed. Ilse Grubrich-Simitis. Frankfurt am Main: S. Fischer, 1985. 83–128.]
Gummere, Francis B. *The Beginnings of Poetry*. New York: Macmillan, 1901.
Habermas, Jürgen. "Walter Benjamin: Consciousness-Raising or Rescuing Critique." In *Philosophical-Political Profiles*. Translated by Frederick G. Lawrence. Cambridge, MA: MIT Press, 1983. 129–163. ["Bewusstmachende oder rettende Kritik – die Aktualität Walter Benjamins." In *Zur Aktualität Walter Benjamins*. Ed. Siegfried Unseld. Frankfurt am Main: Suhrkamp, 1972. 175–223.]
Haeckel, Ernst. *Generelle Morphologie der Organismen*. Berlin: Reimer, 1866.
Haeckel, Ernst. *The History of Creation*. Translated by Sir E. Ray Lankester. New York: Appleton. 1876. [*Natürliche Schöpfungsgeschichte*. Berlin: Reimer, 1868.]
Haeckel, Ernst. *The Evolution of Man: A Popular Exposition on the Principal Points of Ontogeny and Phylogeny*. New York: D. Appleton, 1897. [*Anthropogenie oder Entwickelungsgeschichte des Menschen*. Leipzig: W. Engelmann, 1874.]
Haeckel, Ernst. *The Riddle of the Universe: At the Close of the Nineteenth Century*. Translated by Joseph McCabe. New York: Harper and Brothers, 1900. [*Die Welträtsel. Gemeinverständliche Studie über Monistische Philosophie*. 10th ed. Leipzig: A. Kröner, 1909.]
Hahn, Marcus. *Gottfried Benn und das Wissen der Moderne*. Göttingen: Wallstein, 2011.

Hahn, Marcus. "Zusammenfliessende Eichhörnchen. Über Lucien Lévy-Bruhl und die Ethnologie-Rezeption Robert Musils." In Beil et al., *Medien, Technik, Wissenschaft*, 47–72.
Hake, Thomas. *'Gefühlserkenntnisse und Denkerschütterungen.' Robert Musils 'Nachlaß zu Lebzeiten.'* Bielefeld: Aisthesis, 1998.
Hall, G. Stanley. *Adolescence: Its Psychology and Its Relations to Physiology, Anthropology, Sociology, Sex, Crime, Religion, and Education.* 2 vols. New York: Appleton, 1904.
Hamacher, Werner. "Die Geste im Namen. Benjamin und Kafka." In *Entferntes Verstehen. Studien zu Philosophie und Literatur von Kant bis Celan.* Frankfurt am Main: Suhrkamp, 1998. 280–323.
Hamacher, Werner. "The Word *Wolke* – If It Is One." In *Benjamin's Ground. New Readings of Walter Benjamin.* Ed. Rainer Nägele. Detroit: Wayne State, 1988. 147–176.
Hansen, Miriam. "Benjamin and Cinema. Not a One-Way Street." In *Benjamin's Ghosts: Interventions in Contemporary Literary and Cultural Theory.* Ed. Gerhardt Richter. Stanford: Stanford University Press, 2002. 41–74.
Hartlaub, Gustav Friedrich. *Der Genius im Kinde. Zeichnungen und Malversuche begabter Kinder.* Breslau: F. Hirt, 1922.
Hartzell, Richard E. "The Three Approaches to the 'Other' State in Musil's *Mann ohne Eigenschaften*." *Colloquia germanica* 10 (1976–1977): 204–219.
Heidegger, Martin. "The Origin of the Work of Art." In *Off the Beaten Track*. Translated by Julian Young and Kenneth Haynes. Cambridge: Cambridge University Press, 2002. 1–56. [*Der Ursprung des Kunstwerks.* Stuttgart: Philipp Reclam, 1962.]
Heinz, Andreas. *Anthropologische und evolutionäre Modelle in der Schizophrenieforschung.* Berlin: VWB, 2002.
Heinz, Jutta. "Grenzüberschreitung im Gleichnis. Liebe, Wahnsinn und 'andere Zustände' in Robert Musils *Mann ohne Eigenschaften*." In *Grenzsituationen. Wahrnehmung, Bedeutung und Gestaltung in der neueren Literatur.* Ed. Dorothea Lauterbach, Uwe Spörl, and Uli Wunderlich. Göttingen: Vandenhoeck & Ruprecht, 2002. 235–256.
Henzler, Stefan. "Der Handlungscharakter der Sprache bei Karl Bühler und Bronislaw Malinowski." In *Betriebslinguistik und Linguistikbetrieb. Akten des 24. Linguistischen Kolloquiums, Universität Bremen, 4.–6. September 1989*, Vol. 1. Ed. Eberhard Klein, Françoise Pouradier Duteil, and Karl Heinz Wagner. Tübingen: Niemeyer, 1991. 81–88.
Herbert, Christopher. "Frazer, Einstein, and Free Play." In *Prehistories of the Future: The Primitivist Project and the Culture of Modernism.* Ed. Elazar Barkan and Ronald Bush. Stanford: Stanford University Press, 1995. 133–159.
Herder, Johann Gottfried. *Abhandlung über den Ursprung der Sprache.* Stuttgart: Philipp Reclam, 1985.
Hesse, Hermann. *Klingsor's Last Summer*. Translated by Richard and Clara Winston. New York: Farrar, Straus, and Giroux, 1970. [*Erzählungen*. Vol. 4, *Klingsors letzter Sommer.* Frankfurt am Main: Suhrkamp, 1986.]
Hesse, Hermann. *Demian: The Story of Emil Sinclair's Youth.* Translated by Damion Searls. New York: Penguin, 2013. [*Demian. Die Geschichte einer Jugend.* Frankfurt am Main: Suhrkamp, 2002.]
Heydebrand, Renate von. "Zum Thema Sprache und Mystik in Robert Musils Roman 'Der Mann ohne Eigenschaften.'" *Zeitschrift für deutsche Philologie* 82 (1963): 249–271.

Heydebrand, Renate von. *Die Reflexionen Ulrichs in Robert Musils Roman 'Der Mann ohne Eigenschaften.'* Münster: Aschendorff, 1966.
Hildenbrandt, Vera. *Europa in Alfred Döblins Amazonas-Trilogie. Diagnose eines kranken Kontinents.* Göttingen: Vandenhoeck & Ruprecht, 2011.
Hirn, Yrjö. *The Origins of Art. A Psychological and Sociological Inquiry.* London: Macmillan, 1900.
Hirschkop, Ken. *Linguistic Turns, 1890–1950. Writing on Language as Social Theory.* Oxford: Oxford University Press, 2019.
Hochstätter, Dietrich. *Sprache des Möglichen. Stilistischer Perfektionismus in Robert Musils 'Mann ohne Eigenschaften.'* Frankfurt am Main: Athenäum, 1972.
Hoerl, Erich. *Sacred Channels: The Archaic Illusion of Communication.* Translated by Nils F. Schott. Amsterdam: Amsterdam University Press, 2018.
Hofmannsthal, Hugo von. "Das Gespräch über Gedichte." In *Gesammelte Werke*. Vol. 7, *Erzählungen, Erfundene Gespräche und Briefe, Reisen*. Ed. Bernd Schoeller. Frankfurt am Main: Fischer, 1979. 495–510.
Hofmannsthal, Hugo von. "Philosophie des Metaphorischen [1894]." In *Der Streit um die Metapher. Poetologische Texte von Nietzsche bis Handke*. Ed. Klaus Müller-Richter and Arturo Larcati. Darmstadt: Wissenschaftliche Buchgesellschaft, 1998. 45–48.
Honold, Alexander. *Die Stadt und der Krieg. Raum- und Zeitkonstruktion in Robert Musils Roman 'Der Mann ohne Eigenschaften.'* Munich: Fink, 1995.
Hornbostel, Erich von. "Geburt und erste Kindheit der Musik." *Jahrbuch für musikalische Volks- und Völkerkunde* 7 (1973): 9–17.
Hsu, Francis L. K. "Rethinking the Concept 'Primitive.'" *Current Anthropology* 5.3 (1964): 169–178.
Hui, Alexandra. "Origin Stories of Listening, Melody and Survival at the End of the Nineteenth Century." In *Music and the Nerves, 1700–1900*. Ed. James Kennaway. Basingstoke: Palgrave Macmillan, 2014. 170–190.
Huizinga, Johan. *Homo Ludens: A Study of the Play-element in Culture.* Translated by Richard Francis Carrington Hull. London: Routledge & Kegan Paul, 1949.
Humboldt, Alexander von. *Reise in die Äquinoktial-Gegenden des Neuen Kontinents*. Ed. Ottmar Ette. Frankfurt am Main: Insel, 1991.
Hyman, Stanley Edgar. *The Tangled Bank: Darwin, Marx, Frazer and Freud as Imaginative Writers.* New York: Atheneum, 1962.
Jacobowski, Ludwig. *Anfänge der Poesie. Grundlegung zu einer realistischen Entwickelungsgeschichte der Poesie.* Dresden: E. Pierson's Verlag, 1891.
Jaensch, Erich Rudolph. "Die Völkerkunde und die eidetische Tatsachenkreis." *Zeitschrift für Psychologie* 91 (1923): 88–111.
Jaeger, Siegfried. "The Origin of the Diary Method in Developmental Psychology." In *Contributions to a History of Developmental Psychology: International William T. Preyer Symposium*. Ed. Georg Eckardt, Wolfgang G. Bringmann, and Lothar Sprung. Berlin: Mouton, 1985. 63–74.
Jaeger, Siegfried. "Origins of Child Psychology: William Preyer." In *The Problematic Science. Psychology in Nineteenth-Century Thought*. Ed. William R. Woodward and Mitchell G. Ash. New York: Praeger Publishers, 1982. 300–321.

Jakob, Michael. "Von der 'Frau ohne Eigenschaften' zum 'Mann ohne Eigenschaften.' Anmerkungen zu Clarisse." In *Robert Musils 'Kakanien' – Subjekt und Geschichte*. Ed. Josef Strutz. Munich: Fink, 1987. 116–133.

Jander, Simon. "Die Ästhetik des essayistischen Romans. Zum Verhältnis von Reflexion und Narration in Musils *Der Mann ohne Eigenschaften* und Brochs *Hugenau oder die Sachlichkeit*." *Zeitschrift für deutsche Philologie* 123.4 (2004): 527–548.

Janssen, Sandra. "Neurasthenie oder Psychasthenie? Gottfried Benns Selbstdiagnose als psychiatriegeschichtliches und erkenntnistheoretisches Problem der Rönne-Novellen." In *Neurasthenie. Die Krankheit der Moderne und die Moderne Literatur*. Ed. Maximilian Bergengruen, Klaus Müller-Wilde, and Caroline Pross. Freiburg im Breisgau: Rombach, 2010. 259–285.

Janssen, Sandra. *Phantasmen. Imagination in Psychologie und Literatur 1840–1939. Flaubert – Cechov – Musil*. Göttingen: Wallstein, 2013.

Jensen, Johannes Vilhelm. "Wälder." In *"Die Welt ist tief": Novellen von Johannes V. Jensen*. Berlin: Fischer, 1912.

Jensen, Johannes Vilhelm. *Das verlorene Land*. Translated by Julia Koppel. Berlin: S. Fischer, 1920.

Jung, Carl Gustav. "The Significance of Constitution and Heredity in Psychology." Translated by R.F.C. Hull. In *Collected Works of C. Jung*. Vol. 8, *Structure and Dynamics of the Psyche*. 2nd ed. Princeton, NJ: Princeton University Press, 1972. 107–113. ["Die Bedeutung von Konstitution und Vererbung für die Psychologie." *Die Medizinische Welt* 3.47 (1929).]

Jung, Carl Gustav. *Structure and Dynamics of the Psyche*. Translated by R.F.C. Hull. Princeton: Princeton University Press, 1975. ["Die Struktur der Seele." In *Gesammelte Werke*. Vol. 8, *Die Dynamik des Unbewussten*. Ed. Marianne Niehus-Jung, Lena Hurwitz-Eisner, Franz Riklin, et al. Zurich: Rascher, 1967. 161–183.]

Jung, Carl Gustav. *Symbols of Transformation*. Translated by Gerhard Adler and R.F.C. Hull. Princeton, NJ: Princeton University Press, 2014. [*Wandlungen und Symbole der Libido*. Munich: Deutscher Taschenbuch Verlag, 1995.]

Kafka, Gustav. *Handbuch der vergleichenden Psychologie*. 2 vols. Munich: Reinhardt, 1922.

Kant, Immanuel. *Critique of the Power of Judgment*. Translated by Paul Guyer. Cambridge: Cambridge University Press, 2000. [*Werkausgabe*. Vol. 10, *Kritik der Urteilskraft*. Ed. Wilhelm Weischede. Frankfurt am Main: Suhrkamp, 1994.]

Kappeler, Florian. *Situiertes Geschlecht. Organisation, Psychiatrie und Anthropologie in Robert Musils Roman "Der Mann ohne Eigenschaften."* Munich: Fink, 2012.

Kassung, Christian. *Entropie-Geschichten. Robert Musils 'Der Mann ohne Eigenschaften' im Diskurs der modernen Physik*. Munich: Fink, 2001.

Kaufmann, Doris. "Kunst, Psychiatrie und 'schizophrenes Weltgefühl' in der Weimarer Republik. Hans Prinzhorns Bildnerei der Geisteskranken." In *Kunst und Krankheit. Studien zur Pathographie*. Ed. Matthias Bormuth, Klaus Podoll, and Carsten Spitzer. Göttingen: Wallstein, 2007. 57–72.

Kaufmann, Doris. "Zur Genese der modernen Kulturwissenschaft. 'Primitivismus' im transdisziplinären Diskurs des frühen 20. Jahrhunderts." In *Wissenschaften im 20. Jahrhundert. Universitäten in der modernen Wissenschaftsgesellschaft*. Ed. Jürgen Reulecke and Volker Roelcke. Stuttgart: Steiner, 2008. 41–53.

Kaufmann, Doris. "Pushing the Limits of Understanding': The Discourse on Primitivism in German *Kulturwissenschaften*, 1880–1930." *Studies in History and Philosophy of Science* 39 (2008): 434–443.
Kaufmann, Doris. "Die Entdeckung der 'primitiven Kunst.' Zur Kulturdiskussion in der amerikanischen Anthropologie um Franz Boas, 1890–1940." In *Kulturrelativismus und Antirassismus. Der Anthropologe Franz Boas (1858–1942)*. Ed. Hans-Walter Schmuhl. Bielefeld: transcript, 2009. 211–230.
Kaufmann, Doris. "'Primitivismus': Zur Geschichte eines semantischen Feldes 1900–1930." In *Literarischer Primitivismus*. Ed. Nicola Gess. Berlin: De Gruyter, 2013. 93–124.
Kaufmann, Sebastian. *Ästhetik des 'Wilden.' Zur Verschränkung von Ethno-Anthropologie und ästhetischer Theorie 1750–1850. Mit einem Ausblick auf die Debatte über 'primitive' Kunst um 1900*. Basel: Schwabe, 2020.
Keller, Gottfried. *Green Henry*. Translated by A. M. Holt. New York: Grove, 1960. [*Der Grüne Heinrich*. Stuttgart: G. J. Göschen, 1879.]
Khatib, Sami. "Barbaric Salvage. Benjamin and the Dialectics of Destruction." *parallax* 24.2 (2018): 125–158.
Kirchdörfer-Bossmann, Ursula. *"Eine Pranke in den Nacken der Erkenntnis." Zur Beziehung von Dichtung und Naturwissenschaft im Frühwerk Gottfried Benns*. St. Ingbert: Röhrig, 2003.
Klages, Ludwig. *Vom kosmogonischen Eros*. Munich: G. Müller, 1922.
Klaue, Magnus. "Aufbauende Zerstörung. Metapherntheorie und Sprachmetaphorik im Werk von Fritz Mauthner." *Sprachkunst. Beiträge zur Literaturwissenschaft* 37 (2006): 30–49.
Klee, Paul. *The Diaries of Paul Klee: 1898–1918*. Ed. Felix Klee. Berkeley: University of California Press, 1964.
Kramer, Fritz. *Verkehrte Welten. Zur imaginären Ethnographie des 19. Jahrhunderts*. Frankfurt am Main: Syndikat, 1977.
Krause, Marcus, and Nicolas Pethes, eds. *Literarische Experimentalkulturen. Poetologien des Experiments im 19. Jahrhundert*. Würzburg: Königshausen + Neumann, 2005.
Krause, Robert. *Abstraktion – Krise –Wahnsinn. Die Ordnung der Diskurse in Robert Musils Roman 'Der Mann ohne Eigenschaften.'* Würzburg: Ergon, 2008.
Krause, Robert. "'Man könnte die Geschichte der Grenzen schreiben.' Moosbruggers wildes Denken und die Kultur des Okzidents." *Musil-Forum 31. Studien zur Literatur der klassischen Moderne: Musil und die Fremdheit der Kultur*. Ed. Norbert C. Wolf and Rosmarie Zeller (2009/2010): 23–39.
Kretschmer, Ernst. *A Text-book of Medical Psychology*. Translated from the 10th edition by E. B. Strauss. London: Hogarth, 1952. [*Medizinische Psychologie*. 10th ed. Stuttgart: Georg Thieme, 1950.]
Kretschmer, Ernst. *Medizinische Psychologie*. 12th ed. Stuttgart: Georg Thieme, 1963.
Kronfeld, Arthur. "Der künstlerische Gestaltungsvorgang in psychiatrischer Beleuchtung." *Klinische Wochenschrift* 4.1 (1925): 29–30.
Kubin, Alfred. *The Other Side: A Fantastic Novel*. Translated by Denver Lindley. London: Victor Gollancz, 1969. [*Die andere Seite. Ein phantastischer Roman*. Munich: G. Mueller, 1909.]
Kühn, Herbert. *Die Kunst der Primitiven*. Munich: Delphin Verlag, 1923.
Kühne, Jörg. *Das Gleichnis. Studien zur inneren Form von Robert Musils Roman 'Der Mann ohne Eigenschaften.'* Tübingen: Niemeyer, 1968.

Kuhn, Reinhard. *Corruption in Paradise. The Child in Western Literature.* Hannover, NH: University Press of New England, 1982.

Kuhn, Thomas Samuel. *The Structure of Scientific Revolutions.* Chicago: University of Chicago Press, 1970. [*Die Struktur wissenschaftlicher Revolutionen.* Frankfurt am Main: Suhrkamp, 1995.]

Kunsterziehung. Ergebnisse und Anregungen der Kunsterziehungstage in Dresden, Weimar und Hamburg. Leipzig: Voigtländer, 1906.

Kuper, Adam. *The Invention of Primitive Society: Transformations of an Illusion.* London: Routledge, 1988.

Kuper, Adam. *The Reinvention of Primitive Society. Transformation of a Myth.* London: Routledge, 2005.

Kurasava, Fuyuki. *The Ethnological Imagination: A Cross-Cultural Critique of Modernity.* Minneapolis: University of Minnesota Press, 2004.

Lakoff, George, and Mark Johnson. *Metaphors We Live By.* Chicago: University of Chicago Press, 1980.

Lamprecht, Karl. "Einführung in die Ausstellung von parallelen Entwicklungen in der bildenden Kunst." In *Kongress für Ästhetik und Allgemeine Kunstwissenschaft, Berlin 7.–9. Oktober 1913.* Stuttgart: Enke, 1914. 75–78.

Lange, Konrad von. *Die bewußte Selbsttäuschung als Kern des künstlerischen Genusses.* Leipzig: Veit, 1895.

Lange, Konrad von. *Das Wesen der Kunst. Grundzüge einer illusionistischen Kunstlehre.* Berlin: Grote'sche Verlagsbuchhandlung, 1907.

Largier, Niklaus. "Mystik als Medium. Robert Musils 'Möglichkeitssinn' im Kontext." In *Intermedien. Zur kulturellen und artistischen Übertragung.* Ed. Alexandra Kleihues, Barbara Naumann, and Edgar Pankow. Zurich: Chronos, 2010. 401–411.

Latour, Bruno. *We Have Never Been Modern.* Translated by Catherine Porter. Cambridge, MA: Harvard University Press, 1993.

Leavis, Frank Raymond. *Two Cultures? The Significance of C.P. Snow.* London: Chatto & Windus, 1962.

Leeb, Susanne. *Die Kunst der Anderen: "Weltkunst" und die anthropologische Konfiguration der Moderne.* Berlin: b-books, 2015.

Lehmann, Hans-Thies. "Eine unterbrochene Darstellung. Zu Walter Benjamins Idee des Kindertheaters." In *Szenarien von Theater und Wissenschaft.* Ed. Christel Weiler and Hans-Thies Lehmann. Berlin: Theater der Zeit, 2003. 181–203.

Lehmann, Hartmut. *Die Entzauberung der Welt. Studien zu Themen von Max Weber.* Göttingen: Wallstein, 2009.

Lemke, Anja. "Zur späteren Sprachphilosophie. 'Lehre vom Ähnlichen.' 'Über das mimetische Vermögen.' 'Probleme der Sprachsoziologie. Ein Sammelreferat.'" In *Benjamin-Handbuch.* Ed. Burkhardt Lindner. Stuttgart: Metzler, 2006. 643–653.

Lemke, Anja. *Gedächtnisräume des Selbst. Walter Benjamins 'Berliner Kindheit um neunzehnhundert.'* Würzburg: Königshausen + Neumann, 2008.

Lemke, Sieglinde. *Primitivist Modernism: Black Culture and the Origins of Transatlantic Modernism.* Oxford: Oxford University Press, 1998.

Leonhard, Rudolf. *Das Wort.* Berlin: Graetz, 1932.

Lethen, Helmut. "Masken der Authentizität. Der Diskurs des 'Primitivismus' in Manifesten der Avantgarde." In *Manifeste: Intentionalität*. Ed. Hubert van den Berg and Ralf Grüttemeier. Amsterdam: Brill, 1998. 227–258.

Leucht, Robert, and Susanne Reichlin. "'Ein Gleichgewicht ohne festen Widerhalt, für das wir noch keine rechte Beschreibung gefunden haben.' Robert Musils 'anderer Zustand' als Ort der Wissensübertragung." In Beil et al., *Medien, Technik, Wissenschaft*, 289–322.

Lévi-Strauss, Claude. *The Elementary Structures of Kinship*. Translated by James Harle Bell and John Richard von Sturmer. Boston: Beacon, 1969. [*Les structures élémentaires de la parenté*. Paris: Presses universitaires de France, 1949.]

Lévi-Strauss, Claude. *The Savage Mind*. Chicago: University of Chicago Press, 1966. [*La pensée sauvage*. Paris: Plon, 1962.]

Lévi-Strauss, Claude. *Totemism*. Translated by Rodney Needham. Boston: Beacon, 1963. [*Le Totémisme aujourd'hui*. Paris: Presses universitaires de France, 1962.]

Levinstein, Siegfried. *Das Kind als Künstler*. Leipzig: Voigtländer, 1905.

Lévy-Bruhl, Lucien. *How Natives Think*. Translated by Lilian A. Clare. New York: Washington Square Press, 1966. [*Les fonctions mentales dans les sociétés inférieures*. Paris: F. Alcan, 1910.]

Lévy-Bruhl, Lucien. *The Notebooks on Primitive Mentality*. Translated by Lucien Lévy-Bruhl. New York: Harper & Row Publishers, 1975. [*Les carnets de Lucien Lévy-Bruhl*. Paris: Presses universitaires de France, 1949.]

Li, Victor. *The Neo-Primitivist Turn: Critical Reflections on Alterity, Culture, and Modernity*. Toronto: University of Toronto Press, 2006.

Li, Victor. "Primitivism and Postcolonial Literature." In *The Cambridge History of Postcolonial Literature*. Ed. Ato Quayson. Cambridge: Cambridge University Press, 2012. 982–1005.

Liederer, Christian. *Der Mensch und seine Realität. Anthropologie und Wirklichkeit im poetischen Werk des Expressionisten Robert Müller*. Würzburg: Königshausen + Neumann, 2004.

Lienhardt, Godfrey. "Modes of Thought." In *The Institutions of Primitive Society. A Series of Broadcast Talks*. Ed. Edward E. Evans-Pritchard, Raymond Firtz, John Layard, et al. Oxford: Blackwell, 1959. 95–107.

Lindner, Burkhardt. "Engel und Zwerg. Benjamins geschichtsphilosophische Rätselfiguren und die Herausforderung des Mythos." In *'Was nie geschrieben wurde, lesen.'* Ed. Lorenz Jäger and Thomas Regehly. Frankfurter Benjamin-Vorträge. Bielefeld: Aisthesis, 1992. 236–266.

Lindner, Burkhardt. "Das *Passagen-Werk*, die *Berliner Kindheit* und die Archäologie des 'Jüngstvergangenen.'" In *Studien zu Benjamin*. Ed. Jessica Nitsche and Nadine Werner. Berlin: Kadmos, 2016. 217–242.

Lindner, Martin. *ICH schreiben im falschen Leben*. PressBooks (Online Publ.): Habilitation Uni Passau, 1998.

Lombroso, Cesare. *Criminal Man*. Translated by Mary Gibson and Nicole Hahn Rafter. Durham, NC: Duke University Press, 2006. [*L'uomo delinquente*. Editions 1–5. Torino: Fratelli Bocca, 1876–1896.]

Lombroso, Cesare, and Gina Lombroso-Ferrero. *Criminal Man According to the Classification of Cesare Lombroso*. New York: G. P. Putnam's Sons, 1911.

Lombroso, Cesare. *The Man of Genius*. London: Walter Scott, Ltd., 1896. [*Genio e follia*. Milano: Brigola, 1872.]

Lönker, Fred. "Der Fall Moosbrugger. Zum Verhältnis von Psychopathologie und Anthropologie in Robert Musils *Der Mann ohne Eigenschaften*." *Jahrbuch der Deutschen Schillergesellschaft* 47 (2003): 280–302.
Lorenz, Matthias N. *Distant kinship. Entfernte Verwandtschaft: Joseph Conrads* Heart of Darkness *in der deutschen Literatur von Kafka bis Kracht*. Stuttgart: Metzler, 2017.
Lubrich, Oliver. "Welche Rolle spielt der literarische Text im postkolonialen Diskurs?" *Archiv für das Studium der neueren Sprachen und Literaturen* 1 (2005): 16–39.
Lurje, Walter. *Mystisches Denken, Geisteskrankheit und moderne Kunst*. Stuttgart: Püttmann, 1923.
MacGregor, John. *The Discovery of the Art of the Insane*. Princeton: Princeton University Press, 1989.
Magris, Claudio. "Musil und die Nähte der Zeichen." In *Philologie und Kritik. Klagenfurter Vorträge zur Musilforschung*. Ed. Wolfgang Freese. Munich: Fink, 1981. 177–193.
Malinowski, Bronislaw. *Argonauts of the Western Pacific. An Account of Native Enterprise and Adventure in the Archipelagoes of Melenesian New Guinea*. London: George Routledge & Sons, 1922.
Malinowski, Bronislaw. "Magic, Science, and religion." In *Magic, Science, and Religion, and Other Essays*. Boston: Beacon Press, 1948. 1–71.
Malinowski, Bronislaw. "An Ethnographic Theory of the Magical Word." In *Coral Gardens and Their Magic*. Vol. 2, *The Language of Magic and Gardening*. Bloomington: Indiana University Press, 1965. 211–250.
Malinowski, Bronislaw. "The Problem of Meaning in Primitive Languages." In Ogden, Charles Kay, and Ivor Armstrong Richards, *The Meaning of Meaning. A Study on the Influence of Language upon Thought and of the Science of Symbolism*. San Diego: Harcourt Brace, 1989. 296–336.
Mallarmé, Stéphane. "Ballets." Translated by Evlyn Gould. *Performing Arts Journal*, 15.1 (1993): 106–110. ["Ballets." In *Divigations*. Ed. Eugène Fasquelle. Paris: Charpentier, 1897. 171–179.]
Mann, Thomas. *Reflections of a Nonpolitical Man*. Translated by Walter D. Morris. NY: New York Review Books Classics, 2021. [*Betrachtungen eines Unpolitischen*. Berlin: Fischer, 1918.]
Marett, Robert Ranulph. "Preanimistic Religion [1900]." In *The Threshold of Religion*. London: Methuen, 1914. 1–28.
Marr, Nikolaus. "Über die Entstehung der Sprache." *Unter dem Banner des Marxismus* 1.3 (1926): 558–599.
Martin, Ronald E. *The Languages of Difference: American Writers and Anthropologists Reconfigure the Primitive, 1878–1940*. Newark, DE: University of Delaware Press, 2005.
Mauss, Marcel, and Henri Hubert. *A General Theory of Magic*. Translated by Robert Brain. London: Routledge, 1972. [*Esquisse d'une théorie générale de la magie*. Paris: Presses universitaires de France, 1902.]
Mauthner, Fritz. "Kinderpsychologie." In *Wörterbuch der Philosophie*. Leipzig: Meiner, 1923. Vol. 2, 213–228.
Mauthner, Fritz. *Beiträge zu einer Kritik der Sprache*. Vol. 2, *Zur Sprachwissenschaft*. Vienna: Böhlau, 1999.
Mawson, Douglas. *The Home of the Blizzard. Being the Story of the Australasian Antarctic Expedition, 1911–1914*. London: W. Heinemann, 1915.

McLaughlin, Kevin. "Benjamins Barbarism." *The Germanic Review* 81.1 (2006): 4–20.
Meijers, Anthonie, and Martin Stingelin. "Konkordanz zu den wörtlichen Abschriften und Übernahmen von Beispielen und Zitaten aus Gustav Gerber: 'Die Sprache als Kunst' (Bromberg 1871) in Nietzsches Rhetorik-Vorlesung und in 'Ueber Wahrheit und Lüge im aussermoralischen Sinne.'" *Nietzsche-Studien. Internationales Jahrbuch für die Nietzsche-Forschung* 17 (1986): 350–368.
Menke, Bettine. *Sprachfiguren. Name-Allegorie-Bild nach Walter Benjamin*. Munich: Fink, 1991.
Menninghaus, Winfried. *Schwellenkunde. Walter Benjamins Passage des Mythos*. Frankfurt am Main: Suhrkamp, 1986.
Menninghaus, Winfried. "Walter Benjamins Diskurs der Destruktion." *Studi germanici* 29 (1991): 293–312.
Menninghaus, Winfried. *Walter Benjamins Theorie der Sprachmagie*. Frankfurt am Main: Suhrkamp, 1995.
Menninghaus, Winfried. *Wozu Kunst? Ästhetik nach Darwin*. Berlin: Suhrkamp, 2011.
Meumann, Ernst. *Einführung in die Ästhetik der Gegenwart*. Leipzig: Quelle & Meyer, 1908.
Mikkelsen, Ejnar. *Frozen Justice*. London: Gyldendal, 1922. [*Norden For Lov og Ret*. Copenhagen: Gyldendal, 1920.]
Miller, Gerlinde F. *Die Bedeutung des Entwicklungsbegriffs für Menschenbild und Dichtungstheorie bei Gottfried Benn*. New York: Peter Lang, 1990.
Moebius, Stephan. *Die Zauberlehrlinge. Soziologiegeschichte des Collège de Sociologie (1937–1939)*. Konstanz: UVK, 2006.
Mommsen, Wolfgang J. *Max Weber and German Politics: 1890–1920*. Translated by Michael S. Steinberg. Chicago: Chicago University Press, 1984. [Mommsen, Wolfgang J. *Max Weber und die deutsche Politik 1890–1920*. Tübingen: Mohr, 1974.]
Montesquieu, Charles de Secondat. *Persian and Chinese Letters*. Translated by Mr. Floyd. London: J. and R. Tonson, 1762. [*Lettres persanes*. Amsterdam: Brunel, 1721.]
Moreau, Paul. *De l'homicide commis par les enfants*. Paris: Asselin, 1882.
Mülder-Bach, Inka. *Robert Musil: Der Mann ohne Eigenschaften. Ein Versuch über den Roman*. Munich: Hanser, 2013.
Mülder-Bach, Inka. "Der Fall Moosbrugger." In *Was der Fall ist. Casus und lapsus*. Ed. Inka Mülder-Bach and Michael Ott. Paderborn: Wilhelm Fink, 2014. 145–166.
Mülder-Bach, Inka. "Allegorie und Gleichnis im 'Mann ohne Eigenschaften.'" In *Allegorie. DFG-Symposion 2014*. Ed. Ulla Haselstein. Berlin: De Gruyter, 2016. 273–302.
Müller, Hans Peter. "Wissenschaft als Beruf." In *Max Weber Handbuch*. Ed. Hans-Peter Müller and Steffen Sigmund. Stuttgart: Metzler, 2020. 259–264.
Müller, Max [Friedrich]. "Metaphor." In *Lectures on the Science of Language*. London: Longman, Green, Longman, Roberts, and Green, 1864. 334–484. [*Vorlesung über die Wissenschaft der Sprache. Für das deutsche Publikum bearbeitet von Carl Böttger*. Autorisierte Ausgabe, 2 vols. Leipzig: G. Mayer, 1863–1866.]
Müller, Robert. "Abbau der Sozialwelt." In *Kritische Schriften*. Vol. 2, *Schriften, 1917–1919*. Ed. Ernst Fischer. Paderborn: Igel, 1995. 356–362.
Müller, Robert. *Das Inselmädchen*. Munich: Roland, 1919.
Müller, Robert. "Was erwartet Österreich von seinem jungen Thronfolger?" In *Gesammelte Essays*. Ed. Michael M. Schardt. Paderborn: Igel, 1995. 5–81.
Müller, Robert. *Briefe und Verstreutes*. Paderborn: Igel, 1997.

Müller, Robert. *Tropen. Der Mythos der Reise. Urkunden eines deutschen Ingenieurs. Herausgegeben von Robert Müller. Anno 1915*. Ed. Günter Helmes. 3rd ed. Hamburg: Igel Verlag Literatur & Wissenschaft, 2010.
Müller, Thomas R. "Genie und Wahnsinn." *Soziale Psychiatrie* 141.3 (2013): 11–13.
Müller-Lyer, Franz Carl. *Phasen der Kultur und Richtlinien des Fortschritts. Soziologische Überblicke*. Munich: Lehmann, 1915.
Müller-Richter, Klaus, and Arturo Larcati. *'Kampf der Metapher!' Studien zum Widerstreit des eigentlichen und uneigentlichen Sprechens. Zur Reflexion des Metaphorischen im philosophischen und poetologischen Diskurs*. Vienna: VÖAW, 1986.
Müller-Schöll, Nikolaus. "Nachahmbarkeit. Zur Theorie des Gestischen als eines Theaters der Spur." In *Das Theater des 'konstruktiven Defaitismus.' Lektüren zur Theorie eines Theaters der A-Identität bei Walter Benjamin, Bertolt Brecht und Heiner Müller*. Frankfurt am Main: Stroemfeld, 2002. 139–174.
Müller-Tamm, Jutta. *Abstraktion als Einfühlung. Zur Denkfigur der Projektion in Psychophysiologie, Kulturtheorie, Ästhetik und Literatur der frühen Moderne*. Freiburg: Rombach, 2005.
Müller-Tamm, Jutta. "Die Denkfigur als wissensgeschichtliche Kategorie." In *Wissens-Ordnungen. Zu einer historischen Epistemologie der Literatur*. Ed. Nicola Gess and Sandra Janssen. Berlin: De Gruyter, 2014. 100–120.
Munro, Thomas. *Evolution in the Arts and Other Theories of Culture History*. Cleveland, OH: Cleveland Museum of Art, 1967.
Murphy, Ruth. "Kipling's Just So Stories: The Recapitulative Child and Evolutionary Progress." In *The Child Savage, 1890–2010: From Comics to Games*. Ed. Elisabeth Wesseling. Farnham: Ashgate, 2016. 39–54.
Musil, Robert. *Gesammelte Werke in neun Bänden (GW)*. Ed. Adolf Frisé. Reinbek bei Hamburg: Rowohlt, 1978.
Musil, Robert. *The Man Without Qualities*. Translated by Sophie Wilkins. 2 vols. New York: Vintage, 1996. [*Der Mann ohne Eigenschaften*. Ed. Adolf Frisé. Reinbek bei Hamburg: Rowohlt, 1995.]
Musil, Robert. "Robert Müller." In *KA*, Vol. 12, 9–10.
Musil, Robert. "Aus der Begabungs- und Vererbungsforschung." In *KA*, Vol. 10, *Wissenschaftliche Veröffentlichungen*, s.p.
Musil, Robert. "Grigia." In *Five Women*. Translated by Eithne Wilkins and Ernst Kaiser. Boston: Godine, 1999. 17–56.
Musil, Robert. "Bücher und Literatur." In *GW*, Vol. 8, *Essays und Reden*, 1170–1180.
Musil, Robert. "The Mathematical Man." In Musil, *Precision and Soul*, 39–43.
Musil, Robert. "Commentary on a Metaphysics." In Musil, *Precision and Soul*, 54–58.
Musil, Robert. "Sketch of What the Writer Knows." In Musil, *Precision and Soul*, 61–65.
Musil, Robert. "Psychology and Literature." In Musil, *Precision and Soul*, 65–67.
Musil, Robert. "Helpless Europe." In Musil, *Precision and Soul*, 116–133.
Musil, Robert. "Mind and Experience." In Musil, *Precision and Soul*, 134–149. ["Geist und Erfahrung." In *KA*, Vol. 12, *Essays*, 841–858.]
Musil, Robert. "Toward a New Aesthetic." In Musil, *Precision and Soul*, 193–208. ["Ansätze zu neuer Ästhetik [1925]." In *GW*, Vol. 8, 1137–1154.]
Musil, Robert. "Literati and Literature [1931]." In Musil, *Precision and Soul*, 70–89.

Musil, Robert. *Precision and Soul: Essays and Addresses.* Translated by Burton Pike and David S. Luft. Chicago: University of Chicago Press, 1994. [*GW*, Vols. 8 and 9]

Musil, Robert. *Klagenfurter Ausgabe. Kommentierte Edition sämtlicher Werke, Briefe und nachgelassener Schriften. Mit Transkriptionen und Faksimiles aller Handschriften (KA).* Ed. Walter Fanta, Klaus Amann, and Carl Corino. Klagenfurt: Drava, 2009.

Nägele, Rainer. "Von der Ästhetik zur Poetik. Brecht, Benjamin und die Poetik der Zäsur." In *Lesarten der Moderne. Essays.* Eggingen: Isele, 1998. 98–122.

Nerlich, Brigitte, and David D. Clarke. "Mind, Meaning and Metaphor: The Philosophy and Psychology of Metaphor in 19th-Century Germany." *History of the Human Sciences* 14.2 (2001): 39–61.

Nietzsche, Friedrich. *The Will to Power.* Translated by Anthony M. Ludovici. Mineola: Dover, 2019. [*Der Wille zur Macht. Versuch einer Umwerthung aller Werthe (Studien und Fragmente).* Leipzig: C.G. Naumann, 1901.]

Nietzsche, Friedrich. "Philosophy During the Tragic Age of the Greeks [1873]." Translated by Maximilian A. Mügge. In *The Complete Works of Friedrich Nietzsche.* Vol. 2.2, *Early Greek Philosophy and Other Essays.* Ed. Oscar Levy. New York: Macmillan, 1911. 71–170. ["Die Philosophie im tragischen Zeitalter der Griechen." In *Kritische Studienausgabe.* Vol. 1. Ed. Giorgio Colli and Mazzino Montinari. Munich: Deutscher Taschenbuch Verlag, 1988. 801–872.]

Nietzsche, Friedrich. "On Truth and Lie in a Nonmoral Sense." In *On Truth and Untruth.* Translated by Taylor Carman. New York: Harper, 2010. 15–50. ["Ueber Wahrheit und Lüge im außermoralischen Sinne." In *Sämtliche Werke. Kritische Studienausgabe.* Vol. 1, *Die Geburt der Tragödie [...].* Ed. Giorgio Colli and Mazzino Montinari. Munich: Deutscher Taschenbuch Verlag, 1988. 873–890.]

Nietzsche, Friedrich. *Human, All Too Human.* Translated by Helen Zimmern and Paul V. Cohn. New York: Prometheus, 2009. [*Sämtliche Werke. Kritische Studienausgabe.* Vol. 2, *Menschliches, Allzumenschliches I und II.* Ed. Giorgio Colli and Mazzino Montinari. Munich: Deutscher Taschenbuch Verlag, 1988.]

Nübel, Birgit. *Robert Musil – Essayismus als Selbstreflexion der Moderne.* Berlin: De Gruyter, 2006.

Oehlschläger, Claudia. *Abstraktionsdrang. Wilhelm Worringer und der Geist der Moderne.* Munich: Fink, 2005.

Oesterreich, Konstantin Traugott. "Die Entfremdung der Wahrnehmungswelt und die Depersonalisation in der Psychasthenie. Ein Beitrag zur Gefühlspsychologie." *Journal für Psychologie und Neurologie* 7.6 (1906): 253–276; 8.1–2 (1906): 61–97; 8.5 (1906): 220–237; 9.1–2 (1907): 15–53.

Oesterreich, Konstantin Traugott. *Die Phänomenologie des Ich in ihren Grundproblemen.* Leipzig: Barth, 1910.

Ogden, Charles K., and Ivor A. Richards. *The Meaning of Meaning: A Study of the Influence of Language upon Thought and of the Science of Symbolism.* San Diego: Harcourt Brace, 1989.

Ostermann, Eberhard. "Das wildgewordene Subjekt. Christian Moosbrugger und die Imagination des Wilden in Musils *Mann ohne Eigenschaften*." *Neophilologus* 89 (2005): 605–623.

Otis, Laura. *Organic Memory. History and the Body in the Late Nineteenth and Early Twentieth Centuries.* Lincoln, NE: University of Nebraska Press, 1994.

Paget, Richard. *Human Speech: Some Observations, Experiments, and Conclusions as to the Nature, Origin, Purpose and Possible Improvement of Human Speech.* London: Kegan, 1930.
Paget, Richard. "L'évolution du langage." In Ernst Cassirer, Leo Jordan, Henri Delacroix, et al. *Psychologie du langage.* Paris: Alcan, 1933. 93–100.
Pan, David. *Primitive Renaissance. Rethinking German Expressionism.* Lincoln, London: University of Nebraska Press, 2001.
Parr, Rolf. "Exotik, Kultur, Struktur. Tangenten dreier Perspektiven bei Claude Levi-Strauss." *kultuRRevolution. Zeitschrift für angewandte Diskurstheorie* 32–33 (1995): 22–28.
Payne, Harry C. "Malinowski's Style." *Proceedings of the American Philosophical Society* 125.6 (1981): 416–440.
Payne, Philip. "Musil erforscht den Geist eines anderen Menschen – zum Porträt Moosbruggers im *Mann ohne Eigenschaften.*" *Literatur und Kritik* 11.106–107 (1976): 389–404.
Pelmter, Andrea. *"Experimentierfeld des Seinkönnens" – Dichtung als "Versuchsstätte." Zur Rolle des Experiments im Werk Robert Musils.* Würzburg: Königshausen + Neumann, 2008.
Perloff, Marjorie. "Tolerance and Taboo: Modernist Primitivisms and Postmodernist Poetics." In *Prehistories of the Future. The Primitivist Project and the Culture of Modernism.* Ed. Elazar Barkan and Ronald Bush. Stanford: Stanford University Press, 1995. 339–356.
Pethes, Nicolas. *Mnemographie. Poetiken der Erinnerung und Destruktion nach Walter Benjamin.* Tübingen: Niemeyer, 1999.
Pethes, Nicolas. "Die Transgression der Codierung. Funktionen gestischen Schreibens (Artaud, Benjamin, Deleuze)." In *Gestik. Figuren des Körpers in Text und Bild.* Ed. Margreth Egidi, Oliver Schneider, and Matthias Schöning. Tübingen: Narr, 2000. 299–314.
Pethes, Nicolas. "Literatur- und Wissenschaftsgeschichte. Ein Forschungsbericht." *Internationales Archiv für Sozialgeschichte der deutschen Literatur* 28.1 (2003): 181–231.
Pethes, Nicolas. "Vom Einzelfall zur Menschheit. Die Fallgeschichte als Medium der Wissenspopularisierung in Recht, Medizin und Literatur." In *Popularisierung und Popularität.* Ed. Gereon Blaseio, Hedwig Pompe, and Jens Ruchatz. Cologne: DuMont Literatur und Kunst Verlag, 2005. 63–92.
Pethes, Nicolas. *Zöglinge der Natur. Der literarische Menschenversuch des 18. Jahrhunderts.* Göttingen: Wallstein, 2007.
Pethes, Nicolas. "Versuchsobjekt Mensch. Gedankenexperimente und Fallgeschichten als Erzählformen des Menschenversuchs." In *Experiment und Literatur. Themen, Methoden, Theorien.* Ed. Michael Gamper. Göttingen: Wallstein, 2010. 361–383.
Piaget, Jean. *Le langage et la pensée chez l'enfant.* Paris: Delacheux & Niestlé, 1923.
Piaget, Jean. *Judgment and Reasoning in the Child.* Translated by Marjorie Warden. London: Routledge, 2002. (*Le jugement et le raisonnement chez l'enfant.* Paris: Delacheux & Niestlé, 1924.]
Piaget, Jean. *The Child's Conception of the World.* Translated by Joan and Andrew Tomlinson. Lanham, MD: Littlefield Adams, 1989. *[La représentation du monde chez l'enfant.* Paris: Presses universitaires de France, 1926.]
Piaget, Jean. *The Child's Conception of Physical Causality.* New York: Harcourt Brace & Company, 1930. [*La causalité physique chez l'enfant.* Paris: F. Alcan, 1927.]

Piaget, Jean. "Psychoanalysis in Its Relations with Child Psychology." In *The Essential Piaget: An Interpretive Reference and Guide*. Ed. Howard E. Gruber and J. Jacques Vonèche. New York: Basic Books, 1977. 55–62.

Pietsch Pentecost, Gislind Erna. *Clarisse. Analyse der Gestalt in Robert Musils Roman 'Der Mann ohne Eigenschaften.'* Ph.D. Dissertation: Purdue University, 1990.

Pressler, Günter Karl. *Vom mimetischen Ursprung der Sprache. Walter Benjamins Sammelreferat Probleme der Sprachsoziologie im Kontext seiner Sprachtheorie*. Frankfurt am Main: Peter Lang, 1992.

Preuss, Karl Theodor. *Die geistige Kultur der Naturvölker*. Leipzig: Teubner, 1914.

Preyer, William T. *Mental Development in the Child*. Translated by H. W. Brown. New York: Appleton, 1894. [*Die Seele des Kindes. Beobachtungen über die geistige Entwicklung des Menschen in seinen ersten Lebensjahren*. Leipzig: Th. Grieben, 1882.]

"Primitiv, der bzw. das Primitive." In *Historisches Wörterbuch der Philosophie*. Vol. 7. Ed. Joachim Ritter and Karlfried Gründer. Darmstadt: Wissenschaftliche Buchgesellschaft, 1971–2007. 1318.

"primitiv." In Kluge, Friedrich. *Etymologisches Wörterbuch der deutschen Sprache*. Berlin: De Gruyter, 1995. 647.

"Primitive." In *Wörterbuch der Völkerkunde*. Founded by Walter Hirschberg. Berlin: Reimer, 1999. 295.

Prinzhorn, Hans. *Artistry of the Mentally Ill*. Translated by Eric von Brockdorff from the 2nd edition. New York: Springer, 1972. [*Bildnerei der Geisteskranken. Ein Beitrag zur Psychologie und Psychopathologie der Gestaltung*. 2nd edition. Berlin: Springer, 1968.]

Raulet, Gérard. "Mimesis. Über anthropologische Motive bei Walter Benjamin – Ansätze zu einer anthropologischen kritischen Theorie." *Deutsche Zeitschrift für Philosophie* 64.4 (2016): 581–602.

Recki, Birgit. *Cassirer. Grundwissen Philosophie*. Stuttgart: Reclam, 2013.

Reckwitz, Andreas. "Vom Künstlermythos zur Normalisierung kreativer Prozesse." In *Kreation und Depression*. Ed. Christoph Menke and Juliane Rebentisch. Berlin: Kulturverlag Kadmos, 2010. 98–117.

Rehding, Alexander. "The Quest for the Origins of Music in Germany circa 1900." *Journal of the American Musicological Society* 53.2 (2000): 345–385.

Reichle, Ingeborg. "Vom Ursprung der Bilder und den Anfängen der Kunst. Zur Logik des interkulturellen Bildvergleichs um 1900." In *Image match. Visueller Transfer, "Imagescapes" und Intervisualität in globalen Bildkulturen*. Ed. Martina Baleva, Ingeborg Reichle, and Oliver Lerone Schultz. Paderborn: Fink, 2012. 131–150.

Reinhardt, Ursula. *Religion und moderne Kunst in geistiger Verwandtschaft. Robert Musils Roman "Der Mann ohne Eigenschaften" im Spiegel christlicher Mystik*. Marburg: Elwert, 2003.

Reis, Gilbert. "Eine Brücke ins Imaginäre. Gleichnis und Reflexion in Musils Der Mann ohne Eigenschaften." *Euphorion. Zeitschrift für Literaturgeschichte* 78 (1984): 143–159.

Reschke, Renate. "Barbaren, Kult und Katastrophen. Nietzsche bei Benjamin. Unzusammenhängendes im Zusammenhang gelesen." In *Aber ein Sturm weht vom Paradiese her. Texte zu Walter Benjamin*. Ed. Michael Opitz and Erdmut Wizisla. Leipzig: Reclam, 1992. 303–341.

Rhodes, Colin. *Primitivism and Modern Art*. London: Thames & Hudson, 1994.

Richter, Dieter. *Das fremde Kind. Zur Entstehung der Kindheitsbilder des bürgerlichen Zeitalters*. Frankfurt am Main: Fischer, 1987.
Richter, Sandra. *A History of Poetics. German Scholarly Aesthetics and Poetics in International Context, 1770–1960*. Berlin: De Gruyter, 2010.
Riedel, Wolfgang. "'What's the difference?' Robert Müllers *Tropen* (1915)." In *Schwellen. Germanistische Erkundungen einer Metapher*. Ed. Nicholas Saul, Daniel Steuer, Frank Möbus, and Birgit Illner. Würzburg: Königshausen + Neumann, 1999. 62–76.
Riedel, Wolfgang. "Archäologie des Geistes. Theorien des wilden Denkens um 1900." In *Das schwierige neunzehnte Jahrhundert*. Ed. Jürgen Barkhoff, Gilbert Carr, and Roger Paulin. Tübingen: Niemeyer, 2000. 467–485.
Riedel, Wolfgang. "Robert Musil: *Der Mann ohne Eigenschaften*." In *Lektüren für das 21. Jahrhundert. Schlüsseltexte der deutschen Literatur von 1200 bis 1990*. Ed. Dorothea Klein and Sabine M. Schneider. Würzburg: Königshausen + Neumann, 2000. 265–285.
Riedel, Wolfgang. "Arara ist Bororo oder die metaphorische Synthesis." In *Anthropologie der Literatur. Poetogene Strukturen und ästhetisch-soziale Handlungsfelder*. Ed. Rüdiger Zymner and Manfred Engel. Paderborn: Mentis, 2004. 220–242.
Riedel, Wolfgang. "Endogene Bilder. Anthropologie und Poetik bei Gottfried Benn." In *Poetik der Evidenz. Die Herausforderung der Bilder in der Literatur um 1900*. Ed. Helmut Pfotenhauer, Wolfgang Riedel, and Sabine Schneider. Würzburg: Königshausen + Neumann, 2005. 163–202.
Riedel, Wolfgang. "Wandlungen und Symbole des Todestriebs. Benns Lyrik im Kontext eines metapsychologischen Gedankens." In *Sigmund Freud und das Wissen der Literatur*. Ed. Peter-André Alt and Thomas Anz. Berlin: De Gruyter, 2008. 101–120.
Riemann, Hugo. *Handbuch der Musikgeschichte*. Leipzig: Breitkopf & Hartel, 1904.
Rilke, Rainer Maria. "Unvollendete Elegie 'Lass dir, daß Kindheit war.'" In *Werke*. Vol. 2. Frankfurt am Main: Insel, 1987. 457–459.
Ringelnatz, Joachim. *Geheimes Kinder-Spiel-Buch*. Potsdam: Kiepenheuer, 1924.
Ringelnatz, Joachim. *The Secret-Games-for-Children Book*. Translated by Andrew Lee. Carlsruhe: A. Lee, 1989.
Robertson, Ritchie. "Musil and the 'Primitive' Mentality." In *Robert Musil and the Literary Landscape of his Time*. Ed. Hannah Hickman. Salford: Salford University Press, 1991. 13–33.
Robertson, Ritchie. "Everyday Transcendence? Robert Musil, William James, and Mysticism." *History of European Ideas* 43.3 (2017): 262–272.
Rolf, Eckard. *Symboltheorien. Der Symbolbegriff im Theoriekontext*. Berlin: De Gruyter, 2006.
Rose, Wolfgang, Petra Fuchs, and Thomas Beddies. *"Psychopathie": Die urbane Moderne und das schwierige Kind. Berlin 1918–1933*. Vienna: Böhlau, 2016.
Rossetti, Gina M. *Imagining the Primitive in Naturalist and Modernist Literature*. Columbia, MO: University of Missouri Press, 2006.
Rubin, William, ed. *"Primitivism" in 20th Century Art. Affinity of the Tribal and the Modern*. New York: Museum of Modern Art, 1984. [*Primitivismus in der Kunst des zwanzigsten Jahrhunderts*. Munich: Prestel, 1984.]
Said, Edward. *Orientalism*. New York: Vintage, 1994.
Schilder, Paul. *Wahn und Erkenntnis*. Berlin: Springer, 1918.
Schiller, Friedrich. "On Naïve and Sentimental Poetry." Translated by Julias A. Elias. In *German Aesthetic and Literary Criticism: Winckelmann, Lessing, Hamann, Herder,*

Schiller, and Goethe. Ed. H.S. Nisbet. Cambridge: Cambridge University Press, 1985. 177–232. ["Über naive und sentimentalische Dichtung." In *Werke in drei Bänden*. Vol. 2. Ed. Herbert G. Göpfert and Gerhard Fricke. Frankfurt am Main: Büchergilde Gutenberg, 1992. 540–607.]

Schiller, Friedrich. *On the Aesthetic Education of Man, in a Series of Letters*. Edited and translated by Elizabeth M. Wilkenson and L.A. Willoughby. Oxford: Clarendon Press, 1987. [*Über die ästhetische Erziehung des Menschen* [1796]. In *Sämtliche Werke in sechs Bänden*. Vol. 5. Essen: Phaidon, 1984. 570–669.]

Schiller, Friedrich. "What Means, and for What Purpose Do We Study, Universal History?" In *Complete Works in Two Volumes*. Vol. 2. Edited and translated by Charles J. Hempell. Philadelphia: Kohler, 1861. 346–352. ["Was heißt und zu welchem Ende studiert man Universalgeschichte?" In *Sämtliche Werke in sechs Bänden*. Vol. 6, *Säkularausgabe*. Essen: Phaidon, 1984. 9–23.]

Schleuning, Peter. *Die Freie Fantasie. Ein Beitrag zur Erforschung der klassischen Klaviermusik*. Göttingen: A. Kümmerle, 1973.

Schmarsow, August. "Kunstwissenschaft und Völkerpsychologie." *Zeitschrift für Ästhetik und allgemeine Kunstwissenschaft* 2.3 (1907): 305–339.

Schmidt, Erich. "Die Anfänge der Literatur und die Literatur der primitiven Völker." In Erich Schmidt, Adolf Erman, and Carl Bezold. *Die orientalischen Literaturen*. Leipzig: Teubner, 1924. 1–27.

Schmitz-Emans, Monika. "Artikel Metapher (Langtext)." Formerly published in *Basislexikon Literaturwissenschaft* (last accessed: 3 June 2009) hosted by the Komparatistik Abteilung at the University of Bochum. Now accessible via: https://docplayer.org/25246668-Metapher-autorin-monika-schmitz-emans.html.

Schneider, Manfred. *Der Barbar. Endzeitstimmung und Kulturrecycling*. Munich: Hanser, 1997.

Schneider, Sabine. "Das Leuchten der Bilder in der Sprache. Hofmannsthals medienbewusste Poetik der Evidenz." *Hofmannsthal Jahrbuch* 11 (2003): 209–248.

Scholem, Gershom. "Walter Benjamin." In *On Jews and Judaism in Crisis: Selected Essays*. Ed. Werner J. Dannhauser. Philadelphia: Paul Dry, 2012. 172–197.

Scholem, Gershom. *Walter Benjamin: The Story of a Friendship*. Translated by Harry Zohn. New York: NYRB, 1981. [*Walter Benjamin – die Geschichte einer Freundschaft*. Frankfurt am Main: Suhrkamp, 1975.]

Scholem, Gershom. *Walter Benjamin und sein Engel*. Frankfurt am Main: Suhrkamp, 1992.

Scholz, Friedrich. *Die Charakterfehler des Kindes. Eine Erziehungslehre für Haus und Schule*. Leipzig: E. H. Mayer, 1891.

Schraml, Wolfgang. *Relativismus und Anthropologie. Studien zum Werk Robert Musils und zur Literatur der zwanziger Jahre*. Munich: Eberhard, 1994.

Schulz, Kerstin. "'Als wäre mein Mund so fern von mir wie der Mond' – Das Gleichnis als Denkbild in Robert Musils Roman Der Mann ohne Eigenschaften." In *Denkbilder. Wandlungen literarischen und ästhetischen Sprechens in der Moderne*. Ed. Ralph Köhnen. Frankfurt am Main: Peter Lang, 1996. 119–139.

Schurtz, Heinrich. *Urgeschichte der Kultur*. Leipzig, Vienna: Bibliographisches Institut, 1900.

Schüttpelz, Erhard. *Die Moderne im Spiegel des Primitiven. Weltliteratur und Ethnologie (1870–1960)*. Munich: Fink, 2005.

Schwab, Gabriele. *Imaginary Ethnographies. Literature, Culture, and Subjectivity*. New York: Columbia University Press, 2012.

Schwarz, Thomas. *Robert Müllers Tropen. Ein Reiseführer in den imperialen Exotismus.* Heidelberg: Synchron, 2006.
Schwarz, Thomas. "Robert Müllers Tropen (1915) als neurasthenisches Aufschreibesystem." In *Neurasthenie. Die Krankheit der Moderne und die moderne Literatur.* Ed. Maximilian Bergengruen, Klaus Müller-Wille, and Caroline Pross. Freiburg: Rombach, 2010. 39–155.
Schwennsen, Anja. "Kunst und Mythos zwischen Präsenz und Repräsentation: Cassirers Begriff des mythischen Denkens in literaturwissenschaftlicher Perspektive." In *Zwischen Präsenz und Repräsentation. Formen und Funktionen des Mythos in theoretischen und literarischen Diskursen.* Ed. Bent Gebert and Uwe Mayer. Berlin: De Gruyter, 2014. 205–225.
Sello, Katrin. "Zur 'Fabrikation der Fiktionen.'" In *Carl Einstein. Die Fabrikation der Fiktionen.* Ed. Sibylle Penkert. Reinbek bei Hamburg: Rowolt, 1973. 345–373.
Shelton, Marie-Denise. "Primitive Self. Colonial Impulses in Michel Leiris's *L'Afrique fantôme.*" In *Prehistories of the Future. The Primitivist Project and the Culture of Modernism.* Ed. Elazar Barkan and Ronald Bush. Stanford: Stanford University Press, 1995. 326–338.
Shuttleworth, Sally. *The Mind of the Child: Child Development in Literature, Science and Medicine, 1840–1900.* Oxford: Oxford University Press, 2010.
Sighele, Scipio. *Psychologie des Auflaufs und der Massenverbrechen.* Dresden: Reissner, 1897.
Simmel, Georg. "The Individual Law." In *The View of Life: Four Metaphysical Essays with Journal Aphorisms.* Translated by John A. Y. Andrews and Donald N. Levine. Chicago: University of Chicago Press, 2010. 99–154. ["Der Begriff und die Tragödie der Kultur." In *Das individuelle Gesetz. Philosophische Exkurse.* Ed. Michael Landmann. Frankfurt am Main: Suhrkamp, 1987. 116–147.]
Snow, Charles Percy. *The Two Cultures.* Cambridge: Cambridge University Press, 1959.
Sombart, Werner. *Händler und Helden.* Leipzig: Duncker & Humblot, 1915.
Specht, Benjamin. "'Verbindung finden wir im Bilde.' Die Metapher in und zwischen wissenschaftlichen Disziplinen im späten 19. Jahrhundert." In *Metaphorologien der Exploration und Dynamik 1800/1900. Historische Wissenschaftsmetaphern und die Möglichkeiten ihrer Historiographie.* Ed. Gundhild Berg, Martina King, and Reto Rössler. Hamburg: Meiner, 2018. 41–60.
Stern, Clara, and William Stern. *Die Kindersprache. Eine psychologische und sprachtheoretische Untersuchung.* Darmstadt: Wissenschaftliche Buchgesellschaft, 1965 (1907).
Stern, William. *Psychology of Early Childhood: Up to the Sixth Year of Age.* Translated by Anna Barwell. New York: Henry Holt, 1924. [*Psychologie der frühen Kindheit bis zum sechsten Lebensjahr.* Leipzig: Quelle & Meyer, 1914.]
Stockhammer, Robert. *Zaubertexte. Die Wiederkehr der Magie und die Literatur, 1810–1945.* Berlin: Akademie, 2000.
Stocking, George W. Jr. "'Cultural Darwinism' and 'Philosophical Idealism' in E. B. Tylor." In *Race, Culture and Evolution: Essays in the History of Anthropology.* New York: The Free Press, 1968. 91–109.
Storch, Alfred. *The Primitive Archaic Forms of Inner Experiences and Thought in Schizophrenia: A Genetic and Clinical Study of Schizophrenia.* Translated by Clara

Willard. New York: Nervous and Mental Disease Publishing Company, 1924. [*Das archaisch-primitive Erleben und Denken der Schizophrenen.* Berlin: Springer, 1922.]

Stumpf, Carl. *The Origins of Music.* Translated by David Trippett. Oxford: Oxford University Press, 2012. [*Die Anfänge der Musik.* Leipzig: Barth, 1911.]

Sulloway, Frank J. *Freud, Biologist of the Mind.* London: Burnett, 1979.

Sully, James. *Studies of Childhood.* London: Longmans Green, 1896.

Taussig, Michael. *Shamanism, Colonialism, and the Wild Man. A Study in Terror and Healing.* Chicago: University of Chicago Press, 1991.

Tenbruck, Friedrich H. "Das Werk Max Webers." In *Kölner Zeitschrift für Soziologie und Sozialpsychologie* 69, Suppl. 1 (2017): 375–413.

Tewilt, Gerd-Theo. *Zustand der Dichtung. Interpretationen zur Sprachlichkeit des 'anderen Zustands' in Robert Musils 'Der Mann ohne Eigenschaften.'* Münster: Aschendorff, 1990.

Thurnwald, Richard. *Forschungen auf den Salomo Inseln und dem Bismarck Archipel.* Berlin: D. Reimer, 1912.

Thurnwald, Richard. "Psychologie des Primitiven Menschen." In Kafka, Gustav. *Handbuch der vergleichenden Psychologie.* Vol. 1, *Die Entwicklungsstufen des Seelenlebens.* Munich: Reinhardt, 1922. 147–322.

Thurnwald, Richard. "Primitives Denken." In *Reallexikon der Vorgeschichte.* Vol. 10. Ed. Max Ebert. Berlin: De Gruyter, 1927/28. 294–316.

Tiedemann, Dietrich. "Beobachtungen über die Entwickelung der Seelenfähigkeit bei Kindern." *Hessische Beiträge zur Gelehrsamkeit und Kunst* 2 (1787): 313–333; 486–502.

Todorov, Tzvetan. *The Conquest of America: The Conquest of the Other.* Translated by Richard Howard. Norman, OK: University of Oklahoma Press, 1999. [*La Conqête de l'Amerique. La question de l'autre.* Paris: Le Seuil, 1982.]

Torgovnick, Marianna. *Gone Primitive: Savage Intellects, Modern Lives.* Chicago: University of Chicago Press, 1990.

Treiber, Hubert. "Zur 'Logik des Traumes' bei Nietzsche. Anmerkungen zu den Traumaphorismen aus 'Menschliches, Allzumenschliches.'" *Nietzsche-Studien* 23 (1994): 1–41.

Tylor, Edward Burnett. *Primitive Culture: Researches into the Development of Mythology, Philosophy, Religion, Art, and Custom.* 2 vols. London: Murray, 1871.

Tzara, Tristan. "Negerlieder." In *DADA Zürich. Texte, Manifeste, Dokumente.* Ed. Karl Riha. Stuttgart: Reclam 1995. 104–105.

Unger, Erich. *Das Problem der mythischen Realität. Eine Einleitung in die Goldbergsche Schrift: "Die Wirklichkeit der Hebräer."* Berlin: David, 1926.

Vatan, Florence. *Robert Musil et la question anthropologique.* Paris: PUF, 2000.

Vatan, Florence. "'Und auch die Kunst sucht Wissen.' Robert Musil und literarische Erkenntnis." In *Aisthesis und Noesis. Zwei Erkenntnisformen vom 18. Jahrhundert bis zur Gegenwart.* Ed. Hans Adler and Lynn L. Wolff. Munich: Wilhelm Fink, 2013. 113–129.

Verworn, Max. "Kinderkunst und Urgeschichte." *Korrespondenz-Blatt der Deutschen Gesellschaft für Anthropologie, Ethnologie und Urgeschichte* 27 (1907): 6–17.

Verworn, Max. *Zur Psychologie der primitiven Kunst.* Jena: Fischer, 1907.

Verworn, Max. *Die Anfänge der Kunst. Ein Vortrag.* Jena: Fischer, 1909.

Verworn, Max. *Ideoplastische Kunst. Ein Vortrag.* Jena: Fischer, 1914.

Vico, Giambattista. *Prinzipien einer neuen Wissenschaft über die gemeinsame Natur der Völker.* Hamburg: Meiner, 1990.
Vierkandt, Alfred. *Naturvölker und Kulturvölker. Ein Beitrag zur Sozialpsychologie.* Leipzig: Duncker & Humblot, 1896.
Vischer, Friedrich Theodor. "The Symbol." Translated by Holly A. Yanacek. *Art in Translation* 7.4 (2015): 417–448. ["Das Symbol." In *Kritische Gänge.* Vol. 4. Ed. Robert Vischer. Munich: Meyer and Jesser, 1922 (1887). 420–456.]
Vygotsky, Lev Semenovich. *Thought and Language.* Translated by Alex Kozulin. Cambridge, MA: The MIT Press, 1986. [*Myshlenie y rech.* Moscow: State Social-Economic Pub., 1934.]
Wallaschek, Richard. *Primitive Music: An Inquiry into the Origin and Development of Music, Songs, Instruments, Dances, and Pantomimes of Savage Races.* London: Longmans, Greens, and Co., 1893.
Warburg, Aby. *Schlangenritual. Ein Reisebericht.* Berlin: Wagenbach, 1998.
Weber, Max. "Science as a Vocation." In *The Vocation Lectures: "Science as a Vocation, Politics as a Vocation."* Translated by Rodney Livingstone. Ed. David Owen and Tracy B. Strong. Indianapolis: Hackett, 2004. 1–31. ["Wissenschaft als Beruf." In *Gesamtausgabe.* Ed. Horst Baier et al. Vol. 17.1, *Schriften und Reden.* Ed. Wolfgang Mommsen and Wolfgang Schluchter. Tübingen: J.C.B. Mohr, 1992. 71–111.]
Weber, Samuel. "Citability – of Gesture." In *Benjamin's –abilities.* Cambridge, MA: Harvard University Press, 2008. 95–114.
Weber, Samuel. "Violence and Gesture: Agamben Reading Benjamin Reading Kafka Reading Cervantes." In *Benjamin's –abilities.* Cambridge, MA: Harvard University Press, 2008. 195–210.
Weigel, Sigrid. *Entstellte Ähnlichkeit. Walter Benjamins theoretische Schreibweise.* Frankfurt am Main: Fischer, 1997.
Weiler, Bernd. *Die Ordnung des Fortschritts. Zum Aufstieg und Fall der Fortschrittsidee in der "jungen" Anthropologie.* Bielefeld: transcript, 2006.
Weimar, Klaus. "Die Begründung der Literaturwissenschaft." In *Literaturwissenschaft und Wissenschaftsforschung.* Ed. Jörg Schönert. Stuttgart: Metzler, 2000. 135–149.
Weingart, Brigitte. "Verbindungen, Vorverbindungen. Zur Poetik der 'Partizipation' (Lévy-Bruhl) bei Musil." In Beil et al., *Medien, Technik, Wissenschaft,* 19–46.
Wellershoff, Dieter. *Gottfried Benn, Phänotyp dieser Stunde. Eine Studie über den Problemgehalt seines Werkes.* Cologne: Kiepenheuer & Witsch, 1986.
Wenusch, Monica. *"… ich bin eben dabei, mir Johannes V. Jensen zu entdecken…" Die Rezeption von Johannes V. Jensen im deutschen Sprachraum.* Vienna: Praesens Verlag, 2016.
Werkmeister, Sven. *Kulturen jenseits der Schrift. Zur Figur des Primitiven in Ethnologie, Kulturtheorie und Literatur um 1900.* Munich: Fink, 2010.
Werner, Heinz. *Die Ursprünge der Metapher.* Leipzig: W. Englemann, 1919.
Werner, Heinz. *Die Ursprünge der Lyrik. Eine entwicklungspsychologische Untersuchung.* Munich: Reinhardt, 1924.
Werner, Heinz. *Einführung in die Entwicklungspsychologie.* Leipzig: Barth, 1926.
Werner, Heinz. *Grundfragen der Sprachphysiognomik.* Leipzig: Barth, 1932.
Wesseling, Elisabeth, ed. *The Child Savage, 1890–2010: From Comics to Games.* Farnham: Ashgate, 2016.

Willemsen, Roger. *Das Existenzrecht der Dichtung. Zur Rekonstruktion einer systematischen Literaturtheorie im Werk Robert Musils.* Munich: Fink, 1984.

Willemsen, Roger. "Dionysisches Sprechen: Zur Theorie einer Sprache der Erregung bei Musil und Nietzsche." *Deutsche Vierteljahrsschrift für Literaturwissenschaft und Geistesgeschichte* 60 (1986): 104–135.

Wittmann, Barbara. "Johnny-Head-in-the-Air in America. Aby Warburg's Experiment with Children's Drawings." In *New Perspectives in Iconology: Visual Studies and Anthropology.* Ed. Barbara Baert, Ann-Sophie Lehmann, and Jenke Van den Akkerveken. Brussels: ASP, 2012. 120–142.

Wittmann, Barbara. *Bedeutungsvolle Kritzeleien. Eine Kultur- und Wissensgeschichte der Kinderzeichnung, 1500–1950.* Zurich: Diaphanes, 2018.

Wolf, Norbert Christian. "Salto rückwärts in den Mythos? Ein Plädoyer für das 'Taghelle' in Musils profaner Mystik." In *Profane Mystik? Andacht und Ekstase in Literatur und Philosophie des 20. Jahrhunderts.* Ed. Wiebke Amthor, Hans R. Brittnacher, and Anja Hallacker. Berlin: Weidler, 2002. 255–268.

Wolf, Norbert Christian. *Kakanien als Gesellschaftskonstruktion. Robert Musils Sozioanalyse des 20. Jahrhunderts.* Vienna: Böhlau, 2011.

Wolf, Norbert Christian. "Das wilde Denken und die Kunst. Hofmannsthal, Musil, Bachelard." In *Poetik des Wilden. Wolfgang Riedel zum 60. Geburtstag.* Ed. Jörg Robert and Friederike F. Günther. Würzburg: Königshausen + Neumann, 2012. 335–365.

Wolf, Norbert Christian. "Wahnsinn als Medium poet(olog)ischer Reflexion. Musil mit/gegen Foucault." *Deutsche Vierteljahrsschrift für Literaturwissenschaft und Geistesgeschichte* 88 (2014): 46–94.

Worringer, Wilhelm. *Abstraction and Empathy: A Contribution to the Psychology of Style.* Translated by Michael Bullock. Chicago: Ivan R. Dee, 1997. [*Abstraktion und Einfühlung. Ein Beitrag zur Stilpsychologie.* Munich: R. Piper, 1911.]

Wübben, Yvonne. *Verrückte Sprache. Psychiater und Dichter in der Anstalt des 19. Jahrhunderts.* Konstanz: UVK, 2012.

Wulffen, Erich. *Psychologie des Verbrechens.* Berlin: Langenscheidt, 1908.

Wulffen, Erich. *Gauner- und Verbrechertypen.* Berlin: Langenscheidt, 1910.

Wulffen, Erich. *Sexualverbrecher.* Berlin: Langenscheidt, 1910.

Wulffen, Erich. *Das Kind. Sein Wesen und seine Entartung.* Berlin: Langenscheidt, 1913.

Wundt, Wilhelm. *Elements of Folk Psychology: Outlines of a Psychological History of the Development of Mankind.* Translated by Edward Leroy Schaub. London: Allen & Unwin, 1916. [*Elemente der Völkerpsychologie. Grundlinien einer psychologischen Entwicklungsgeschichte der Menschheit.* Leipzig: A. Kröner, 1912.]

Wundt, Wilhelm. "Die Zeichnungen des Kindes und die zeichnende Kunst der Naturvölker." In *Festschrift Johannes Volkelt zum 70. Geburtstag.* Munich: C.H. Beck, 1918. 1–24.

Zenk, Volker. *Innere Forschungsreisen. Literarischer Exotismus in Deutschland zu Beginn des 20. Jahrhunderts.* Oldenburg: Igel, 2003.

Zumbusch, Cornelia. *Wissenschaft in Bildern, Symbol und dialektisches Bild in Aby Warburgs Mnemosyne-Atlas und Walter Benjamins Passagen-Werk.* Berlin: Akademie, 2004.

Index

Adorno, Theodor Wiesengrund 18, 352, 354, 359 f.
Aebli, Hans 100
Agamben, Giorgio 326, 344, 346, 348
Albers, Irene 60, 74
Anacker, Regine 228
Anz, Thomas 96, 160 f., 251
Asman, Carrie 334
Aue, Maximilian 278
Auerbach, Erich 9

Bach, Carl Philipp Immanuel 265
Bachelard, Gaston 24, 62 f., 208
Balázs, Bela 242
Balint, Michael 130
Basu, Priyanka 143, 144, 147, 150
Bataille, Georges 167, 173
Beard, Philip H. 278
Beddies, Thomas 85
Behrens, Heike 104
Benjamin, Walter 18, 29, 85–87, 93, 96, 166, 173, 194, 200, 303–340, 342–357, 359 f.
Benn, Gottfried 18, 21, 29, 128, 205, 209, 228–235, 359 f.
Bergson, Henri 177, 209
Beuy, Joseph 160
Biese, Alfred 11, 154, 184, 186, 194–200
Bilang, Karla 26
Binswanger, Ludwig 113, 162
Bleuler, Eugen 98, 130 f., 136, 251–257, 261, 263, 267, 270, 273, 297 f.
Bloch, Ernst 14, 19–22, 234, 310, 348, 359
Blome, Eva 2, 213
Blumenberg, Hans 9
Boas, Franz 48, 68, 148, 185
Boas, George 75, 168
Böhme, Hartmut 17
Bohrer, Karl Heinz 14
Bonacchi, Silvia 251, 255, 258, 301
Borges, Jorge Luis 19
Bovet, Pierre 78

Bracken, Christopher 18
Brandstetter, Gabriele 9
Braun, Wilhelm 269
Brecht, Bertholt 324, 334, 349–351, 355 f.
Breton, André 14 f.
Breuer, Josef 114
Bronfen, Elisabeth 14
Brüggemann, Heinz 304, 308
Bücher, Karl 150, 152, 153 f.
Bühler, Charlotte 108
Bühler, Karl 79 f., 89, 94 f., 170, 184, 338, 342 f.
Burckhardt, Jacob 135
Burk, Karin 330
Buschendorf, Bernhard 195
Büssgen, Antje 230

Caillois, Roger 95
Cassirer, Ernst 10, 184–186, 192 f., 198–201, 238, 241, 316, 335–337
Cavanaugh, John C. 77, 103
Cendrars, Blaise 26
Chagall, Marc 165
Cheng, Joyce S. 12, 27
Clarke, David D. 195
Clifford, James 25, 67 f., 71–73
Columbus, Christopher 66
Conrad, Joseph 71, 209 f., 213–215
Cooper, Frederick 1
Crapanzano, Vincent 48

Dacqué, Edgar 21, 207–209, 231
Därmann, Iris 17, 36, 62, 358
Darwin, Charles 103, 150, 159
Derrida, Jacques 22, 62, 358
Dessoir, Max 144
Deutsch, Werner 104
Dietrich, Stephan 215, 217, 225 f.
Dilthey, Wilhelm 158, 161, 177, 187, 195
Döblin, Alfred 18 f., 29, 359
Durkheim, Emile 36, 47–52, 54, 56 f., 59, 87, 167, 243

Einstein, Carl 14f., 18, 26, 156, 173
Etherington, Ben 1, 12f., 21, 28
Evans-Pritchard, Edward Evan 43

Fabian, Johannes 3, 41f., 61, 358
Fanta, Walter 292
Ferenczi, Sándor 123–127, 232f.
Fiedler, Konrad 345
Fischer, Bernhard 234
Fittler, Doris 309, 316f., 323, 325
Flechsig, Paul 229
Fleck, Ludwik 64
Foucault, Michel 4, 33f., 146, 201
Frank, Michael C. 41, 215
Franke, Anselm 2, 7f., 19
Frazer, James George 44–47, 49, 57, 69–71, 73f., 90f., 110, 117
Freud, Sigmund 62, 80f., 84, 96–98, 100, 110, 114–132, 134, 136, 138, 169, 172f., 180, 184, 250, 261, 290, 296, 308f., 345
Friedländer, Salomo 308
Frobenius, Leo 34, 69, 79, 238
Fry, Roger 26
Fuchs, Petra 85

Gardian, Christoph 47, 115, 136, 227
Gauguin, Paul 26, 243
Geertz, Armin 43
Geertz, Clifford 67f., 71, 73
Geisenhanslüke, Achim 114
Geissler, Peter 115
Genette, Gérard 182, 197
Georg, Eugen 114, 207f., 231
Gerber, Gustav 187, 195
Gess, Heinz 135
Gess, Nicola 33, 85, 129, 135, 143, 167, 173, 176, 236f.
Geulen, Eva 305, 329
Giddens, Anthony 1
Gisi, Lucas Marco 33, 216, 223
Giuriato, Davide 305, 311, 321, 333
Gockel, Bettina 110, 113, 160–162
Goebel, Eckhart 285
Goethe, Johann Wolfgang 107, 161, 171
Goldstein, Kurt 343
Götze, Carl 97

Gould, Stephen Jay 66, 76, 78, 86, 126f., 134
Greenblatt, Stephen 66, 217
Grill, Genese 271, 282
Groos, Karl 78, 81f., 84, 86, 88f., 92–97, 99, 103, 106–108, 170–172, 173, 237, 245, 307, 333
Grosse, Ernst 23, 143, 147–149, 152, 154
Gruber, Gernot 150
Grubrich-Simitis, Ilse 121f., 123, 124, 127
Gummere, Francis B. 143, 154

Habermas, Jürgen 346, 348
Haeckel, Ernst 6, 76–78, 124, 126–128, 148, 152f., 168, 177, 206, 211f., 228, 334
Hahn, Marcus 209, 228, 230f., 237
Hake, Thomas 248
Hall, Stanley 78, 82f., 106
Hamacher, Werner 318, 344, 348
Hansen, Miriam 345
Hartlaub, Gustav Friedrich 11, 97, 144, 168f.
Hartzell, Richard E. 278
Heidegger, Martin 7, 146
Heinz, Andreas 115
Heinz, Jutta 269, 296
Henzler, Stefan 185
Herbert, Christopher 69
Herder, Johann Gottfried 185, 341
Hesse, Hermann 135f.
Hildenbrandt, Vera 19
Hirschkop, Ken 184, 200
Hochstätter, Dietrich 294
Hoerl, Erich 55f.
Hoffmann, E.T.A. 139
Hofmannsthal, Hugo von 176f., 186, 197, 222
Holert, Tom 2, 7f., 19
Honold, Alexander 274
Hornbostel, Erich von 223, 238, 243, 297, 301
Hsu, Francis L. K. 22, 61, 358
Hubert, Henri 48f., 59, 243
Huch, Friedrich 165
Hui, Alexandra 150
Huizinga, Johan 173f.

Humboldt, Alexander von   238, 244, 336
Hyman, Stanley Edgar   69

Jacobowski, Ludwig   143, 152f.
Jaensch, Erich Rudolph   237, 239–242
Jakob, Michael   274
Jakobson, Roman   183
Jander, Simon   299f.
Janssen, Sandra   288
Jaspers, Karl   113
Jensen, Johannes V.   210, 213, 215, 243
Johnson, Mark   9
Jung, Carl Gustav   21, 110, 115f., 118, 130–136, 177, 206, 231, 290, 296

Kafka, Franz   325, 334, 348–351
Kandinsky, Wassily   26
Kant, Immanuel   157, 175
Kappeler, Florian   237, 281, 285, 301
Kassung, Christian   271
Kaufmann, Doris   113, 143, 148, 160, 162, 164
Kaufmann, Sebastian   33, 143f., 147
Khatib, Sami   311
Kirchdörfer-Boßmann, Ursula   228
Klages, Ludwig   21, 114, 177, 238, 290
Klaue, Magnus   193
Klee, Paul   12, 26, 173
Koch-Grünberg, Theodor   214
Kokoschka, Oskar   165
Kramer, Fritz   35f.
Kraus, Karl   309, 312–314, 356
Krause, Marcus   248
Krause, Robert   269
Kretschmer, Ernst   10f., 115, 130, 136–140, 180f., 237–239, 241f., 250–256, 258, 261f., 267–269, 272–274, 293
Kronfeld, Arthur   163–165
Krueger, Felix   110
Kubin, Alfred   13, 15, 165, 244
Kühn, Herbert   144, 154–156, 162
Kuhn, Reinhard   80
Kuhn, Thomas Samuel   64
Kühne, Jörg   296
Kuper, Adam   22, 35, 61, 73, 358
Kurasava, Fuyuki   8

Lacoue-Labarthe, Philippe   167
Lakoff, George   9
Lamarck, Jean-Baptiste de   76, 124f., 127, 206, 212
Lange, Konrad   92, 94, 171
Larcati, Arturo   188
Largier, Niklaus   282
Latour, Bruno   1
Leavis, Frank Raymond   23
Leeb, Susanne   2, 11, 27, 145f., 160
Lehmann, Hans-Thies   330
Lehmann, Hartmut   8, 17
Leiris, Michel   74
Lemke, Anja   333, 335, 346
Lemke, Sieglinde   2
Leonhard, Rudolf   340, 342f.
Lethen, Helmut   156
Leucht, Robert   282
Lévi-Strauss, Claude   22, 61f., 326, 358
Levinstein, Siegfried   106
Lévy-Bruhl, Lucien   36, 51–60, 72f., 87, 98, 100–102, 178, 180f., 185, 236f., 239, 241f., 244, 269, 271, 285, 296, 317, 320, 335–338
Li, Victor   22, 35, 73
Liederer, Christian   211–213, 215, 223–228
Lienhardt, Godfrey   60, 358
Lindner, Burkhardt   313, 352
Lindner, Martin   104
Lombroso, Cesare   83–85, 161, 172
Lönker, Fred   269
Löwith, Karl   17
Lubrich, Oliver   65
Lukács, Georg   14
Lurje, Walter   165f.

MacGregor, John   140, 160
Magris, Claudio   269, 271
Malinowski, Bronislaw   67–73, 182–184, 186, 192, 197–199, 201
Mallarmé, Stéphane   345–347
Mann, Thomas   16
Marc, Franz   26
Marett, Robert R.   47, 315
Marr, Nikolai   337f., 344, 350
Martin, Ronald E.   117, 131
Marx, Karl   21

Mauss, Marcel 48 f., 59, 243
Mauthner, Fritz 103, 184–186, 190–194, 197, 199 f.
Mawson, Douglas 243
McLaughlin, Kevin 311
Meijers, Anthonie 187
Menke, Bettine 311, 348
Menninghaus, Winfried 311–313, 325, 340, 347
Mikkelsen, Ejnar 243
Miller, Gerlinde F. 228, 233
Moebius, Stephan 60, 167, 357
Mommsen, Wolfgang 18
Moreau, Paul 83
Mülder-Bach, Inka 288, 294
Müller, Friedrich Max 176
Müller, Hans-Peter 16
Müller, Max 198
Müller, Robert 13, 18, 29, 128, 205, 209–214, 216–218, 222 f., 225, 227–229, 234, 236, 243 f., 246–248, 287 f., 300, 324, 360
Müller, Thomas R. 161
Müller-Lyer, Franz Carl 237 f., 241, 244
Müller-Richter, Klaus 188
Müller-Schöll, Nikolaus 349
Müller-Tamm, Jutta 9, 35, 146, 156, 213, 216, 220, 223 f., 226, 228
Munro, Thomas 143
Murphy, Ruth 87
Musil, Robert 13, 23, 29, 85, 236–253, 255, 258–260, 262 f., 268 f., 271, 274–282, 284 f., 288–292, 294, 296–298, 300–302, 304, 306, 308, 359 f.

Nägele, Rainer 348 f.
Nancy, Jean-Luc 167
Nerlich, Brigitte 195
Nietzsche, Friedrich 161, 167, 172, 184, 186–190, 192 f., 195, 197–200, 225, 258, 260, 274, 281 f., 293 f., 308, 338
Nübel, Birgit 300

Oehlschläger, Claudia 156
Oesterreich, Konstantin Traugott 251, 255, 258 f., 267, 279 f., 290–292
Ogden, Charles K. 182, 200

Ostermann, Eberhard 269, 275, 278, 281 f., 288
Otis, Laura 127 f., 134

Paget, Richard 338 f.
Pan, David 117, 308
Parr, Rolf 62
Pauly, August 124
Payne, Harry C. 71
Payne, Philip 292
Pelmter, Andrea 248
Perloff, Marjorie 74
Pestalozzi, Johann Heinrich 103
Peters, Sibylle 9
Pethes, Nicolas 23, 25, 80, 248, 251, 313 f., 338, 348, 355
Piaget, Jean 10, 79, 95, 97–102, 106, 115, 320
Picasso, Pablo 26
Pietsch Pentecost, Gislind Erna 251, 264
Pressler, Günter Karl 335
Preuss, Karl Theodor 47, 49, 58, 110, 136 f., 238, 315 f., 337
Preyer, William T. 77, 90, 94, 103 f., 106 f., 170
Prinzhorn, Hans 144, 162–164

Rathenau, Walter 285
Raulet, Gérard 320
Recki, Birgit 199
Rehding, Alexander 146, 149, 151
Reichle, Ingeborg 143, 147
Reichlin, Susanne 282
Reinhardt, Ursula 282
Reis, Gilbert 300 f.
Reschke, Renate 311
Rhodes, Colin 27
Ribot, Théodule 89
Richards, Ivor Armstrong 182 f., 200
Richter, Dieter 80
Richter, Sandra 158
Ricoeur, Paul 67
Riedel, Wolfgang 4, 11, 177, 201, 216, 226, 230, 232, 284 f., 295 f., 301
Riegl, Alois 148
Riemann, Hugo 151
Rilke, Rainer Maria 96

Ringelnatz, Joachim 85, 308
Robertson, Ritchie 237, 283
Rose, Wolfgang 85
Rossetti, Gina M. 2
Rousseau, Jean-Jacques 4
Rubin, William 25–27
Ruskin, John 168

Said, Edward 64–67
Scherer, Wilhelm 159
Schilder, Paul 110f., 113, 115, 136
Schiller, Friedrich 3–6, 33, 96, 171, 205, 228, 230, 235
Schmarsow, August 148
Schmidt, Erich 143, 153
Schmitz-Emans, Monika 186
Schneider, Manfred 311
Schneider, Sabine 177
Scholem, Gershom 303, 314f., 318, 322
Scholz, Friedrich 82
Schraml, Wolfgang 237, 246
Schulz, Kerstin 295
Schurtz, Heinrich 148
Schüttpelz, Erhard 1, 17, 26, 28, 36
Schwarz, Thomas 210–212, 214, 216f., 221f., 226, 228
Schwennsen, Anja 185, 200
Seidmann-Freud, Tom 330
Sello, Katrin 15
Sepp, Hans 281
Shelton, Marie-Denise 74
Shuttleworth, Sally 77, 80
Sighele, Scipio 82
Simmel, Georg 185
Snow, Charles Percy 23
Sombart, Werner 16
Specht, Benjamin 176, 187, 195
Spencer, Herbert 150, 159
Spinner, Samuel 1
Steinen, Karl von den 10, 53
Stern, Clara 104, 179–182, 339
Stern, William 77, 79, 81, 84, 89, 93–96, 103–108, 172, 179–182, 326, 339
Stingelin, Martin 187
Stockhammer, Robert 184
Stocking, George W. Jr. 71
Storch, Alfred 111–115, 136, 161, 164

Stumpf, Carl 143, 150
Sulloway, Frank J. 126f.
Sully, James 77–80, 89–94, 105–108, 118, 167f., 170, 179f.

Taine, Hyppolite 103
Taussig, Michael 36
Tenbruck, Friedrich H. 8, 16
Tewilt, Gerd-Theo 269, 296
Thurnwald, Richard 10, 180, 238
Tiedemann, Dietrich 103
Todorov, Tzvetan 73
Torgovnick, Marianna 1, 27f., 43
Turner, William 69
Tylor, Edward Burnett 10, 34, 36–47, 57, 60, 62, 159, 179, 181, 183, 187, 206, 315
Tzara, Tristan 26

Unger, Erich 231

Vatan, Florence 237, 246, 300
Verworn, Max 144, 155
Vico, Giambattista 185, 193, 196, 237
Vierkandt, Alfred 36, 46, 110, 144, 238
Vischer, Friedrich Theodor 159, 184, 186, 194f., 197–200, 296
von Heydebrand, Renate 237
Vygotsky, Lev Semenovich 87, 326

Wallaschek, Richard 144, 150f.
Walser, Robert 325
Warburg, Aby 53
Weber, Max 7–9, 12, 15–18, 359
Weber, Samuel 349
Weiler, Bernd 33
Weimar, Klaus 152
Wellershoff, Dieter 231
Werkmeister, Sven 1, 4, 28, 33–35, 39, 43, 68f., 178, 214, 218, 220, 223, 226, 237, 246, 248
Werner, Heinz 6, 79, 114, 143, 153, 180, 340–344
Wesseling, Elisabeth 75
Willemsen, Roger 237, 273, 278, 296
Wittmann, Barbara 2, 75, 79, 97, 103, 106, 144f., 168, 173

Wolf, Norbert Christian 237, 251, 281f., 285, 288, 300
Worringer, Wilhelm 156, 162
Wübben, Yvonne 160

Wundt, Wilhelm 35f., 42, 46f., 49, 58, 110, 136, 144, 178f., 187, 196, 315, 339
Wyneken, Gustav 8, 304f.

Zenk, Volker 210
Zumbusch, Cornelia 348

www.ingramcontent.com/pod-product-compliance
Lightning Source LLC
Chambersburg PA
CBHW031750220426
43662CB00007B/345